Food Texture
Measurement and Perception

Andrew J. Rosenthal, BTech, MSc, PhD, MIFST
Senior Lecturer in Food Science
School of Biological and Molecular Sciences
Oxford Brookes University
Oxford, United Kingdom

A Chapman & Hall Food Science Book

AN ASPEN PUBLICATION®
Aspen Publishers, Inc.
Gaithersburg, Maryland
1999

The author has made every effort to ensure the accuracy of the information herein. However, appropriate information sources should be consulted, especially for new or unfamiliar procedures. It is the responsibility of every practitioner to evaluate the appropriateness of a particular opinion in the context of actual clinical situations and with due considerations to new developments. The author, editors, and the publisher cannot be held responsible for any typographical or other errors found in this book.

Aspen Publishers, Inc., is not affiliated with the American Society of Parenteral and Enteral Nutrition.

Library of Congress Cataloging-in-Publication Data
Rosenthal, Andrew J.
Food texture: measurement and perception/Andrew J. Rosenthal.
p. cm.
Includes bibliographical references (p.) and index.
ISBN 0-8342-1238-2
1. Food texture. I. Title.
TX531.R595 1999
664'.07—dc21
98-45264
CIP

Orders: (800) 638-8437
Customer Service: (800) 234-1660

About Aspen Publishers • For more than 35 years, Aspen has been a leading professional publisher in a variety of disciplines. Aspen's vast information resources are available in both print and electronic formats. We are committed to providing the highest quality information available in the most appropriate format for our customers. Visit Aspen's Internet site for more information resources, directories, articles, and a searchable version of Aspen's full catalog, including the most recent publications: **http://www.aspenpublishers.com**

Aspen Publishers, Inc. • The hallmark of quality in publishing
Member of the worldwide Wolters Kluwer group.

Editorial Services: Kathy Litzenberg
Library of Congress Catalog Card Number: 98-45264
ISBN: 0-8342-1238-2

Printed in the United States of America
1 2 3 4 5

To Harry who, maybe without knowing, planted and
fed the seed of food science in my mind.
Thank you.

Table of Contents

Contributors

Arjen Bot, PhD
Unilever Research Vlaardingen
Vlaardingen
The Netherlands

Dick W. de Bruijne, Technical PhD
Unilever Research Vlaardingen
Vlaardingen
The Netherlands

Bogdan J. Dobraszczyk, BSc, MSc, PhD
Department of Food Science and Technology
The University of Reading
Reading, United Kingdom

Michael Edwards, BSc, PhD
Microscopist
Campden and Chorleywood Food Research Association
Gloucestershire, United Kingdom

Marion L. Greaser, PhD
Muscle Biology Laboratory
Department of Animal Sciences
College of Agricultural and Life Sciences
University of Wisconsin
Madison, Wisconsin
United States

M. Robin Heath, BDS, PhD, FDS, RCS (Engineering)
Department of Prosthetic Dentistry
The Royal London School of Medicine and Dentistry
London, United Kingdom

David Kilcast, BSc, PhD
Head of Sensory and Consumer Science
Leatherhead Food Research Association
Leatherhead, United Kingdom

Albert M. Pearson, PhD
Formerly of the Department of Animal Sciences
Oregon State University
Corvalis, Oregon
United States

Jon F. Prinz, BChD, MMedSc, PhD
Department of Prosthetic Dentistry
The Royal London School of Medicine and Dentistry
London, United Kingdom

Andrew J. Rosenthal, BTech, MSc, PhD, MIFST
Senior Lecturer in Food Science
School of Biological and Molecular Sciences
Oxford Brookes University
Oxford, United Kingdom

Jo Smewing, BSc, MPhil, MIFST
Applications Manager
Stable Micro Systems Ltd.
Surrey, United Kingdom

Andrew C. Smith, BSc, PhD
Head of Biomaterials Science Group
Institute of Food Research
Norwich, United Kingdom

Julian F. V. Vincent, MA, PhD, DSc
Centre for Biomimetics
The University of Reading
Reading, United Kingdom

Ton van Vliet, MSc, PhD
Department of Food Technology and Nutrition Sciences
Wageningen Agricultural University
Wageningen
The Netherlands

Preface

Food texture is essentially a human experience arising from our interaction with food and its structure and behavior when handled. To understand our response to food structure and its breakdown involves a matrix of disciplines, from aspects of physiology and psychology of perception, to physical and chemical studies of the composition and structure of food materials and their behavior when deformed or stressed.

The original concept behind this book was to take a holistic view of food texture, starting with the determination of food texture, its perception in the mouth, and its measurement by both sensory and instrumental methods, and to examine the relation between those methods. The book has been divided into two sections. The first deals with perception of food texture and techniques for its measurement. The second focuses on individual groups of food commodities. Having said this, the first section is interlaced with appropriate food-related examples, which helps to reinforce the applied nature of the subject.

Chapter 1 introduces the topic, examining how instrumental and sensory approaches can complement each other in determining the texture of food. Throughout the book we have given instrumental and sensory methods equal weight, thus countering the notion that sensory evaluation techniques are less reliable than instrumental tests. The second chapter deals with oral perception and the eating process. It deals with the physiology of the mouth and related muscles; it follows the breakdown path of food, exploring the mechanisms of texture perception in the mouth. Techniques that are used to evaluate texture from a sensory perspective are reviewed in Chapter 3. While some of these techniques are used in other areas of sensory evaluation, Dr. Kilcast includes some valuable examples showing techniques that have been applied to the evaluation of food texture. In Chapter 4, Dr. van Vliet characterizes the various types of foods in terms of their rheological properties and describes suitable testing procedures to measure such properties. Chapter 5, which examines methods to evaluate the structure and material properties of foods, examines how these relate to food texture. The authors adopt the view that foods are *materials* and while many foods behave viscoelastically, their structure may be evaluated with the armory of tests developed by material scientists to evaluate engineering materials.

The last five chapters of the book focus on the texture of selected foods. Each of the commodities is approached in a common format, starting with the structure and how it affects the texture from both sensory and rheological perspectives. Some consideration is also given to any textural changes that occur during common processing operations and storage. Since most of the sensory and rheological techniques are covered in the first five chapters of the book, we have tried to avoid repetition; however, where a specialized technique was developed for just one commodity, it is dealt with in the appropriate commodity chapter.

Chapter 6 covers starch-based foods including bread, crisp breads, cakes, breakfast cereals, and extruded snack foods. With so many of these foods being solid foam structures, the chapter examines the strength of model systems with a cellular structure and compares these to real starch-based foamed foods. Chapter 7 deals with fabricated fat-based foods, including water in oil emulsions, such as margarine, halvarines, chocolate, low-fat and zero-fat spreads; and oil in water emulsions, such as mayonnaise and dressings. Chapter 8 examines flesh foods and deals extensively with the structure of muscle and meat along with many factors that influence meat texture prior to and after slaughter. Chapter 9 covers the texture of vegetables and fruit, starting with the plant cell structure and examining how it is affected by processing. Hydrocolloids (Chapter 10), though only occasionally eaten on their own, are frequently introduced into processed foods to provide structure in the form of a thickener, producing a viscous paste or a solid gel. The chapter reviews the variety of hydrocolloids used in processed foods and identifies the unique textural characteristics of each.

Inevitably, with a multi-authored book, there are small amounts of overlap among the chapters. The different approaches and varied styles of writing, however, help reinforce the material.

Sadly, Professor Albert Pearson passed away before the book was published.

Acknowledgments

During the planning and preparation of this book I was fortunate enough to work with a number of publishing professionals. I would like to make a special mention of Rose Gilliver and Pam Cowburn, whose encouragement and persistence were probably the two greatest factors in commissioning the authors and collecting the manuscripts. Thanks also to Jack Bruggeman, Kathy Litzenberg, and Shana Wagger, who transformed the manuscripts into this finished book.

I am grateful to my colleagues, Jeya Henry and Neil Heppell, for their interest and encouragement, and to my boss, David Beadle, for treating it all as a scholarly activity.

I am especially grateful to Sheila (as well as our Bearded Collies, Omri, Ishshah, and Halva) for encouraging me, tolerating me, and lightening my mood when necessary.

Andrew J. Rosenthal
BTech, MSc, PhD, MIFST

CHAPTER 1

Relation between Instrumental and Sensory Measures of Food Texture

Andrew J. Rosenthal

Derived from the Latin *textura*, which means a weave, *texture* originally was taken to refer to the structure, feel, and appearance of fabrics. It was not until the 1660s that it started to be used to describe "the constitution, structure or substance of *anything* with regard to its constituents, formative elements" (*Oxford English Dictionary*, 1989). Various attempts to define *food texture* have culminated in some international agreement with the development of the international standard ISO 5492 (International Organization for Standardization [1992]), which deals with the vocabulary used for sensory evaluation. This defines texture as "All the mechanical, geometrical and surface attributes of a product perceptible by means of mechanical, tactile and, where appropriate, visual and auditory receptors." Clearly, food texture is about perception, making it above all other things a human experience. It is about our perception of a foodstuff that originates in that product's structure and how the product behaves when handled and eaten. Furthermore, it incorporates *all* the attributes (mechanical, geometric, and surface) of the food, suggesting that the experience of texture is one of many stimuli working in combination.

Bearing such a definition in mind, is it reasonable to presume that such a complex array of interactions of stimuli could be measured instrumentally? This is a moot point, for while many samples can be subjected to assorted instrumental testing techniques, on what basis shall we acknowledge that the results produced have any relation to human perception? Prior to the 1940s, it was generally considered that sensory measurements of food texture were purely subjective and as such generally unreliable. Variation in individuals as well as variability of any one person from day to day seemed to make the sensory analysis of food an art and not worthy of serious scientific study. It was inconceivable that an individual's response could be anything other than personal, hedonic, and prejudiced by that person's beliefs and biases. Since our scientific ethos was founded on reproducibility, most serious researchers were persuaded to rely on instrumental testing techniques carried out under standardized conditions. Such techniques were considered to be reliable, with relatively small inherent variation or error. An attitude developed in which instrumental tests were referred to as "objective" while sensory work was denigrated as being "subjective."

In these early years of texture measurement, the study of texture focused on the efforts of the rheologist, who measured flow and deformation of food materials. In his discussion paper "Is Rheology Enough for Food Texture Measurement?" Bourne (1975) suggested that rheological measurements, which often focus on a single large deformation, resulting in the sample's breaking into pieces, were inadequate in defining food texture. When an individual eats a foodstuff, the sample is chewed beyond this initial breakdown, and the stimuli that result form part of the overall texture sensation. While the initial bite is an important aspect of texture, so too are the subsequent bites, the viscosity, stickiness, and consistency of the food as it mixes with saliva. Equally important are aspects of the food's appearance, the mechanical properties, and sounds that occur when it is handled, cut, and eaten. Clearly, rheology is not enough to explain all the rich and complex aspects of texture that are experienced by humans.

Attitudes about the objectivity of sensory research began to change shortly after the Second World War. The U.S. Army had made a considerable investment in developing nutritious rations for its troops, only to find that many of them did not appreciate what they were being offered to eat. A focused research program was developed by the U.S. Quartermasters Corps to look at issues of food acceptability and choice. Development of controlled sensory testing procedures, the separation (in people's minds) between "sensory" and "affective" tests, and advances in multivariate statistical techniques that were made possible by the application of relatively powerful computers all helped to bring the perception of sensory testing into scientific respectability.

Without a doubt, texture plays a key role in our appreciation of food. Our perception of food texture often constitutes a criterion by which we judge its quality and is frequently an important factor in whether we select an item or reject it. We squeeze and prod fruits and cheese to gauge their ripeness and tilt bottles to estimate the viscosity of their contents. Texture can be expressed in the sounds that foods make when handled, to the extent that we listen to foods to estimate their quality—for example, listening to the sound of a melon when it is tapped. Familiarity with a product brings knowledge about how its texture and behavior change during processing and storage. With continued contact with a specific food, certain individuals have developed great expertise, to the extent that they become expert judges of that product's quality. Such individuals (e.g., master craftspeople) often determine the texture of foods with empirical test methods such as prodding, tapping, or squeezing. These so-called "foreman's finger tests" have become established measures of quality for some products and may even set market prices. Surely, if an individual can assess texture by prodding, then a machine that prods and measures the force involved will give a reproducible measure of that aspect of texture and therefore quality. By applying suitable calibration, we are able to create machines that might provide information that formerly only our expert could offer.

Having accepted the idea that a machine might take the place of a human to assess food texture, we must remember that before all else, food texture is essentially a human experience that arises from our interaction with food. It is worthwhile to be cautious when considering physical test procedures, for data can be collected by subjecting any material to any procedure, but results from the test do not necessarily mean anything in terms of texture. Such a comment is not intended to disqualify instrumental tests that may not relate directly to human perception, for such procedures may be valid for all kinds of other reasons. For example, viscosity of a liquid is frequently measured to gain an idea of resistance to flow and therefore the pumping requirements to push the fluid down a pipe. However, as we will see later, under some conditions instrumental measurements of viscosity bear little relation to the experience as perceived in the mouth.

THE PROCESS OF FOOD TEXTURE PERCEPTION

Human perception of food has been described as a cyclic process that starts with an anticipation originating primarily from visual cues but also flavored by our prior experiences. Various aspects of appearance, such as color, size, and shape, as well as aspects of structure (e.g., openness), preempt our physical interaction with the food (Kramer, 1973). Though not always associated with texture perception, visual cues provide a gauge of viscosity (Shama, Parkinson & Sherman, 1973) and the "wobbly" behavior of semisolid, jellylike foods. Our participation leads to manual manipulation either directly or with tools (e.g., cutting with a knife). Visually perceived changes in the food with handling add to our impressions of the food's texture. Even before the food is in the mouth, we have gathered a substantial amount of knowledge about the food's texture from visual, tactile, and even auditory stimuli.

Initial perception in the mouth (i.e., without biting) is at a relatively low shear rate. Two categories of sensation have been identified: those due to touch, which occur regardless of any shearing, and those that require a small amount of deformation. With no shear at all, we gather impressions about the food's homogeneity, such as the presence, size, and shape of particles or air cells. At slightly higher shear rates caused by movement of the tongue, the food is deformed and flows. Under these conditions, characteristics such as elasticity, stickiness to the palate, and viscous behavior are perceived (Sherman, 1969).

During the first few chews, much of the structure is broken. Brittle materials fracture, fibrous materials are torn, and the food is kneaded and mixed with saliva to form a coherent bolus (Heath, 1991). During these chewing cycles, a high degree of shearing is achieved, and a wide variety of textural characteristics are perceived. Attributes that relate to physical makeup (e.g., hardness-softness) and deformation and breakdown (e.g., brittleness, plasticity, crispness, and sponginess) are detected (Sherman, 1969). Movement of the jaw during subsequent chewing cycles of the so-called "mush phase" are more regular than in the early stages of mastication. During this period, saliva is secreted into the bolus as it is further kneaded prior to swallowing. Textural attributes perceived are those that relate to the particular nature (e.g., smoothness, coarseness, powderiness, lumpiness, and pastiness), consistency (e.g., creaminess, wateriness), and adhesion to the palate (e.g., stickiness).

Hutchings and Lillford (1988) developed a model to explain the breakdown path to which a food is subjected before it is ready for swallowing. They reinforced the idea that the breakdown path is a dynamic process occurring over a period of time and postulated that the key attributes that affect the process are the degree of lubrication and the structure that the food possesses. During mastication, the structure is broken down by mechanical action. Lubrication is due in part to the release of saliva during chewing but also arises from the composition of the food in terms of moisture and fat, both of which act as lubricants. Changes in structure and lubrication occur until a threshold is reached, at which point the material is swallowed.

After swallowing, we perceive a residual masticatory impression that arises from the remains of disintegrated food and any mouth-coating materials. Such attributes include meltdown properties on the palate, greasiness, gumminess, and stringy sensations (Sherman, 1969).

Since perception is a cyclic process, the information gathered during the handling, biting, chewing, and swallowing feeds back into our anticipation of the next portion.

COMPARISON OF HARDWARE FOR INSTRUMENTAL AND SENSORY MEASUREMENT

Many stimuli contribute to our perception of texture, including visual and auditory cues as well as those related to touch and movement. Visual and auditory cues are gathered through specialized sense organs—the eyes and ears. In contrast, the sensors of material characteristics are spread throughout the body, sometimes being categorized as those sensitive to touch (*somasthesis*) and movement/position (*kinesthesis*). Various skin organelles have been identified, such as Pacinian corpuscles, Meissner's corpuscles, Ruffini endings, and free nerve endings. While some of these organelles have been associated with the perception of certain attributes (e.g., Pacinian corpuscles with perception of pressure), it is often thought that the law of specific nerve endings does not apply to skin senses and that all types of nerve endings contribute to our general perception of texture. In addition to the tactile sense organelles in the hard and soft palate, tongue, gums, and periodontal membrane surrounding the teeth, there are vitally important nerve endings in the oral muscles and joints. Signals from these nerves provide information on jaw position, muscle tension, and length.

In comparison to the sensing apparatus of the human body, instrumental testing devices rely on transducers to convert material and physical measurements into visual or electrical outputs that can be either observed directly or fed into data-recording/processing equipment. Instrumentation depends on the type of test being performed, but for mechanical testing, it often involves strain gauges and load cells to measure forces and position or movement detectors. Crucial to the type of test being applied is the geometry of the test cell and how the sample is held. Tests carried out on solid and viscoelastic materials are often done under compression, shear, torsion, or tension (see Chapter 4 of this volume).

Successful transducers usually have a linear response that, through calibration with standards, can represent defined physical characteristics in terms of absolute units. In contrast, human perception is governed by psychophysical phenomena, which tend to be nonlinear. For example, Weber's law states that the magnitude of a "just noticeable difference" is proportional to the intensity of the stimulus already present. Thus, the body adapts to the forces that are exerted, being most sensitive when small forces are applied. Moreover, the response to a stimulus is most noticeable when a change in that stimulus occurs; if the stimulus is held constant, the response lessens with time. The overall consequence is that we are receptive to relative differences and changes but rather poor at identifying any absolute measures.

Factors like temperature are frequently influential on rheological behavior and consequently most apparatus for measurement of such parameters will be accurately thermostated. Typical instrumental methodology would involve introducing the sample to the device and leaving it to reach a steady temperature before the physical test is applied. But this approach is quite different from what occurs in the mouth during eating. While the temperature within the center of the body is fairly constant at 37°C, the mouth is normally a few degrees below that. Food that is introduced is rarely at the same temperature, and there follows a brief change in the temperature of the food, which may lead to a change in its physical behavior. A well-known example of how this temperature change affects the texture of food is provided by chocolate. As with most triglyceride material, cocoa butter can be crystallized in a variety of polymorphic forms. The desirable V crystal has a relatively sharp melting point around 33.5°C (Schlichter-Aronhime & Garti, 1988). When chocolate at around 20°C is introduced to the mouth, the temperature starts to rise and the fat crystals start to melt in the mouth, giving rise to the sensation of dissolving and melting away. Similar sensations are important in ice cream and margarines (see Chapter 7 this volume).

In addition to thermal melting, the presence of saliva within the mouth leads to a dissolution of water-soluble materials. About 1.5 litres of saliva are secreted each day. Saliva is a dilute non-Newtonian liquid that contains digestive enzymes (α amylase) as well as a variety of unusual proteins and polypeptides. Histatins, for example, are a group of polypeptides found only in saliva that have exceptional antibacterial properties (Schenkels, Veerman, & Amerongen, 1995). Saliva acts as a lubricant as well as a solvent, allowing food materials to be effectively dissolved and broken down during mastication. Most rheological test equipment does not have any facility to introduce a solvating lubricant onto the sample during testing.

The speed of movement of the jaw and the tongue within the mouth is a critical factor in our perception of food texture. Tornberg et al. (1985) showed that when chewing meat, the jaw can move at between 200 and 400 cm/min in contrast to typical instrumental testing machines, which operate around at 20 cm/min. Bearing in mind that only a few liquid foods behave as Newtonian, the others will exhibit shear dependent viscosity, and in such a situation the apparent viscosity will depend on the rate at which the liquid is being sheared. Shama and Sherman (1973) found that shear rates in the mouth are applied in a range from 0.1 to 1000 s^{-1}. Since most foods are non-Newtonian, their apparent viscosity will depend on the shear rate applied. It is therefore necessary to match shear rates applied in an instrumental test with those that might be experienced in the mouth, or the measured viscosity may not equate to what would be perceived orally. Shama and Sherman showed that the shear rate applied in the mouth actually depends on the viscosity of the food, such that low-viscosity foods receive relatively high shear rates while high-viscosity foods tend to be sheared more slowly (Figure 1–1). Consequently, some iteration is needed to find appropriate shear rates that measure the apparent viscosity as it would be perceived in the mouth, and the appropriate shear rate may vary from one food to another.

Takahashi and Nakazawa (1991) studied the swallowing process by introducing pressure transducers into the palate of volunteers. Using a laboratory viscometer at shear rates that occur in the mouth, they determined the viscosity of a variety of carboxymethyl cellulose solutions. Different volumes of these solutions (<25 cm^3) were then offered to the volunteers, and the retaining time that the samples were in the mouth, palatal pressure during swallowing, and work involved in swallowing were monitored. The swallowing pressure changed only from about 0.1 to 0.2 mPa over a viscosity range of 10 to 10,000 mPa•s. The time that the sample was retained in the mouth and the work involved in swallowing remained almost constant up to a critical viscosity of 1 Pa•s, above which both the time retained and the work increased mark-

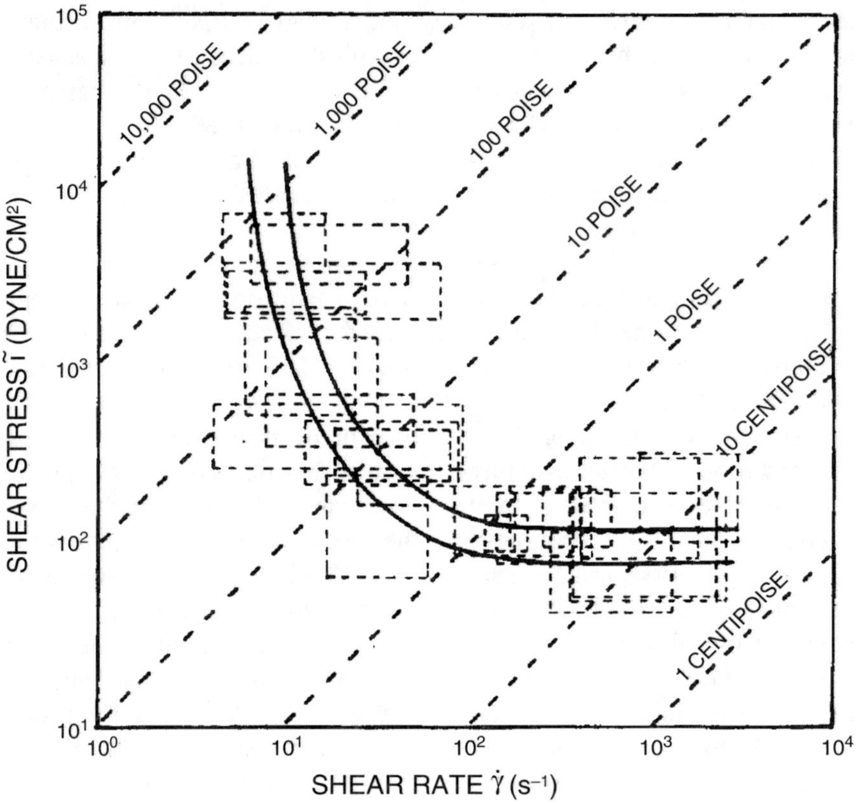

Figure 1–1 Bounds of Shear Stress and Shear Rate Associated with Oral Evaluation of Viscosity. *Source:* Reprinted with permission from F. Shama and P. Sherman, Identification Stimuli Controlling the Sensory Evaluation of Viscosity: II Oral Methods, in *Journal of Texture Studies*, Vol. 4, pp. 111–118, © 1973.

edly. The effect of sample volume varied with solution viscosity, low-viscosity liquids being swallowed in one deglutition (<15 cm^3) while high-viscosity liquids were swallowed in several small volumes.

The action of mastication and secretion of saliva combine in the breakdown path until the bolus is suitable for swallowing (Hutchings & Lillford, 1988). Clearly, this is a time-dependent process and one in which the nature of the food, and hence its texture, is changing. Few instrumental tests consider this kind of broad time frame; in fact, most focus on the first or the first and second bites. While a considerable amount of the structural breakdown may occur during these early parts of mastication, other sensory attributes experienced closer to the time of swallowing frequently do not get evaluated.

At this stage, it is worth returning to the ISO definition of food texture. A key aspect of food texture is that it can arise from a *combination* of physical properties; sensations such as wetness have been attributed to the simultaneous stimuli of cold and pressure, while stickiness is due to the sequence of a gentle pressure followed immediately by pulling on the skin (Szczesniak, 1990). Despite these multifarious sensations, few texture-measuring instruments focus on more than one physical property at a time; this has resulted in investigators' examining different physical aspects in isolation, each yielding a facet of the experience but none an equivalence to the entire sensation. For example, Dacremont, Colas, and Sauvageot (1991) measured the sounds traveling to the ear during eating. By placing a conventional microphone adjacent to the ear as well as a miniature contact microphone inside the ear, they were able to distinguish between the sounds that are conducted through the bone and those that travel as waves through the air. Such an approach allows some distinction to be made between sounds that originate from different types of food. Crispy foods (described as *croustillant* by the French authors) produce high-pitched sounds with frequencies in excess of 5 kHz. Such sounds travel predominantly through the air. In contrast, low-frequency crackly and crunchy sounds (described as *craquant* and *croquant* by the authors) travel predominantly through the bone. Distinction between crackly and crunchy sounds is postulated to be due to crunchy foods' creating vibrations that stimulate touch sensors (Dacremont, 1995). Vickers (1988) reviewed the research into acoustical properties of crisp foods in relation to their sensory properties. She pointed out that acoustic data alone provide only a partial explanation of sensory crispness. By combining acoustic data with force-deformation failure tests in multiple regression, one can obtain a better correlation with sensory crispness for some foods. It could of course be argued that examining each physical phenomenon individually allows a separation of the relative contribution that each has to the overall experience.

APPROACHES TO INSTRUMENTAL MEASUREMENT OF TEXTURE

Scott-Blair (1958) categorized the instrumental techniques used to measure food texture into three groups:

1. *empirical tests*, which measure something physical under well-defined conditions
2. *imitative tests*, which attempt to simulate the conditions to which the material is subjected in the mouth
3. *fundamental tests*, which measure well-defined physical properties such as viscosity or elastic modulus

Advances in medical instrumentation have created a fourth category of techniques that examine the neurophysiology of the eating experience.

The following discussion will predominantly (though not exclusively) focus on techniques used to obtain "textural" information about gelatin gels. While the topic of hydrocolloid gel testing is reviewed more comprehensively in Chapter 10 of this book, the comparison of techniques considered here helps to clarify the nature of the various types of tests that exist.

Empirical Tests

By definition, empirical tests are developed by experimentation and observation, and as such they may lack a rigorous scientific basis. However, this does not disqualify their use, and in some sectors of the food industry, empirical tests act as standards that are used to grade food quality. For example, the Adam's consistometer is used as a standard test to ascertain the quality of apple puree or creamed corn. It consists of a flat horizontal plate with a series of concentric, equally spaced rings etched on the surface. The sample is placed in a short tube positioned over the central ring, and the tube is lifted, allowing the food to spill out and flow radially across the plate. Clearly, the distance covered by the puree in a given time depends on a variety of phenomena such as hydrostatic force, particle size and shape, friction, and surface tension. Consequently, there is no simple scientific explanation for the flow behavior, yet it does relate well to quality of the food product.

Empirical tests have been criticized as usually being specific to particular narrow ranges of products. Each empirical test measures characteristics of the food in an arbitrary way, and since they are uniquely different from each other, they tend not to compare well and cannot be used predictively (Bagley & Christianson, 1987).

The Bloom Gelometer (Bloom, 1925) was once an industry standard for testing gelatin. The importance of the test is typified by the fact that gelatins are both characterized and sold on the basis of their gel strength, which is given in Bloom units (with very weak gelatin gels being around 100 Bloom, while firm gels are in the region of 250 Bloom). While developed to test gelatin gels, the instrument has been applied to other gel-like materials.

One of the original patent drawings of the Bloom Gelometer is shown in Figure 1–2. Itemized parts on the sketch are referred to in the following discussion. The essence of the test is that a cylindrical plunger of fixed weight and 12.7 mm in diameter (Item 10) is placed on the top of the gel (Item 9). The plunger is connected via a rod (Item 11) to a vessel (Item 13) into which lead shot can be gradually added from a dispensing mechanism (Items 33–37). When the force acting down on the gel is enough to force the plunger to penetrate to a depth of 4 mm, the flow of lead shot is halted by an intricate electrical magnetic mechanism (Items 41–55). The Bloom strength of the test sample is obtained from the weight of the plunger, the lead shot–collecting vessel, and the lead shot that was required to force the plunger 4 mm into the surface of the gel.

As with all effective empirical tests, sample preparation and test procedures are methodically specified. In this case, the gelatin is made up to a concentration of 6.67% (w/w) and must be matured at a temperature of 10°C for between 16 and 18 hours prior to testing. Failure to hold a sample long enough can lead to spurious results; for example, a gelatin gel held at 10°C and tested after 5½ and 10 hours can give values that differ by 20% (Szczesniak, 1975).

While the Bloom Gelometer was an industrial standard, the basis of measuring the force required to push a 12.7-mm–diameter plunger 4 mm deep into a gel is very much an arbitrary measure. This begs the question, Does the Bloom Gelometer measure texture? Strictly speaking, no. The Bloom strength of the gel cannot describe the combinations of experience that occur when a gel is eaten. Yet there is undoubtedly a relationship between the Bloom strength of a gel and the texture that it exhibits. Moreover, the Bloom strength is useful in classifying gels for different food uses, ranging from soft gels suitable for children to hard pastilles. In addition to food use, the Bloom Gelometer is important in the standardizing gelatins for other industries, such as adhesives and photographic emulsions.

Imitative Tests

Imitative tests attempt to mimic mastication with some kind of machine that chews the food. The machine is instrumentated to provide measurements of stress and/or strain during the test sequence. In the past, a variety of such tests were created. Some actually incorporated human dentures that rocked over each other, imitating the movement of the jaw. While there is some sense in creating a test cell with a geometry similar to the human mouth, gaining data from the machine depends on other factors such as the type and position of the sensors and the relative motion of the jaws. However, some minor modi-

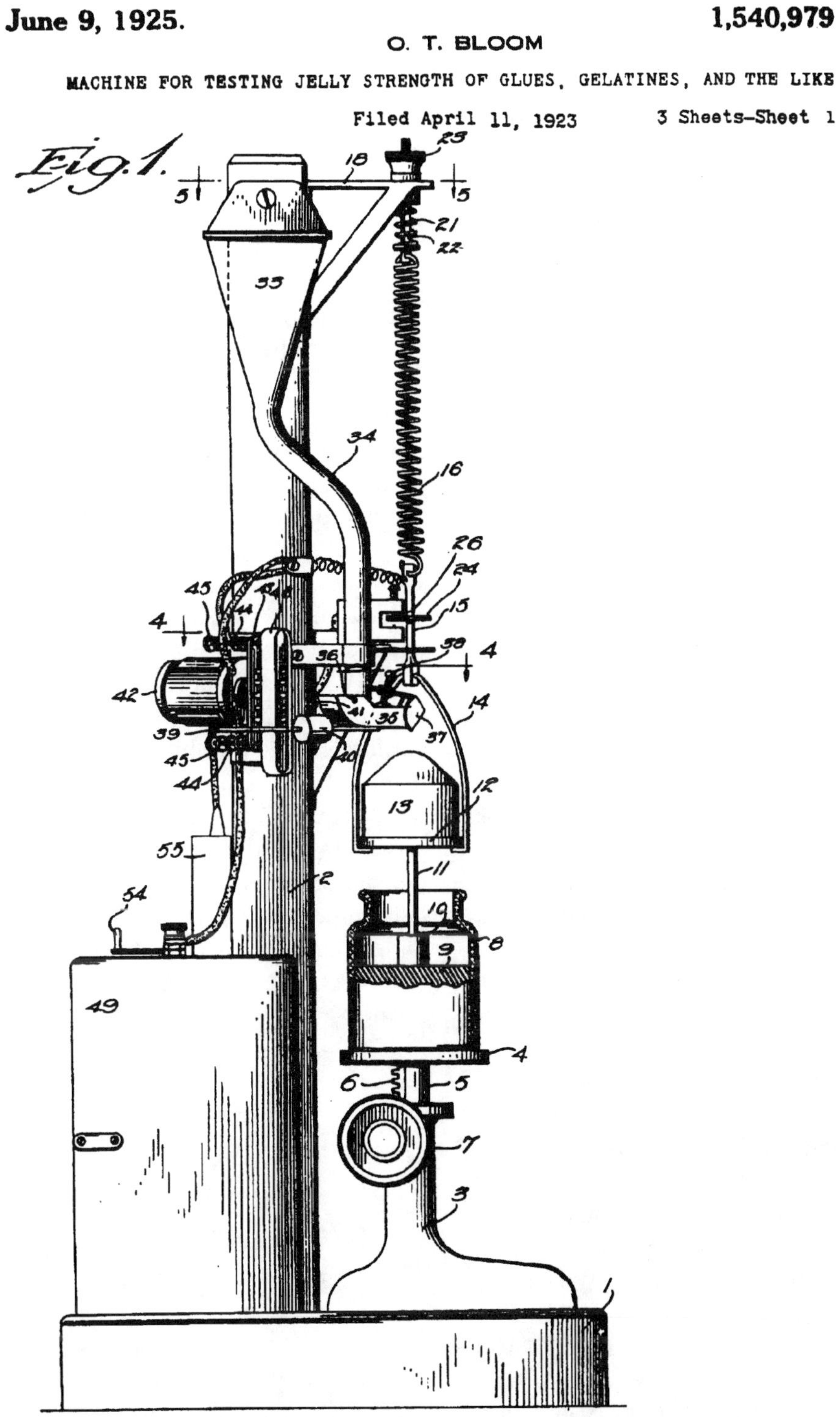

Figure 1–2 The Bloom Gelometer (Based on Original Patent Sketch).

fications, such as replacing the dentures with plungers of known cross-section area so that defined stresses can be applied, allow them to yield valuable data for comparative applications such as quality assurance.

An imitative test that has caught the imagination of many food technologists because it purports to provide standardized values of food texture is *Texture Profile Analysis* (TPA), created at General Foods in the mid-1960s. In a series of papers, Szczesniak and her coworkers defined a variety of textural terms, which are explained in Exhibit 1–1 (Szczesniak, 1963). By using nine trained assessors who were familiar with these terms, various products were first ranked and then rated. For each of the sensory attributes, nine products were selected that were deemed to exist at equal distances from each other and that spanned the sensory continuum. For example, their scale for hardness ranged from cream cheese (with a sensory rating of 1), through hard-boiled egg white, frankfurter, yellow cheese, olives, peanuts, raw carrot, peanut brittle (the candy part), and rock candy (rated as 9). Each standard had defined details of preparation or brand names, as well as details of testing temperature (Szczesniak Brandt, & Friedman, 1963). Standards were therefore established that enabled anyone to consider each of the textural attributes at distinct defined levels.

In addition to these sensory standards, a compressive force-deformation instrument was described (Friedman, Whitney, & Szczesniak, 1963). Based on an instrument that deformed the food via a pivotal motion (resembling the human jaw), the General Foods Texturometer used a flat-ended plunger to contact the food sample. The plunger rocked back and forth at a rate of 108 cm per minute, and the food sample, with a standard height of 0.5 inch (0.0127 m), was deformed by 75%. A two-bite cycle was employed, and the stress that developed in the food sample was measured as the sample was compressed. After this "first bite," the load was removed from the sample and allowed to relax somewhat. As the plunger pulled away from the surface of the sample, any tension due to stickiness was observed. The second bite then compressed the sample again before allowing it to relax for a second time. The resistance during deformation of the food was monitored throughout this two-bite cycle. An idealized stress-strain TPA curve is shown in Figure 1–3. In an attempt to relate the sensorial texture definitions already defined (Exhibit 1–1), a mathematical function was attributed to each term on the basis of the data from the texturometer's stress-strain curve. This was achieved through "careful experimentation and consideration of the important dependent variables" (Friedman et al., 1963, p. 393). Illustrations of how these instrumental measures were determined have been incorporated into Exhibit 1–1. For example, cohesiveness was defined as the ratio of the work required to compress the sample on the second bite to the work required to compress the sample on the first bite. As such, it was determined by comparing the ratio of the area under the two peaks that correspond to the two bites. The proof that the instrument performed in accordance with the sensory panel

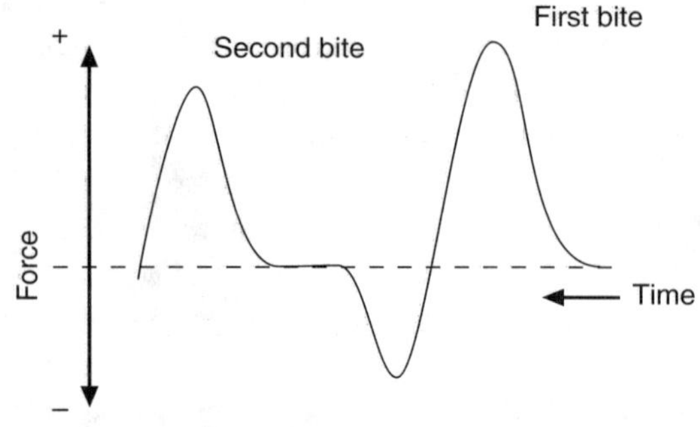

Figure 1–3 Aspects of a Texture Profile Analysis Curve

Exhibit 1–1 Parameters Measured by Texture Profile Analysis

Parameter	*Sensorial Definition*	*Instrumental Definition*
Hardness	Force required to compress a food between the molars.	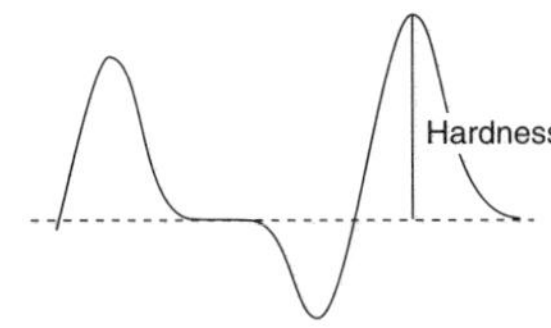
Elasticity	The extent to which a compressed food returns to its original size when the load is removed.	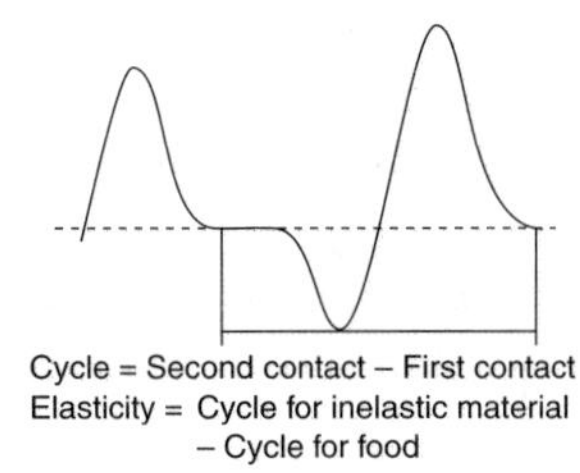
Adhesiveness	The work required to pull the food away from a surface.	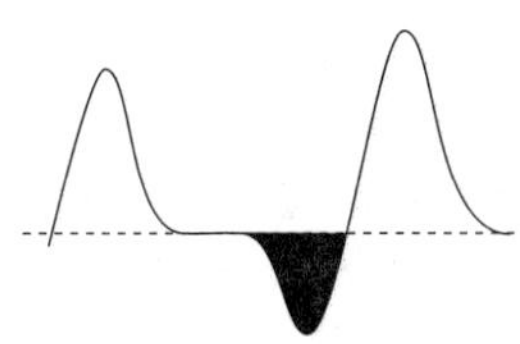
Cohesiveness	The strength of the internal bonds making up the food.	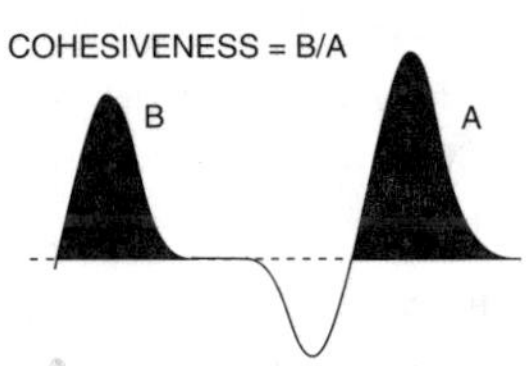
Brittleness	The force at which the material fractures. Brittle foods are never adhesive.	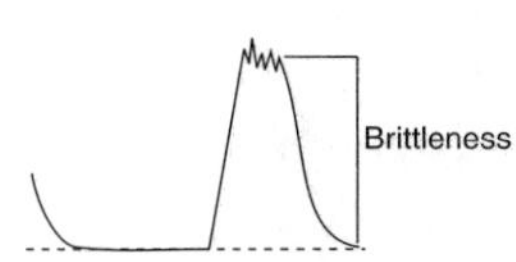
Chewiness	The energy required to chew a solid food until it is ready for swallowing.	= Hardness × Cohesiveness × Elasticity
Gumminess	The energy required to disintegrate a semisolid food so that it is ready for swallowing.	= Hardness × Cohesiveness

was achieved by correlating the sensory rating against the Texturometer output. Best-fit lines or curves were then fitted to the data (Szczesniak et al., 1963).

It is evident that the development of TPA has proved a valuable aid to assessing food texture. However, care should be exercised in accepting the results for purposes other than comparative evaluation. The technique is clearly imitative of what goes on in the mouth. But it should be noted that quite apart from the differences, identified earlier, between instrumental and human testing (e.g., temperature control, saliva), the relationships between some of the sensory characteristics that TPA purports to measure are not linear. For example, between the sensory hardness ratings of 1 and 2 there are about 10 Texturometer units, while between the sensory hardness ratings of 8 and 9 there are about 70 (Szczesniak et al., 1963). Another feature of TPA as measured with the Texturometer is that the instrument has a rocking action that results from the pivotal construction (analogous to the temporomandibular joint). The area of contact between a horizontal-plane sample and the flat-faced plunger will therefore vary, being small at first and then rapidly increasing until it is all in contact with the food.

Despite these shortcomings, the greater availability of other force-deformation compressive testing machines on the market has encouraged transposing TPA to other instruments for comparative tests. For example, Bourne (1966) substituted an Instron Universal Testing Machine and performed TPA to compare pears at different stages of ripeness.

As long as TPA is limited to comparative testing, some of its shortcomings may not matter, but if workers overlook such points and treat the instrumental analogues of these sensory terms as absolute values, the basis becomes risky. Szczesniak and Hall (1975) recognized this potential for abuse and stated that "proper use of the Texturometer is still much of an art since the operator must supply the thinking of which the instrument is not capable" (p. 118).

Szczesniak (1975) used TPA to examine gelatin and carageenan gels. So that some commonality existed between the experimental gels, they were formulated to the same gel strength as measured with a Bloom Gelometer. While the Bloom strengths were the same, differences in TPA hardness and adhesiveness were apparent. An interesting aspect of their study is that TPA measurements were carried out over a range of temperatures. The result of varying the test temperature from 10°C to 20°C was a dramatic drop in the gelatin gel hardness, presumably due to melting. In contrast, the carrageenan gels, which did not melt, showed only a small decrease. Cohesiveness ratings of the carageenan and the gelatin gels appeared almost indistinguishable from each other and hardly changed along the temperature gradient.

Henry, Katz, Pilgrim, and May (1971) carried out TPA on a range of semisolid foods (such as gelatin gels). In addition to the characteristics identified in Exhibit 1–1, they defined five further variables, which related to the material's behavior during the upstroke while the plunger is being pulled away from the sample. In addition to the instrumental testing, they also performed sensory evaluation on the foods. Performing correlation analysis on each of the attributes allowed the identification of four factors that accounted for 95% of the sensory variance. Each of these factors correlated well with specific attributes. For example, 33% of the sensory variance was explained by the factor referred to as "stringy and sticky." Multiple regressions were then performed on the four sensory factors and each of the instrumental measurements. It was found that the sensory factors could be explained by just eight of the instrumental measurements, suggesting that in the case of semisolid foods, much of the data obtained through TPA may well be redundant.

Fundamental Tests

Fundamental tests measure innate physical properties of materials such as their Young's modulus or Poisson's ratio. Such tests are scientifically rigorous, and data are expressed in well-defined scientific units. Moreover, relationships between fundamental properties of materials allow prediction of values for one property based on known values for others (see Chapter 4 of this book).

Coefficient of viscosity is a fundamental property of Newtonian liquids, and its measurement therefore constitutes a fundamental test of food texture. However, with non-Newtonian liquids, comments made earlier in this chapter need to be considered if one is to contemplate drawing parallels with sensations of viscosity in the mouth. In practice, if one is to quote apparent viscosities of non-Newtonian liquids, one needs to define the shear rates that were applied in performing the test. Moreover, exact details of test procedures become crucial if one deals with time-dependent flow behavior. Since thixotropy is a quite common rheological behavior in food systems, some measure of shear history needs to be established prior to instrumental testing for the results to mean anything useful (let alone to provide a fundamental measure of the food's texture).

Saunders and Ward (1953) devised an ingenious and simple method of measuring the elastic modulus of gelatin gels. The Saunders and Ward apparatus (Figure 1–4) consists of a U-tube, one limb of which is approximately 2 cm in diameter and the other limb of which is a capillary. The gel is poured into the wide tube (radius R) and allowed to set in the apparatus. The length of the gel (L) needs to be considerably greater than its radius. The capillary tube (radius a) contains a drop of a low–surface tension, colored liquid, such that any movement of the gel results in the displacement of the drop of liquid.

Once the gel is formed in the U-tube, the entire apparatus can be placed into a temperature-controlled bath, To the open end of the wide tube a known small pressure (P) is applied. Since the gel adheres to the walls of the tube, the stress that results from the applied pressure causes the gel to shear as if it were made up of concentric cylinders: those nearest the center of the tube are displaced the most, whereas those at the tube wall do not move at all. This parabolic deformation is measured by the difference in the height of the dye (h) in the capillary tube. Relating the

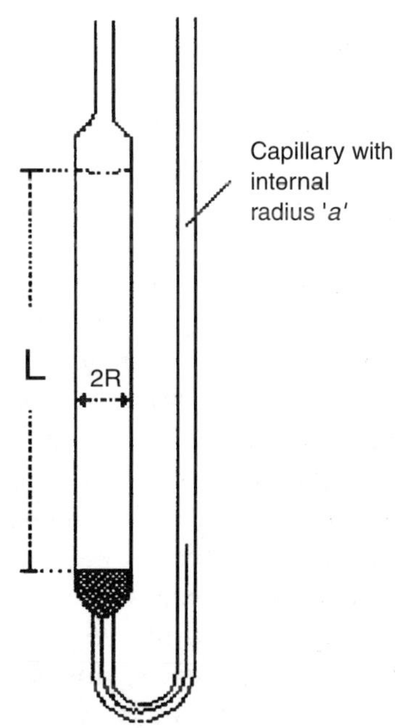

Figure 1–4 Saunders and Ward's Apparatus

applied pressure to the geometry and displacement of the gel allows calculation of the elastic modulus (*G*).

$$G = \frac{PR^4}{8La^2h}$$

Do such fundamental measurements relate to food texture? It has been argued that some aspects of food texture such as food breakdown can be treated essentially as an issue of mechanical failure. Since there are well-established theories that relate mechanical failure of a material when stressed to its fundamental physical properties, it follows that the breakdown behavior of food should be measurable from such characteristics (see, e.g., Chapter 5 of this book). However, while contributing to food texture, mechanical failure represents only a small part of the experience, predominantly during the early stages of mastication.

A knowledge of fundamental physical properties of food can be very valuable for reasons other than establishing food texture—for example, developing packaging material to prevent bruising of fruits.

Physiological Tests

In recent years, considerable interest has been paid to measuring physiological activity during eating. Various techniques have been used to investigate characteristics such as jaw movement, muscle activity (electromyography), and sounds made during mastication (gnathosonics) (Boyar & Kilcast, 1986a). Advances in these areas have been linked to developments in medical technology: for example, early work with electromyography caused considerable discomfort with the use of needle electrodes, but fortunately, adhesive surface electrodes have now taken over for most oral measurements.

Various muscles can be monitored, but the two most commonly used are the masseter and the temporal, both of which are relatively close to the surface of the face and can be felt through the side of the cheek.

Techniques like electromyography have been used to investigate human physiological response to eating gelatin gels. Boyar and Kilcast (1986b) had assessors linked to an electromyograph eat both gelatin and carageenan table jellies. By integrating the electromyograph peak height during chewing, one gains an appreciation of the total muscle activity. In the case of gelatin, the force involved during chewing rose during the first couple of seconds and then began to decline as the gel was dispersed in the saliva, melting and dissolving away. In contrast, the carageenan gels took longer to break up in the mouth.

In another study, electromyography was used to study relatively concentrated gelatin and starch gels formed into various confectionery products such as wine gums. The integrated muscle activity during chewing of a starch:gelatin wine gum mixture (50:50) began to break down rapidly in the mouth. In contrast, the pure gelatin wine gum required an initial rise in integrated muscle activity for about 20 seconds before the gum began to melt and dissolve away. Other gelatin confectionery products were also examined (gelatin chew and whipped gelatin chew), and these were found to behave in the same way (Kilcast & Eves, 1991).

RELATIONSHIP BETWEEN INSTRUMENTAL AND SENSORY MEASURES OF TEXTURE

The advantages and disadvantages of both sensory and instrumental techniques have been discussed, and while these techniques are clearly different, an understanding of the relation between them is necessary to enable correlating one with the other, as is desirable in formulating procedures such as quality control. Many researchers have investigated such relationships, and the approaches taken have been comprehensively reviewed by Kapsalis, Kramer, and Szczesniak (1973); other review articles consider some of the relationships that have been found to exist (e.g., Szczesniak, 1987).

Before considering how sensory and instrumental measures of food texture interact, it is worthwhile to reflect on what is being measured and compared. Bearing in mind that texture can arise from multifarious stimuli and that most instrumental measurements tend to concentrate on one property of the food, we should not assume that there will necessarily be any relationship between an instrumental measurement and the sensory experience.

Some distinction needs to be made between trained and consumer panels of assessors. The former are often composed of individuals who have been selected on the basis of their sensory discrimination and descriptive power. Moreover, they are often trained to develop and fine-tune their power of perception. As such, they may be far more discriminating between subtle nuances of texture than members of the general public. Both trained and consumer panels have a role to play, yet that role is very different. One should not assume that the measurements of texture that come from a trained panel are similar to those perceived by a consumer panel. The consumer may not detect the same detail as the trained assessor, and as such it is probably appropriate that the consumer's opinion be the final arbiter in assessing acceptability. Nevertheless, in the past, some prestigious research institutes have claimed that their trained panel can identify quality characteristics such as acceptability in the same way as the general public.

Concepts of quality such as ripeness frequently involve changes in texture. Identifying correlations between sensory and instrumental tests can prove useful for quality assurance applications.

Texture arises from the mechanical, geometrical, and surface attributes of foods, and since these are themselves dependent on structure and chemical composition, it would seem reasonable to attempt to monitor changes in chemical composition as a measure of food texture and possibly quality. Szczesniak (1973) reviewed some of these indirect measurements of food texture. The power of such chemical techniques can be illustrated by considering changes to banana and plantain during ripening. During ripening, the flesh softens due to the starchy material's being broken down into sweet simple sugars. Alcohol-insoluble solids, a measure of complex carbohydrate, (such as starch, pectin, or cellulose) decrease, making them an excellent index of maturity. Other indirect indices include moisture content of vegetables and free fatty acids in tomatoes.

When considering correlations between sensory and instrumental measurements, it is appropriate to consider the linearity of the psychophysical response. While Weber's law relates to the magnitude of the just noticeable difference, Fechner's law shows that many stimuli bear a log relation to perception. However, the perception of some stimuli does not follow Fechner's law, and other models have been put forward to explain their behavior, such as that of Stevens, which essentially follows a power law, and that of Beidler, which exhibits an S-shaped curve rising to a maximum (Meilgaard, Civille, & Carr, 1991). Fitting the right model to a particular stimulus requires an element of preliminary experimentation since each requires that data be collected in a different way (e.g., magnitude estimation for Stevens' model).

Elejalde and Kokini (1992) examined the psychophysics of in-mouth viscosity perception. Using magnitude estimation, they showed that perception of viscosity followed a power law relationship and correlated well with shear stress ($R^2 = 0.96$) and that shear stress in the mouth is in fact the sensory mechanism for oral evaluation of viscosity, since the slope is very close to unity.

CONCLUSION

Food texture is essentially a human experience arising from our interaction with food—its structure and behavior when handled. To understand our response to food structure and its breakdown involves us in a matrix of disciplines, from aspects of physiology and psychology of perception to physical and chemical studies of the composition and structure of food materials and their behavior when deformed or stressed.

In the past, most of the serious research into food texture was undertaken by rheologists and physical scientists. The sensorial side of the subject was often considered to be subjective and was relegated to psychologists who studied perception. The last 30 years have seen great advances in the techniques and equipment (including computing power) available to both rheologists and sensory scientists. The consequence of these technological advances is a convergence between instrumental physical measurement and techniques of sensory science.

REFERENCES

Bagley, E., & Christianson, D. (1987, March). Measurement and interpretation of rheological properties of foods. *Food Technology, 41*(3), 96–99.

Bloom, O. (1925). Machine for testing jelly strength of glues, gelatins, and the like. U.S. Patent 1540979.

Bourne, M. (1966). Texture profile of ripening pears. *Journal of Food Science, 33*, 223–226.

Bourne, M. (1975). Is rheology enough for food texture measurement? *Journal of Texture Studies, 6*, 259–262.

Boyar, M., & Kilcast, D. (1986a). Food texture and dental science. *Journal of Texture Studies, 17*, 221–252.

Boyar, M., & Kilcast, D. (1986b). Electromyography as a novel method for examining food texture. *Journal of Food Science, 51*(3), 859–860.

Dacremont, C. (1995). Spectral composition of eating sounds generated by crispy, crunchy and crackly foods. *Journal of Texture Studies, 26*(1), 27–43.

Dacremont, C., Colas, B., & Sauvageot, F. (1991). Contribution of air and bone conduction to the creation of sounds perceived during sensory evaluation of foods. *Journal of Texture Studies, 22*(4), 443–456.

Elejalde, C., & Kokini, J. (1992). The psychophysics of pouring, spreading and in-mouth viscosity. *Journal of Texture Studies, 23*, 315–336.

Friedman, H., Whitney, J., and Szczesniak, A. (1963). The Texturometer: A new instrument for objective texture measurement. *Journal of Food Science, 28*, 390–396.

Heath, M. (1991). The basic mechanics of mastication: Man's adaptive success. In J. Vincent & P. Lillford, (Eds.), *Feeding and the texture of food* (pp. 143–166). Cambridge, UK: Cambridge University Press.

Henry, W., Katz, M., Pilgrim, F., and May, A. (1971). Texture of semi-solid foods: Sensory and physical correlates. *Journal of Food Science, 36*, 155–161.

Hutchings, J., & Lillford, P. (1988). The perception of food texture: The philosophy of the breakdown path. *Journal of Texture Studies, 19*, 103–115.

International Organization for Standardization. (1992). *Sensory analysis: Vocabulary* (ISI 5492). Available from the ISO Central Secretariat, Case postale 56, 1211 Geneva 20, Switzerland.

Kapsalis, J., Kramer, A., & Szczesniak, A. (1973). Quantification of objective and sensory texture relations. In A. Kramer & A. Szczesniak (Eds.), *Texture measurement of foods: Psychophysical fundamentals; Sensory, mechanical and chemical procedures and their interrelationships* (pp. 130–160). Boston: D. Reidel.

Kilcast, D., & Eves, A. (1991). Integrating texture and physiology techniques. In J. Vincent & P. Lillford (Eds.), *Feeding and the texture of food* (pp. 167–183). Cambridge, UK: Cambridge University Press.

Kramer, A. (1973). Food texture: Definition, measurement and relation to other food quality attributes. In A. Kramer & A. Szczesniak (Eds.), *Texture measurement of foods: Psychophysical fundamentals. Sensory, mechanical, and chemical procedures and their interrelationships* (pp. 1–9). Boston: D. Reidel.

Meilgaard, M., Civille, B., & Carr, B. (1991). Sensory evaluation techniques. Boston: CRC.

The Oxford English Dictionary. (1989). Oxford, UK: Clarendon.

Saunders, P., & Ward, A. (1953). An absolute method for the rigidity modulus of gelatin gels. In V.G.W. Harrison (Ed.), *Proceedings of the Second International Congress on Rheology* (pp. 284–290). Oxford, UK: Butterworths.

Schenkels, L., Veerman, E.C.I., & Amerongen, A.V.N. (1995). Biochemical composition of human saliva in relation to other mucosal fluids. *Critical Reviews in Oral Biology and Medicine, 6*, 161–175.

Schlichter-Aronhime, J., & Garti, N. (1988). Solidification and polymorphism in cocoa butter and the blooming problems. In N. Garti & K. Sato (Eds.), *Crystallization and polymorphism of fats and fatty acids* (pp. 363–393). New York: Marcel Dekker.

Scott-Blair, G. (1958). Rheology in food research. *Advances in Food Research, 8*, 1–56.

Shama, F., Parkinson, C., & Sherman, P. (1973). Identification stimuli controlling the sensory evaluation of viscosity: I. Non oral methods. *Journal of Texture Studies, 4*, 102–110.

Shama, F., & Sherman, P. (1973). Identification stimuli controlling the sensory evaluation of viscosity: II. Oral methods. *Journal of Texture Studies, 4*, 111–118.

Sherman, P. (1969). A texture profile of foodstuffs based upon well-defined rheological properties. *Journal of Food Science, 34*, 458–462.

Szczesniak, A. (1963). Classification of textural characteristics. *Journal of Food Science, 28*, 385–389.

Szczesniak, A. (1973). Indirect methods of objective texture measurements. In A. Kramer and A. Szczesniak, (Eds.), *Texture measurement of foods: Psychophysical fundamentals. Sensory, mechanical and chemical procedures and their interrelationships* (pp. 109–117). Boston: D. Reidel.

Szczesniak, A. (1975). Textural characterization of temperature sensitive foods. *Journal of Texture Studies, 6*, 139–156.

Szczesniak, A. (1987). Correlating sensory with instrumental texture measurements: An overview of recent developments. *Journal of Texture Studies, 18*, 1–15.

Szczesniak, A. (1990). Psychorheology and texture as factors controlling the consumer acceptance of food. *Cereal Foods World, 35*, 1201–1204.

Szczesniak, A., Brandt, M., & Friedman, H. (1963). Development of standard rating scales for mechanical parameters of texture and correlation between the objective and the sensory methods of texture evaluation. *Journal of Food Science, 28*, 397–403.

Szczesniak, A., & Hall, B. (1975). Application of the General Foods Texturometer to specific food products. *Journal of Texture Studies, 6*, 117–138.

Takahashi, J., & Nakazawa, F. (1991). Effects of viscosity of liquid foods on palatal pressure. *Journal of Texture Studies, 22*, 13–24.

Tornberg, E., Fjelkner-Modig, S., Ruderus, H., Glantz, P., Randow, K., & Stafford, D. (1985). Clinically recorded masticatory patterns as related to the sensory evaluation of meat and meat products. *Journal of Food Science, 50*, 1059–1066.

Vickers, Z. (1988). Instrumental measures of crispness and their correlation with sensory assessment. *Journal of Texture Studies, 19*, 1–14.

CHAPTER 2

Oral Processing of Foods and the Sensory Evaluation of Texture

M. Robin Heath and Jon F. Prinz

In this chapter, we discuss the sensory mechanisms by which texture is evaluated and describe the modulation of mastication in response to the changing texture of food. The perception of texture involves the interaction of food with the teeth, mucosa, and saliva. It varies with each chew as particle size is reduced by comminution and as the bolus is ensalivated. When two or more materials are taken into the mouth, they interact so that the properties of the mixture may differ markedly from the individual foods if eaten in isolation, as with boiled potatoes and gravy. Any attempt at describing the texture of food should therefore take its dynamic nature into account.

The mouth is richly innervated, and sensations that arise during chewing provide much of the pleasure that contributes to appetite. Animals have evolved sensory mechanisms to evaluate palatability by taste and texture in response to the need to evaluate the suitability of potential materials as food. Texture is especially important to herbivores, providing an assay of nutritional content (Hill & Lucas, 1996). Toughness has been shown to be an important cue in food selection: monkeys show a marked preference for tender young (protein-rich) leaves over tougher (high-fiber) mature leaves. Their decision on whether to eat the leaves appears to be made after rubbing the leaves between the fingers and taking a small trial bite; tough old leaves are rejected while less tough, more nutritious leaves are eaten (Choong et al., 1992).

The evaluation of the palatability of food begins even before it is placed in the mouth. If it can be said that "the first bite is made by the eyes," the second is made by the hands as the food is handled. For foods from which a portion must be bitten off (e.g., an apple), the force required to bite off the piece of food provides an assessment of its mechanical properties. Some of this information is derived from periodontal receptors in the incisors (Sim, Lucas, Pereira, & Oates, 1993) but some is also derived from proprioception in the arm.

The most sensitive oral sensors are in the front of the mouth, allowing the food to be rejected easily. But as anyone who has had to give a tablet to a dog will know, once the food item reaches the back of the mouth it is almost invariably swallowed.

In addition to the nutritional assay of food, an assessment of the food's potential to abrade the teeth is important. For many animals, life span is determined by tooth wear, so it is of vital importance that an animal detect and if possible avoid abrasive food items (Utz, 1983, 1986). The wear of teeth is becoming an increasing dental problem as more people keep their natural teeth throughout life.

Agriculture, fire, pottery, and stone tools are relatively recent innovations that have dramatically modified the human diet, and the human masticatory apparatus has, of course, not yet fully adapted to these changes. This must be borne in mind when attempting to interpret human data on chewing. Strategies that evolved to cope with a natural diet of unprocessed plant and animal materials may not be necessary for the softer modern diet of cooked, processed convenience foods. The physiological reserve is generous for processed foods for all but the most elderly edentate people.

MASTICATION

Mastication, the first stage in the digestion of food, is a complex process whereby food taken into the mouth is processed into a form suitable for swallowing. This involves the breakdown of the food into smaller particles, the incorporation of saliva, the agglomeration and shaping of the resultant mixture into a cohesive bolus, and finally the transport of the bolus to the pharynx to prepare for swallowing. Even foods that can be swallowed whole are generally chewed to some extent. For many processed foods, chewing is physiologically unnecessary, but the drive to chew remains. Probably a number of factors give a masticatory drive. Chewing releases flavor and smell, increasing the pleasure to be derived from the food, but no single factor such as particle size, viscosity, or the persistence of taste gives simple correlations with the number of chews made prior to swallowing.

Masticating is effected by voluntary muscles that are normally unconsciously controlled by a wide range of sensory stimuli from the mouth (Takada, Miyawaki, & Tatsuda, 1994), although conscious interference with the reflex patterns is possible. The cerebral cortex has been mapped, and the relative importance of the sensory receptors of the facial and oral area is clear from the proportion of the sensory cortex devoted to it (Penfield & Rasmussen, 1957). However, much of the coordination of mastication by the central nervous system is subcortical, involving the basal ganglia and the brainstem (Lund & Olsen, 1983). The effectiveness of this subconscious control is indicated by those occasions when the unexpected stimuli trigger conscious attention—a mechanism used by manufacturers of soft foods containing hard chips such as chocolate chip ice cream.

MODULATION OF MASTICATION

Mammalian mastication is a cyclical motor activity mediated by a central pattern generator (Lund & Enomoto, 1988). This is modulated by sensory feedback from receptors in the masticatory apparatus that monitor the location and state of the food. The inputs to this control system are the sensations arising in the oral mucosa, periodontium, teeth, muscles, and receptors in the temporomandibular joint (TMJ) (Figure 2–1). The outputs are the control of bite force, mandibular trajectory, soft tissue behavior, and initiation of swallowing. By comparing how different foods are chewed, we can gain an insight into the sensory mechanisms by which the food is assessed.

The masticatory sequence can be divided into three phases: (a) ingestion—transfer of food to between the teeth by the tongue, (b) main sequence—rhythmic chewing in which the food is comminuted and the bolus formed, and (c) clearance and swallowing (Figure 2–2). Texture is evaluated during all three stages. Viewed from the front, each cycle resembles a teardrop, but no two cycles are identical (Figure 2–3).

It is worthwhile to divide the physical properties of food into two broad classifications, intrinsic and extrinsic (Figure 2–4). The intrinsic properties are those such as hardness, strength, and elasticity, which are sensed mainly in terms of force-deformation behavior by the muscles of mastication during the closing stroke. The extrinsic properties are those determined by the surface characteristics, such as the particle size distribution, surface profile, and frictional coefficient, which are sensed by the tongue and mucosa mainly during the opening stroke (Chew & Lucas, 1988). The sensory homunculus illustrates the relative importance of the mouth and its sensory apparatus; although the system works together as a whole, it is possible to ascribe some aspects of sensation to individual components of the system.

Sensations Arising in the Mouth from the Mucosa

The oral mucosa are liberally supplied with sensory nerve endings; these lie beneath the epithelium. The lips are particularly sensitive to temperature, while the ability to make the finest two-point discrimination (1–2 mm) is found on the upper surface of the tip of the tongue (Ringel & Ewanowski, 1965). The superficial lingual sensory nerves are very fast-adapting. Light mechanical stimulation produces a short discharge of pulses but none thereafter despite continued stimulation (Figure 2–5). It is easy to argue that this rapid adaptation is essential for sensory perception of lingual contact during chewing and speech.

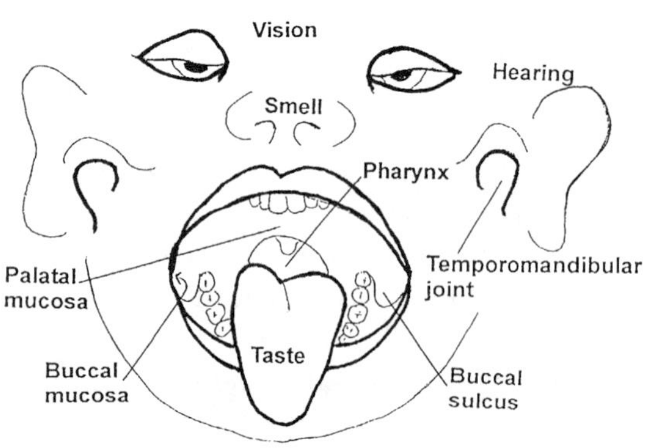

Figure 2–1 Sources of Sensation. Sensation from the eyes, nose, ears, lips, teeth, mucosa, temporomandibular joints, and muscles of mastication all contribute to the perception of texture.

The vermillion border or the outer aspects of the lips are the first part of the stomatognathic system to come into contact with the food. It is the lips that make the initial assessment of the surface roughness of the food and determine the food's temperature.

Although food does not normally enter the labial sulcus, the mucosa here is remarkably sensitive. It has been suggested that the mucosa here has a role in the detection of tannins (Prinz & Lucas, 1996). Although usually described as tasteless, tannins have a profound astringent effect (Green, 1993). It has been shown that tannins increase intraoral friction (Prinz & Lucas, 1996). The pellicle that forms on the teeth would be affected immediately after the first bite; hence, monitoring the friction between the mucosa and the pellicle provides a rapid detection system for tannins.

Food placed on the tip of the tongue is normally transferred to the cheek teeth. En route, some foods are pressed up against the palate, a process that provides additional information on the physical properties of the food. These slow manipulations—"tongue chews" —typically occur with foods that are expected to be soft or for which the texture is unknown (banana, ice cream, and frozen yogurt) (Hiiemae, Thexton, & Crompton, 1978; Prinz & Lucas, 1995).

Parts of large mouthfuls of some foods are "pouched"—that is, they are held buccal to the teeth, awaiting further processing. Stretch receptors in the cheek allow an assessment of

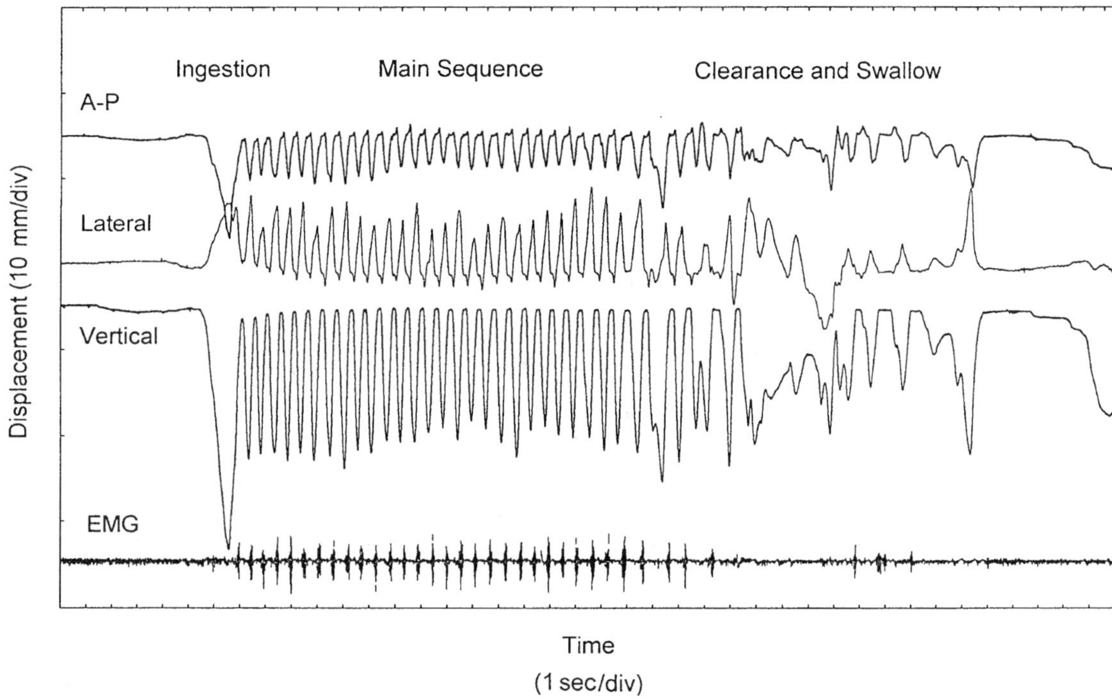

Figure 2–2 Typical Traces of a Full Sequence from Ingestion through the Rhythmic Movements of Main-Sequence Mastication to Swallowing and Clearance, for a Subject Chewing Meat. Mandibular trajectories in anteroposterior, lateral, and vertical planes, together with the electrical activity of the muscles of mastication (EMG), are shown.

the compliance of the pouched material to be made while touch-sensitive receptors relay information on the surface profile and frictional coefficient. Typically, very little food is allowed forward into the labial sulcus, material being held back by the action of modiolar muscles at the angle of the mouth.

Sensations Arising in the Mouth from the Teeth

Perception of the extrinsic and intrinsic characteristics of different foods probably contributes to the subconscious control of the different teeth used to chew through the sequence of one mouthful. The periodontal membrane of the teeth is a very sensitive load detection system (Mioche & Peyron, 1995). Particles as small as 15 μm can be detected between the teeth (Utz, 1986). The widely differing morphologies of the teeth suggest that their functions are different. The position of the bolus during chewing (and hence the teeth used) appears to be influenced by the mechanical properties of the food. With chewing gums, chewing commences on the premolars, but as the wad is softened, the bolus is moved backwards onto the molar teeth.

Small items of food (seeds, etc.) are typically dealt with at the front of the mouth. Not only does the mobility of the tip of the tongue provide fine position control, but it is also the most sensitive area to touch. Oral discrimination of shaped objects, such as stars, squares, and "D" shapes, termed *oral stereognosis*, is best when subjects use the tip of the tongue

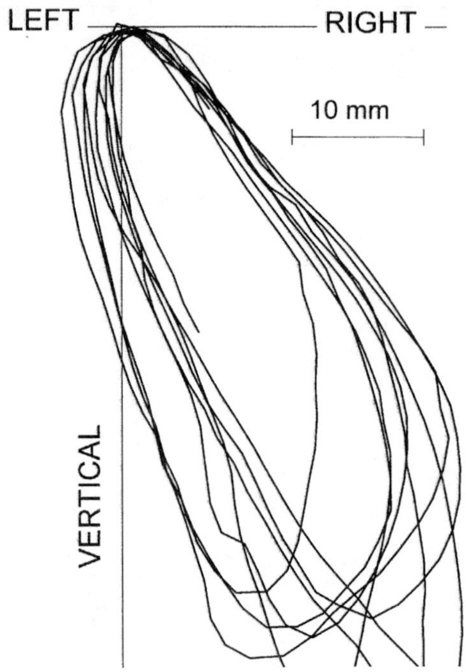

Figure 2–3 Incisal Movement Viewed from in Front, Illustrating Jaw Movement during Five Seconds of Main-Sequence Chewing

and are allowed to feel the test pieces with their incisors (Müller, Heath, Kazazoglu, & Hector, 1993; Öwall & Vorwerk, 1974).

The incisors are typically used to bite off portions of food from a larger piece. This action has been described as "sculpting." When biting objects such as apples, the hand brings the food up to the teeth, and the head is simultaneously brought down, impacting the teeth into the food. The lower jaw is then elevated. The single roots of the anterior teeth are morphologically more favorable for sensitivity to lateral loads than the multirooted postcanine teeth. The intrinsic texture of the food can thus be monitored and subconsciously checked against that expected from extrinsic factors, particularly appearance or quality of known branded foods.

With tougher foods, the canines and premolars are brought into play because these teeth are capable of taking greater lateral loads by virtue of their larger roots. The positioning of the food between the teeth is at least partially dependent on how "hard" the food is: hard foods are placed more anteriorly

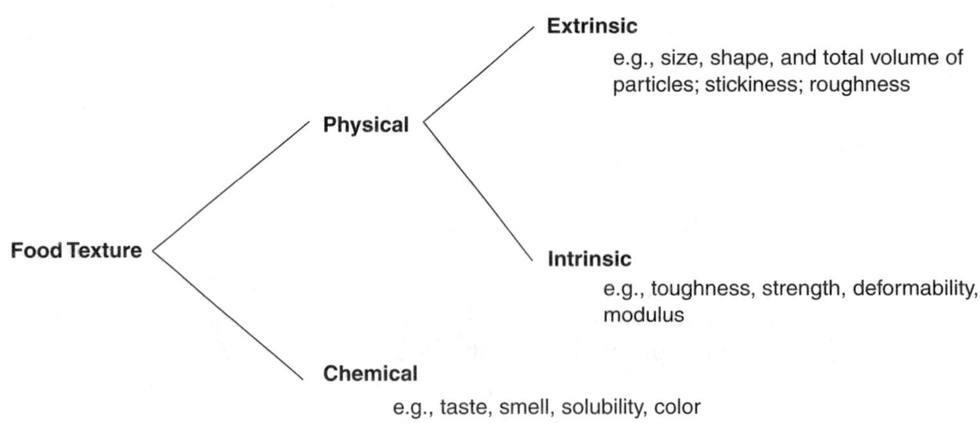

Figure 2–4 Categorization of Perceived Food Characteristics

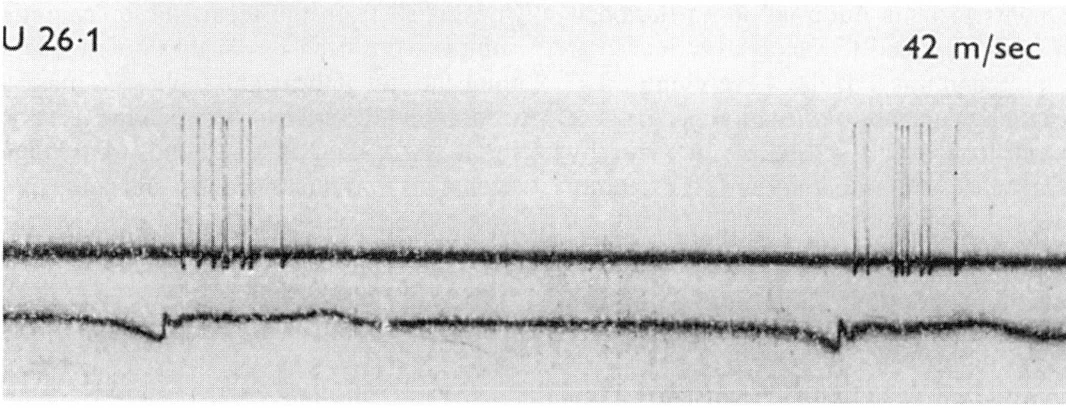

Figure 2–5 Fast Adaptation in Lingual Sensory Nerves. Single sensory nerves in the tongue show a short burst of discharges (upper trace) with an extended light touch (lower trace). The nerve can then respond again to a new stimulus in less than 20 ms of the end of the first stimulus. *Source:* Reprinted with permission from R. Porter, Lingual Mechanoreceptors Actuated by Muscle Twitch, *Journal of Physiology*, Vol. 183, pp. 101–111, © 1966.

and are moved backwards as they fragment, imbibe saliva, and soften. This is because the premolars have a smaller surface area and can exert greater pressures than can the wider molars. This is in spite of the mechanical disadvantage that the premolars have over the molars, which are closer to the joint. The premolars are also particularly favorably placed for foods that require both high force and fine control. It has long been recognized that the food can be well controlled between the tongue and the modiolar muscles (Fish, 1933).

Most herbivores have large molars. These teeth are presumably optimized for the comminution of leaves and seeds. A large surface area also provides efficient softening of viscoelastic foods. Human molars are relatively flat teeth, which with a "natural" diet become flatter as they abrade (Smith, 1984). By virtue of their multiple roots, the molars are able to withstand greater loads than the anterior teeth.

Sensations Arising in the Mouth from the Muscles

Two major groups of muscles are involved in mastication and a third in swallowing. The muscles of mastication are grouped into the opening and closing muscles, but the muscles of facial expression and the tongue also have a significant role in mastication. Sensory nerve endings in both the body of the muscle and the tendons provide proprioceptive feedback on muscle length and tension. This ability to monitor both load and displacement allows physical properties such as the Young's modulus, compliance, and strength to be assessed.

Sensations Arising in the Mouth from the Temporomandibular Joint

Anesthesia of the TMJ has small but measurable effects on chewing patterns (Öwall, 1978). Receptors in the joint's capsule and ligaments provide information on the position of the jaw and the loads experienced at the joint, and this aids the perception of texture. Redundancy of information occurs here as well; load is also sensed by the teeth, by the muscles of mastication and, for denture wearers, by the mucosa.

Sensations Arising in the Mouth from the Ears

The role of the auditory system is often underestimated (Drake, 1963; Vickers, 1980). Sound levels during chewing can be surprisingly high when chewing crisp foods (over 90 dB(A)), but even soft foods such as chewing gum produce detectable sounds (Watt, 1976). Sharp cracking sounds, such as those produced by brittle foods, are sufficient to activate reflex muscle inhibition when replayed via headphones (Basmajian & De Lucia, 1985). This inhibition can also be elicited by a sudden release of tension, as when food fractures, but inhibition occurs only in very tense muscles (Müller et al., 1993). This reflex is commonly thought to prevent damage to the teeth by switching off the muscles before they come into contact. At its most extreme, the antagonistic opening muscles become active (Karkazis, Kossioni, Heath, & van Willigan, 1993).

Hearing is also involved in the detection of abrasives in food, allowing particles as small as 12 μm to be detected, more than an order of magnitude smaller than the mucosal detection threshold. Masking with white noise interferes with the assessment of particle size. Using white noise to mask the dental sounds, Laine and Siirila (1977) showed that the particle size detection threshold increased by a factor of three, indicating just how well sounds are conducted from the teeth to the ears (Prinz, 1997).

Sounds are also produced by soft, malleable foods such as chewing gum as the food "unsticks" from the teeth; these separation sounds were used by Watt (1976) to determine the end of the occlusal phase of mastication. Stick-slip friction between the teeth and the waxy coating of some foods (e.g., runner beans) also produces sounds, and some processed foods have been deliberately engineered to produce sounds, fizzing and popping as they are wetted.

ORAL PROCESSING

Ingestion and the First Bite

Probably the most important phase in the assessment of texture is incision and the first manipulations. Agrawal, Lucas, Prinz, and Bruce (1997) tested a range of 15 foods varying in texture from soft processed cheese to nuts. They were able to show a high correlation of masticatory muscle activity with a mechanical index E/ξ (where E is Young's modulus and ξ is toughness). This index also correlates well with the rate of production of new surface area. In many subjects, this correlation was present even during the first bite, but it was present in all subjects in subsequent cycles.

Main-Sequence Mastication

Once the food has been placed between the cheek teeth, the jaws close and the main chewing sequence starts. The most obvious change to hard foods effected by the teeth is particle size reduction as the food is fractured. The rate of particle size reduction and the resulting particle size distributions are of obvious importance. A simple count of the number of main-sequence chewing strokes prior to swallowing provides a measure of food quality (Tornberg et al., 1985); this method has been suggested as a method for assessing marbling in pork (Harrington, 1962).

Clearance and Swallowing

Clearance is performed by the tongue prior to swallowing. During the clearance phase, the tongue attempts to collect any particles that were not incorporated into the bolus. Clearance is rarely completely successful, and many food particles are left adhering to the tongue, oral mucosa, and teeth; this debris may remain in the mouth for many minutes before it is finally removed.

Swallowing is arbitrarily divided into three stages based on the position of the food bolus (Jenkins, 1977). Stage I is in the mouth and is initiated normally by a subconscious reflex. Stage II is in the pharynx and lasts only about 1 s. Stage III is in the esophagus and ends only at the entrance to the stomach. The initiation of swallowing is supposed to depend only on conditions surrounding Stage I, but information based on experience concerning Stages II and III may well influence this. For example, it is known from fluoroscopic studies of bariumized tablets that certain materials get stuck in the pharynx and esophagus more readily than do others (Hey, Jorgensen, Sorennsen, Hasselbach, & Wamberg, 1982). The peristaltic muscular waves of the pharynx/esophagus may drive softer particles down to the stomach more rapidly than harder particles. Particle size, wetness, and/or plasticity as perceived orally may influence the initiation of swallowing.

Hutchings and Lillford (1988) described two thresholds that need to be satisfied before swallowing is initiated: a food particle size threshold and a lubrication threshold. Although these thresholds can be quantified (Prinz & Lucas, 1995), this "two-threshold" concept appears to have deflected interest from another function of saliva, which is that it helps food particles to adhere together to form a coherent mass—the bolus. This is important in mammals because when food is swallowed it must be coherent enough to pass the airway without loose particles being inhaled. Despite anatomical arrangements unique to mammals that encourage a specified food path, there is a risk of potentially fatal accidents if food is inhaled. This problem is solved to some extent by a very rapid swallowing action, and by formation of a cohesive food mass, the risk of inhaling food is reduced.

Debris

After the major part of a chewed mouthful has been swallowed, some remnants are left, but how much is left after a first swallow varies depending on the foods. Briefly, the forces that hold particles together to form a bolus are due to Stephan adhesion (Bowden & Tabor, 1950), while those attaching particles to the mucosa are due to surface tension (Lucas & Prinz, 1996; Prinz & Lucas, 1997). If the forces that tend to hold particles to the mucosa and teeth exceed those attracting the particles to each other, they will be retained in the mouth, forming debris. The sauces and gravies that are frequent accompaniments to foods increase salivary viscosity; this increases the interparticle adhesion, which results in greater cohesion.

Typically juicy foods such as apples involve subtle intercalated swallows. Some others, including almonds, usually involve more than one clearly defined swallow. Control of these must involve sensory monitoring of the texture of the food, but not consciously because these are part of reflex actions. The sensation that particles are left may provoke clearance movements by the tongue and cheeks and may even get through to consciousness if this clearance is not effective. The extent of conscious or unconscious attention is intriguing. If asked, individuals can detect particles even after the last major swallow (Figure 2–6) but subsequently may have some particles remaining even when they can no longer detect them (Figure 2–7). The sensory detection is particularly dependent on movement of the tongue. There is considerable variation between individuals, even for residual particles with sharp margins. These differences between people's drive for effective clearance are intriguing and probably contribute to the variation in oral health.

However, contrary to intuition, it is the so-called "healthy" foods that tend to be left behind in the mouth. These are foods such as carrot, lettuce, and spinach, which break down slowly, leaving large particles. Other foods, such as fibrous meat and some fruits and vegetables, may become wedged between teeth and are perceived by the periodontal afferent. For denture wearers, some textural sensations are unpleasant. In edentate people, debris collects under the fitting surfaces of dentures (Figure 2–8). All lower complete dentures move during chewing, and it is inevitable that the food debris gets under the surface of the denture. Soft foods are probably irrelevant to comfort, but particles of nut or tomato seeds are undoubtedly very uncomfortable. This may contribute to the exclusion of some foods from the diet of denture wearers (Heath, 1972). Thus, the evaluation of the texture continues long after swallowing for those foods that do not form a cohesive bolus or are incompletely cleared.

CONCLUSION

People have evolved a generously innervated system with which to assess food before committing the mouthful to chewing, much less swallowing. The mouth is also remarkably well adapted for the control and comminution of a wide range of food types. The result is a redundancy of information from the mouth for chewing the cooked and processed foods that make up modern diets; if one sensory input is blocked, another can take over, with the result that chewing patterns are little affected. The extent of this is illustrated by the success of denture wearers, who have lost their periodontal sensation but continue to chew well while simultaneously controlling their dentures.

However, the net sensations from the teeth, mucosa, and mandibular joint all contribute to the monitoring of food quality from first contact on the lips until after swallowing. Subconscious monitoring probably also accounts for the prompt consciousness of any discrepancy between the qualities expected by eye and hand-feel and those perceived intraorally.

The food industry aims to produce foods that are both pleasurable and not too difficult to eat. Pleasure is derived from taste, smell, temperature, and vision. However, the perception of texture also gives pleasure and is the sensation that is the most relevant to a consideration of food structures.

A more detailed understanding of oral function may serve a wide variety of professional interests. This may be applicable in quality control in the food industry, in the development of foods that are both easy to eat and palatable, in the design of dental prostheses, and in the diagnosis and treatment of dysphagia.

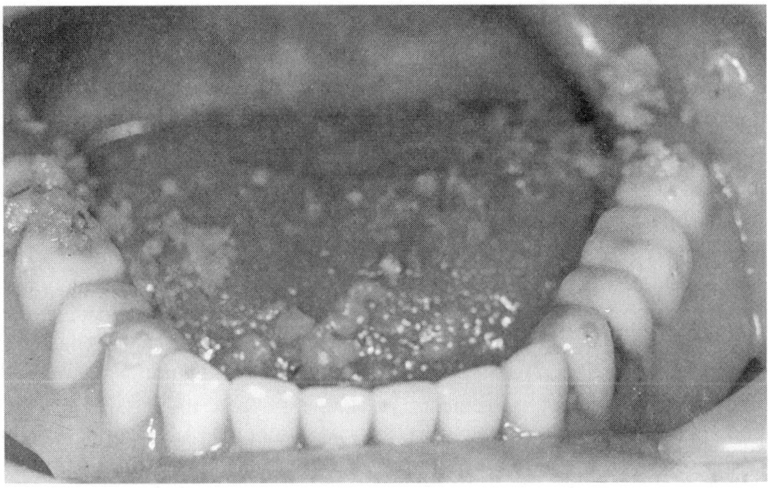

Figure 2–6 Debris in the Mouth of a Denture Wearer after Chewing 5 g of Peanuts. This photograph was taken immediately after the final swallow. Note the extensive debris attached to the mucosa and over the denture. The patient was aware of this debris, which took a considerable time to clear.

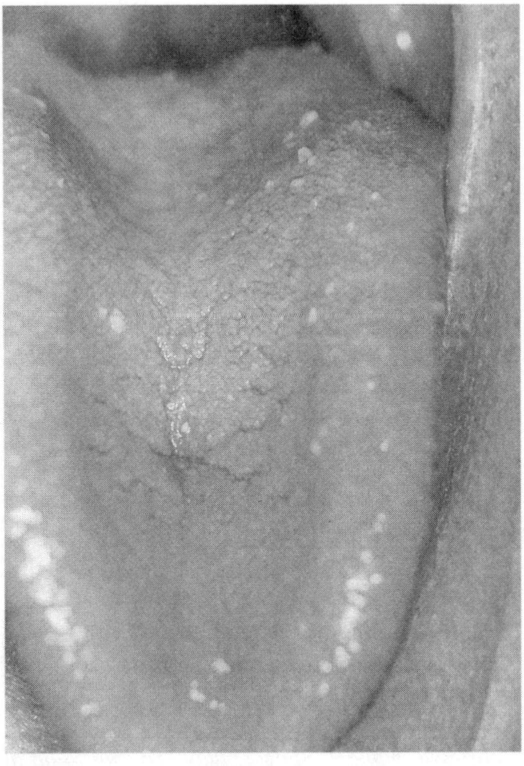

Figure 2–7 The Tongue of the Same Person Pictured in Figure 2–6 after He Cleaned All the Perceptible Particles from His Mouth. Numerous small particles remain.

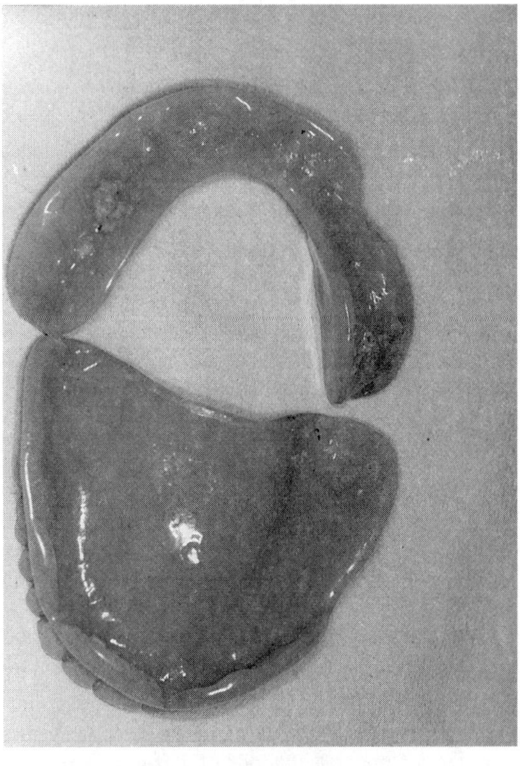

Figure 2–8 These Dentures Show Particles on the Surfaces that Fit the Mucosa after the Person Chewed Peanuts. The discomfort caused by the many particles under the lower denture caused the individual to exclude peanuts from his normal diet.

REFERENCES

Agrawal, K.R., Lucas, P.W., Prinz, J.F., & Bruce, I.C. (1997). Mechanical properties of food responsible for resisting their breakdown in the human mouth. *Archives of Oral Biology, 42*(1), 1–9.

Basmajian, J.V., & De Lucia, C.J. (1985). *Muscles alive*. Baltimore: Williams & Wilkins.

Bowden, F.P., & Tabor, D. (1950). *Friction and the lubrication of solids: Part 1*. London: Oxford University Press.

Chew, C.L., & Lucas, P.W. (1988). Food texture as a factor in the control of masticatory movements. *Journal of Dentition, 16,* 210–214.

Choong, M.F., Lucas, P.W., Ong, J.Y.S., Pereira, B.P., Tan, H.T.W., & Turner, I.M. (1992). Leaf fracture toughness and sclerophylly: Their correlations and ecological implications. *New Phytology, 121*, 597–610.

Drake, B.K. (1963). Food crushing sounds: Comparison of objective and subjective data. *Journal of Food Science, 28,* 233–241.

Fish, E.W. (1933). *Principles of full denture prosthetics* (1st ed.). London: John Bale.

Green, B.G. (1993). Oral astringency: A tactile component of flavor. *Acta Psychologica Amsterdam, 84*(15), 119–125.

Harrington, G. (1962). Chew count as a measure of tenderness of pork with different degrees of marbling. *Journal of Food Science, 27*, 106–110.

Heath, M.R. (1972). Dietary selection by elderly persons related to dental state. *British Dental Journal, 132*, 145–150.

Hey, H., Jorgensen, F., Sorennsen, K., Hasselbach, H., & Wamberg, T. (1982). Oseophageal transit of six commonly used tablets and capsules. *British Medical Journal, 285*, 1717–1719.

Hiiemae, K.M., Thexton, A.J., & Crompton, A.W. (1978). Intra-oral food transport: The fundamental mechanism of feeding. In D.S. Carsol & J.S. McNamara (Eds.), *Muscle adaptation in the craniofacial region* (pp. 181–208). Ann Arbor: University of Michigan Press.

Hill, D.A., & Lucas, P.W. (1996). Toughness and fibre content of major leaf foods of Japanese macaques (*macaca fuscata yakui*) in Yakushima. *American Journal of Primatology, 38,* 221–231.

Hutchings, J.B., & Lillford, P.J. (1988). The perception of food texture: The philosophy of the breakdown path. *Journal of Texture Studies, 19,* 103–115.

Jenkins, G.N. (1977). *The physiology and biochemistry of the mouth*. Oxford, UK: Blackwell.

Karkazis, H.C., Kossioni, A., Heath M.R., & van Willigan, D.J. (1993). Anterior digastric muscle responses to sudden unloading of the mandibular elevator muscles in younger and older adults. *Journal of Oral Rehabilitation, 20,* 433–440.

Laine, P., & Siirila, H.S. (1977). The effect of auditory sense in interocclusal microdiscrimination and size discrimination of persons with natural dentitions and full dentures. *Proceedings of the Finnish Dental Society, 73*(1), 27–31.

Lucas, P.W., & Prinz, J.F. (1996). Mechanisms of bolus formation: Their relevance to chewing and the swallow threshold. *American Journal of Physical Anthropology, 22*(1), 153.

Lund, J.P., & Enomoto, S. (1988). The generation of mastication by the mammalian central nervous system. In A. Cohen, S. Rossignol, & S. Grillner (Eds.), *Neural control of rhythmic movements in vertebrates* (pp. 201–283). New York: John Wiley.

Lund, J.P., & Olsen, K. (1983). The importance of reflexes and their control during jaw movement. *Trends in Neuroscience, 6*, 459–463.

Mioche, L., & Peyron, M.A. (1995). Bite force displayed during assessment of hardness in various texture contexts. *Archives of Oral Biology, 40*, 415–423.

Müller, F., Heath, M.R., Kazazoglu, E., & Hector, M.P. (1993). Contribution of periodontal receptors and food quality to masseter muscle inhibition in man. *Journal of Oral Rehabilitation, 20*, 281–290.

Öwall, B. (1978). Interocclusal perception with anaesthetized and un-anaesthetized temporomandibular joints. *Swedish Dental Journal, 2*, 199–208.

Öwall, B., & Vorwerk, P. (1974). Analysis of a method for testing oral tactility during chewing. *Odontologiska Revy, 25*(1), 1–25.

Penfield, W., & Rasmussen, T. (1957). *The cerebral cortex of man: A clinical study of localization of function*. New York: Macmillan.

Prinz, J.F. (1997). Vibration transmission across the

temporomandibular joint. *Journal of Oral Rehabilitation, 24*, 224–229.

Prinz, J.F, & Lucas, P.W. (1995). Swallow thresholds in human mastication. *Archives of Oral Biology, 40*, 401–403.

Prinz, J.F, & Lucas, P.W. (1996). Salivary lubrication: The effect of tannins. *American Journal of Physical Anthropology, 22*, 190–191.

Prinz, J.F, & Lucas, P.W. (1997). Masticating and swallowing: An optimization model. *Proceedings of the Royal Society of London* B, *264,* 1715–1721.

Ringel, R.L., & Ewanowski, S.J. (1965). Oral perception: I. Two point discrimination. *Journal of Speech and Hearing Research, 8*, 389–397.

Sim, B.J., Lucas, P.W., Pereira B.P., & Oates, C.G. (1993). Mechanical and sensory assessment of the texture of refrigerator stored spring roll pastry. *Journal of Texture Studies, 24*(1), 27–44.

Smith, B.H. (1984). Patterns of molar wear in hunter-gatherers and agriculturists. *American Journal of Physical Anthropology, 63*, 39–56.

Takada, K., Miyawaki, S., & Tatsuda, M. (1994). The effects of food consistency on jaw movement and posterior temporalis and inferior orbicularis oris in children. *Archives of Oral Biology, 39,* 793–805.

Tornberg, E., Fjelkner-Modig, S., Ruderus, H., Glantz, P.O., Randow, K., & Stafford, G.D. (1985). Clinically recorded patterns as related to the sensory evaluation of meat and meat products. *Journal of Food Science, 50*, 1059–1066.

Utz, K.H. (1983). The interocclusal tactile fine sensibility of natural teeth. *Journal of Oral Rehabilitation, 10*, 440–441.

Utz, K.H. (1986). Untersuchungen uber die interokklusale taktile feinsensibilitat naturicher zahne mit hilfe von aluminium-oxid-teilchen. *Deutsche Zahnarztliche Zeitschrift, 41*, 313–315.

Vickers, Z.M. (1980). Relationships of chewing sounds to judgements of crispness and crunchiness. *Journal of Food Science, 47,* 121–124.

Watt, D.M. (1976). Monitoring mastication. *Journal of Dentition, 4*, 271–278.

CHAPTER 3

Sensory Techniques To Study Food Texture

David Kilcast

Eating food should be a pleasurable experience, and in developed countries in which consumers enjoy the choice of a wide range of foods, eating experiences that are not pleasurable will ultimately result in the failure of that product. If, on the other hand, the available range of foods is limited by factors such as nonavailability or poverty, the eating experience is much less important than satisfying hunger and providing adequate nutritive value.

The consequence of examining these extreme circumstances is that we must recognize that in trying to understand what consumers desire from food, no single factor can be considered in isolation from other factors. For some years, psychology researchers have been developing models to understand consumer behavior (e.g., Shepherd & Sparks, 1994). Although there are many possible circumstances under which nonsensory factors such as price and nutritional image can have dominant effects, the sensory characteristics of foods are central to continued purchase of foods.

The sensory evaluation of food is frequently defined by the term *tasting*. A brief moment's thought will reveal that this term is inadequate to describe all the perceptions involved in eating food. When we eat food, we perceive a whole range of different characteristics relating to the appearance, flavor, and texture of the food. Numerous tools are available for investigating the sensory properties of foods, and the information required must be carefully defined if appropriate tools are to be selected. Systematic development of new products will inevitably depend on the use of different tools at different stages of the development cycle.

PRINCIPLES OF SENSORY EVALUATION

Human beings employ a range of senses in perceiving food quality (Exhibit 3–1). The discussion below summarizes these senses briefly. Fuller descriptions can be found in the following references: *appearance* in Hutchings (1994), *odor* in Maruniak (1988), *taste* in Plattig (1988), and *texture* in Bourne (1982) and Brennan (1988).

The Human Senses

The visual senses are of particular importance in generating an initial impression of food quality that often precedes the input from the remaining senses. Indeed, if the appearance of the food creates a negative impact, then the other senses may not come into play at all. The visual sense is often equated

Exhibit 3–1 The Human Senses

Sense	Perception	
Vision	Appearance	
Gustation	Taste	→ Flavor
Olfaction	Odor/aroma	→ Flavor
Chemical/trigeminal	Irritant	→ Flavor
Touch	Texture	
Hearing	(Texture)	

only with color but provides input on many more appearance attributes that can influence food choice, such as size, shape, surface gloss, and clarity. In particular, the visual senses can provide an early, and strong, expectation of the flavor and textural properties of foods.

Taste (gustation) is strictly defined as the response by the tongue to soluble, involatile materials. These have classically been defined as four primary basic taste sensations—salt, sweet, sour, and bitter—although in some countries this list is extended to include sensations such as metallic, astringent, and *umami*, this last sensation associated with monosodium glutamate. The taste receptors are organized groups of cells, known as tastebuds, located within specialized structures called *papillae*. These are located mainly on the tip, sides, and rear upper surface of the tongue. Sweetness is detected primarily on the tip of the tongue, salt and sour on the sides of the tongue, and bitter on the rear of the tongue. Taste stimuli are characterized by the relatively narrow range between the weakest and the strongest stimulants (about 10^4) and are strongly influenced by factors such as temperature and pH (Meilgaard, Civille, & Carr, 1991).

The odor response is much more complex. Odors are detected as volatiles entering the nasal passage, either directly via the nose or indirectly through the retronasal path via the mouth. The odorants are sensed by the olfactory epithelium, which is located in the roof of the nasal cavity. Some 150 to 200 odor qualities have been recognized, and there is a very wide range (about 10^{12}) between the weakest and the strongest stimulants (Meilgaard et al., 1991). The odor receptors are easily saturated, and specific anosmia (blindness to specific odors) is common. It is thought that the wide range of possible odor responses contributes to variety in flavor perception. Both taste and odor stimuli can be detected only if they are released effectively from the food matrix during the course of mastication.

The chemical sense corresponds to a pain response through stimulation of the trigeminal nerve. This effect is produced by chemical irritants such as ginger and capsaicin (from chili), both of which give a heat response, and chemicals such as menthol and sorbitol, which give a cooling response. With the exception of capsaicin, these stimulants are characterized by high thresholds. The combined effect of the taste, odor, and chemical responses gives rise to the sensation generally perceived as flavor, although these terms are often used loosely.

Texture is perceived by the sense of touch and comprises two components: *somesthesis*, a tactile, surface response from skin, and *kin-*

esthesis (or proprioception), a deep response from muscles and tendons. The events contributing to texture perception are shown schematically in Figure 3–1. For many foods, visual stimuli will generate an expectation of textural properties. The touch stimuli themselves can arise from tactile manipulation of the food with the hands and fingers, either directly or through the intermediary of utensils such as a knife or spoon. Oral contact with food can occur through the lips, tongue, palate, and teeth, all of which provide textural information.

The descriptions given above, while appropriate for the individual sensing modalities, fail to take into account their interactive nature, however, shown schematically in Figure 3–2. These interactions have been extensively reviewed by Cardello (1996). Color, which is obviously an important appearance characteristic, can be shown to have an influence on flavor perception. For example, Dubose, Cardello, and Maller (1980) found significant increases in perceived flavor intensity in beverages with increasing color intensity. Textural properties of foods have substantial effects on the perception of flavor, with the effect of viscosity having received particular attention. Cardello (1996) has reviewed the literature, which focuses mainly on the effects of viscosity on taste, with less work reported on odorous volatiles. Sound emission from crisp and crunchy foods has been shown to be of great importance in the perception of their texture (e.g., Vickers, 1991) and to form a basis for the discrimination of crisp and crunchy foods.

The importance of the interaction between the texture of foods and their perceived flavor can be clearly seen if the time course of events during food consumption is considered. This is shown schematically in Figure 3–3. As already indicated, strong expectations of the flavor and texture characteristics can be generated before the food is introduced into the mouth. As food enters the mouth and is either bitten or manipulated between tongue and palate, catastrophic changes occur to the structure of the food that strongly influence the way in which tastants and odorants are released from the food. Of particular importance are temperature increase (cold foods) or decrease (hot foods) and dilution by saliva. Salivary introduction

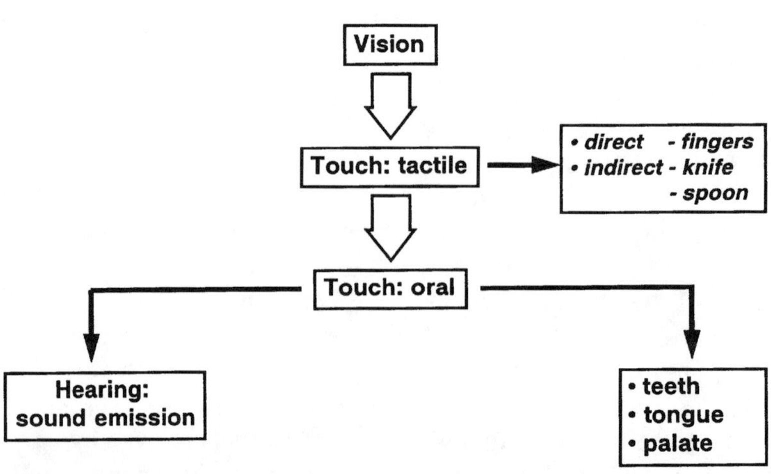

Figure 3–1 Schematic Diagram of the Processes Leading to the Perception of Texture

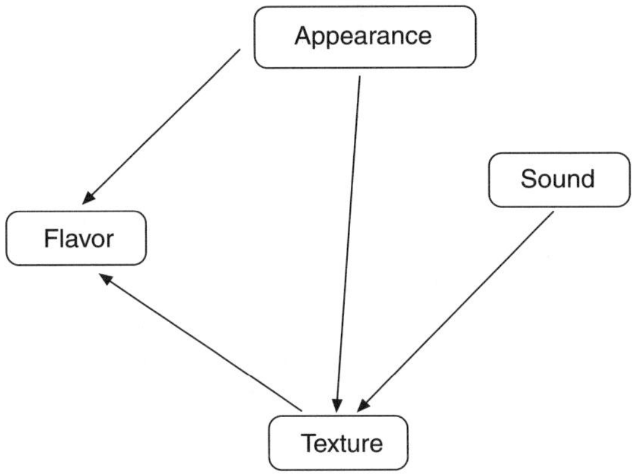

Figure 3–2 Interactive Processes Operating for the Human Senses

also serves to lubricate the food bolus. The factors that influence such release are under active study (e.g., Overbosch, Afterof, & Haring, 1991) and include

- rate and mode of production of new surface
- rate of production of saliva
- dissolution and dispersion of the food
- release of involatile tastants and volatile odorants
- transport of volatiles to the nasal cavity

These processes can result in important effects on perceived flavor and can produce enormous difficulties in ensuring high flavor quality if changes to foods are made that produce different structural characteristics—for example, when reducing the fat content.

Factors Influencing the Quality of Sensory Data

The complex nature of food quality perception creates many difficulties for the sensory analyst, whose primary task is to use human subjects as an instrument to measure the sensory quality of foods. The factors that should be considered in assessing the performance of human subjects in this way are

- *accuracy:* giving the correct answer without systematic error or bias
- *precision:* showing little variation
- *validity:* measuring the correct property (Piggott, 1995)

The wide range of responses of human beings to sensory stimuli is reflected in highly variable (low-precision) sensory data, but the precision can be improved by careful selection of a range of human subjects who can produce a response with lower variability, and by extensive training. Improving accuracy can be achieved by recognizing the various sources of physiological and psychological biases that can influence human subjects (Meilgaard et al., 1991). Some of the most important effects are shown in Tables 3–1 and 3–2.

Other important psychological effects that are more obvious but are nevertheless frequently overlooked are those arising from unwanted interaction between assessors, from lack of motivation, and from extrovert/introvert personality traits that can influence how assessors use assessment systems. A major role of the sensory analyst lies in minimizing the effects of such biases, and consequently improving test accuracy, by careful design of the testing protocol. Of particular importance are the minimization of adaptation, enhancement/suppression, and contrast effects by careful selection of the type and number of samples and the elimination of presentation order effects by using balanced presentation orders. Reduction of random variation can be achieved by careful assessor selection procedures, followed by training and periodic calibration of the panel. Sensory measurement of sensory characteristics is, of course, of far greater validity than any instrumental measurement of sensory characteristics, but

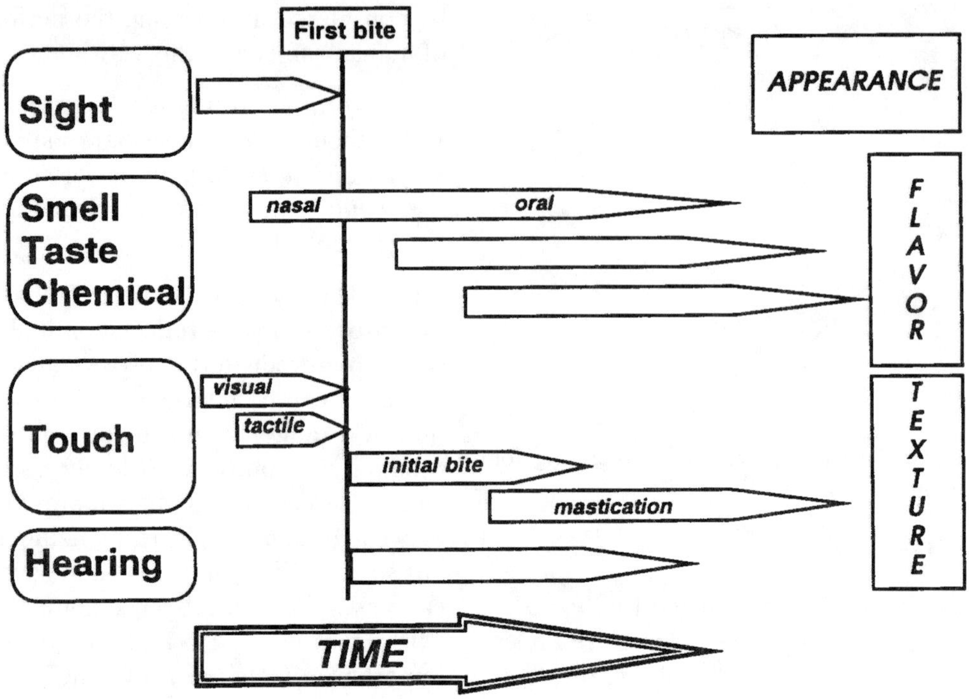

Figure 3–3 Representation of the Time Order of the Different Sensory Modalities

Table 3–1 Physiological Factors Influencing Sensory Response

Name	*Effect*
Adaptation	The decrease in, or change in sensitivity to, a given stimulus as a result of continued exposure to that stimulus (adaptation) or a similar stimulus (cross-adaptation). Increase in response (cross-potentiation) can also occur between different stimuli.
Enhancement	The effect of the presence of one substance increasing the perceived intensity of a second substance.
Suppression	The effect of the presence of one substance decreasing the perceived intensity of a second substance.
Synergy	The effect seen when the perceived combined intensity of two substances is greater than the sum of the individual intensities.

Table 3–2 Psychological Factors Influencing Sensory Responses

Name	*Effect*
Expectation error	Information given or learned about the sample can trigger preconceived ideas and lead assessors to find what they expect to find. This is particularly prevalent in threshold tests in which stimuli are presented in increasing order of intensity but can be observed in many other situations.
Error of habituation	When presented with a series of samples of slowly increasing or decreasing stimuli, assessors tend to continue to give the same response. This can result in different samples being missed, especially in routine quality control testing.
Stimulus error	This error can occur when irrelevant criteria, such as packaging format, influence the observer.
Logical error	This error can occur when two or more characteristics of the samples are associated in the minds of the assessors.
Halo effect	This effect is frequently seen when multiple sample attributes are assessed. The score given to one attribute can influence the score given to a different attribute. This can occur within the same sensory modality, between modalities, or between sensory and acceptability assessments.
Presentation order	There are several effects in this general category, the most important being as follows:
Contrast effect	Presentation of one sample with a higher stimulus intensity than the other samples in the set will give a lower intensity rating to the other samples and can compress the range between them.
Error of central tendency	Responses are biased toward the samples presented in the center of a set. In addition, there is an effect resulting from the reluctance to use scale extremes, which effectively focuses scoring to the central portion of a scale.
Pattern effect	Assessors will seek to use all available clues and to find patterns in presentation orders.
Positional bias	Attitudes and judgments will change over a period of time—for example, satiation, fatigue, or boredom may affect assessors. The first sample presented in a sequence is often seen as abnormal.

nonetheless great care is needed to define the nature of the data required and to select the appropriate experimental procedures.

CLASSIFICATION OF TEXTURAL PROPERTIES

Most studies in which the importance of different sensory modalities to consumer acceptability is investigated conclude that flavor is the most important modality, followed by texture and then appearance (e.g., Moskowitz & Krieger, 1995). Such conclusions do not reflect the enormous efforts that the food industry devotes to designing appealing textural characteristics and maintaining those characteristics in long-term production. Research with consumers in the United States carried out by Szczesniak and Kahn (1971) showed that awareness of texture lies at a subconscious level and that textural properties are taken for granted. If the expectations of texture are violated, however, awareness of textural defects is accentuated, and texture becomes a focal point for criticism and rejection of the food. Expectations are being increasingly recognized as important factors in food choice by consumers (e.g., Cardello, 1994; Vickers, 1991).

Bourne (1982) classified textural characteristics into three groups:

1. *critical:* those foods for which texture is the dominant quality characteristic—for example, meat, potato chips, celery
2. *important:* foods in which texture makes a significant, but not dominant, contribution in comparison with flavor and appearance—for example, fruits, bread, and sugar confectionery
3. *minor:* foods in which texture makes a negligible contribution to overall quality—for example, beverages and thin soups

Such a classification is quite arbitrary but is potentially useful to both product developer and sensory analyst.

Many consumer studies have been carried out on the importance of texture and on the use of textural terms. In a study in which 29 pureed foods were fed to blindfolded subjects, Schiffman (1973) showed a wide range of successful identifications, with young adults of normal weight being able, on average, to correctly identify only 40% of the foods. Apples, strawberries, and fish had success rates of over 78%, whereas foods such as cucumber, lamb, and cabbage had success rates of less than 10%.

Word association studies by Szczesniak and Kleyn (1963) and Szczesniak (1971) showed that awareness of texture depended on factors such as gender (women were more texture conscious than men), experience (people working with food were texture conscious), socioeconomic class (consumers in higher socioeconomic groups were more texture conscious than those in the lower groups), and food type (bland, crispy, and crunchy foods provoked greater texture awareness). Szczesniak and Kahn (1971) also investigated the possibility that there might be innate preferences for specific textures. This was found to be the case for crispy and crunchy textures, but other textures such as soggy, watery, lumpy, sticky, slimy, crumbly, and tough, which gave rise to a sense of lack of control in the mouth, were disliked.

In the study by Szczesniak and Kleyn (1963), 78 descriptive words were used by the 100 respondents. In direct contrast, studies by Yoshikawa, Nishimaru, Tashiro, and Yoshida in Japan (1970a, 1970b, 1970c) showed a much higher level of texture awareness. A total of 140 female college students described the texture of 97 foods and produced 406 descriptive terms. This richer vocabulary reflects the greater variety of textures in Japanese cuisine and greater sensitivity to subtle textural differences. Such studies also, however, point to the substantial difficulties in classifying textural attributes across different cultures and in different languages.

The complex nature of food texture has led to several attempts to design classification schemes to aid the assessment of food texture and to classify specific terms. The most extensive and important early studies were carried out by Szczesniak and colleagues at General Foods in the United States (Brandt, Skinner, & Coleman, 1963; Szczesniak, 1963; Szczesniak, Brandt, & Friedman, 1963). These initial papers formed the basis for the Texture Profile Method, described further below, and the starting point lay in the classification scheme described in the first paper in this sequence (Szczesniak, 1963). Three categories of textural characteristics were defined in this paper:

1. *mechanical characteristics*—relating to the reaction of food to stress
   - primary parameters: hardness, cohesiveness, viscosity, elasticity, adhesiveness
   - secondary parameters: brittleness, chewiness, gumminess
2. *geometrical characteristics*—relating to the size, shape, and orientation of the particles within the food—for example, powdery, gritty, lumpy, flaky, fibrous, cellular, aerated, crystalline
3. *other characteristics*—relating to the perception of the moisture and fat contents of the food, for example, dry, moist, oily

This terminology classification was specific for each product type but in each case was based on underlying rheological characteristics. In practice, the temporal nature of sensory perception was also recognized, and the attributes were divided into initial (first bite), masticatory, and residual (after-swallowing) phases.

An alternative classification scheme was proposed by Sherman (1969) as a consequence of alleged defects in the definitions of several mechanical characteristics, in which three categories were defined.

1. *primary*—fundamental property of the food; geometric characteristics such as particle size, shape, size distribution, air content, cell size, cell size distribution, cell shape
2. *secondary*—derived by a combination of two or more attributes in unknown proportion; rheological characteristics such as viscosity, elasticity, adhesion
3. *tertiary*—subdivided according to the type of process involved
   - mastication: hard, soft; brittle, plastic, crisp; rubbery, spongy; smooth, coarse; powdery, lumpy, pasty; creamy, watery, soggy; sticky, tacky
   - disintegration: greasy, gummy, stringy; melt-down properties
   - nonmasticatory: appearance; sampling and slicing characteristics; spreading, creaming characteristics, pourability

The nonmasticatory characteristics were associated with an initial perceptual phase; the remaining characteristics were associated with initial perception on the palate and mastication (high-shearing) phases. One possible criticism of this classification is that it is based primarily on the rheological properties of the foods rather than on a perceptual basis.

The need to take these classifications further to a more practical level was recognized by Jowitt (1974), who proposed a system of definitions for textural terms. The terms were classified into the following groups:

1. general—*structure, texture, consistency*
2. terms relating to the behavior of the material under stress or strain—for example, *firm, chewy, stringy, rubbery, sticky, glutinous, brittle, crumbly, crisp*
3. terms relating to the structure of the material
   - relating to particle size or shape—for example, *smooth, chalky, gritty, mealy*
   - relating to shape and arrangement of structural elements—for example, *flaky, stringy, aerated, glassy, gelatinous, spongy*
4. terms relating to mouth-feel characteristics—for example, *body, watery, greasy, slimy, mushy, astringent, cooling*

Definitions were provided for each of these terms. The need for terms has been recognized by the standards bodies of different countries, and these terms and definitions have been incorporated into standards (British Standard BS 5098, the equivalent of International Standard ISO 5492; British Standards Institution, 1992). However, the frequent absence of direct translation equivalents of some terms between different languages led to difficulties in securing agreement on the terms, and those listed represent a compromise on which there were dissenting views. The importance of securing an understanding of the use of textural terms across different languages was recognized by Drake (1989), who built up a list of 54 English terms for textural/rheological properties of foods and their equivalents in 22 other languages. Such understanding is particularly important as the operation of the food industry becomes more international.

As a direct consequence of their practical importance, some textural terms have received particular attention and have been the subject of numerous studies (Guinard & Mazzucchelli, 1996). These terms include *firmness* and *hardness*, *crispness* and *crunchiness*, *creaminess*, *smoothness*, viscosity-related properties, *juiciness*, *astringency*, and *carbonation*.

BASIC REQUIREMENTS FOR SENSORY ANALYSIS

The principles that must be considered in satisfying the requirements outlined earlier in this chapter, and the systems that need to be implemented for their practical application, are common to the investigation of all sensory modalities. Only one method has been designed specifically for assessing food texture (see the section "The Texture Profile Method" below), and the remaining methods require only a change in emphasis for the investigation of specific modalities. However, the detailed operation of many sensory tests needs examination and, frequently, modification for the investigation of texture. In this and the following two sections, the discussions will be generalized to all sensory modalities, but with reference to specific uses in textural investigations. More detailed discussions are given in standard texts (e.g., Meilgaard et al., 1991; Muñoz, Civille, & Carr, 1992; Piggott, 1988; Poste, Mackie, Butler, & Larmond, 1991; Stone & Sidel, 1993; Watts, Ylimaki, Jeffery, & Elias, 1989).

In developing and implementing a high-quality sensory evaluation system, a number of interrelated requirements can be defined (Figure 3–4).

Clear Definition of the Objectives of the Sensory Evaluation System

Clear definition of the objectives of the sensory evaluation system is central to the establishment of any system that will be of suf-

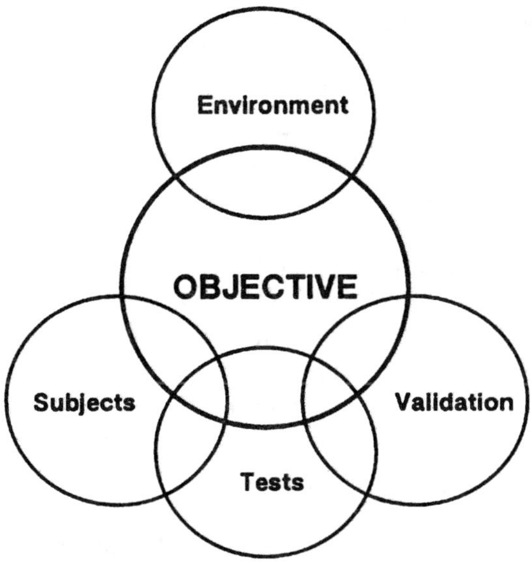

Figure 3–4 Requirements for Formal Sensory Analysis and Their Relationships

ficient accuracy to measure the required parameters with maximum precision and that will be cost-effective. Factors that need to be considered include

- *The primary purpose of the system.* For example, is it to be for research and development use, quality assurance use, or both? If sensory evaluation is needed for both purposes, then different facilities, assessors, and locations may be needed.
- *To whom, and by what route, the sensory data are to be directed and acted upon.* The management structure of the system should be defined, including clear lines of responsibility.
- *The resources that will be available*—facilities, equipment, personnel, and finances for both start-up and ongoing use, including projected improvements.
- *The purpose for which the sensory data are required* (e.g., product optimization, cost reduction, shelf-life determination). Clear identification of such requirements will help define the required resources and the type and frequency of tests.
- *The extent to which the system will be applied* (e.g., to all or part of the product range). The organization of a system in a company producing a single high-volume product is generally more straightforward than in a multiproduct company, especially if there is frequent change in the product lines.
- *Likely frequency of use.* This will be determined by many factors but needs careful consideration in assessing resource requirements.
- *Flexibility to upgrade or adapt for different uses.* In particular, location and design might be chosen to permit future installation of a computerized data acquisition system.
- *Possible need for integration with instrumental data.* Instrumental measurements of flavor and texture are unlikely to replace the need for sensory testing, but specific measurements of quality factors can be particularly useful in quality control operations, provided that they are carefully calibrated against suitable sensory assessments.

Provision of a Dedicated Sensory Testing Environment

A suitable environment is crucial to providing high-quality sensory data. The environment must provide the required working conditions for the assessors as well as a work area for sample preparation and for data analysis. Detailed advice is given in a number of publications (e.g., Stone & Sidel, 1993; British Standard BS 7183, the equivalent of International Standard ISO 8589; British Standards Institution, 1989a). A major requirement for any environment is that there be easy access for the assessors. There are three main components of a sensory evaluation environment: preparation area, testing environment, and booth design.

Preparation Area

The preparation area should be adjacent to the testing area but separated to prevent access by the assessors. Equipment in the area should be appropriate to the products being tested, and specialized equipment may be needed. The size of the area is determined by the amount of work to be undertaken, but an extensive work surface should be available for preparing and organizing the presentation of large sample numbers. Ventilation should be sufficient to clear cooking odors, and equipment and cleaning materials should be odor-free.

Testing Environment

The basic test area should provide individual booths for assessors, but equally important is the availability of a room for round-table discussions, especially for deriving descriptive vocabularies for profiling studies. The test area should have easy accessibility and be located near but separate from the preparation area. The assessors must not enter or leave through the preparation area, as this could bias their judgment. The test area must be easy to reach and away from noise, such as that of main roads, offices, busy corridors, and process machinery.

The temperature must be comfortable for the assessors. Air conditioning should ensure a steady ambient temperature of about 22°C. Good temperature control is particularly important for the assessment of the texture of temperature-sensitive products. There should preferably be a relative humidity of between 45% and 50%, although controlling this precisely may be impractical.

The area must be odor-free. Ventilation is necessary to extract odors, and activated carbon filters may be necessary to treat incoming air supplies. Slight positive air pressure from testing booth to preparation area helps minimize odor transfer. Flooring, wall coverings, and furniture should be odor-free and nonporous.

Neutral colors such as off-white, light gray, cream, and light green have been found to be more restful for the assessor to work in than white, but white or pale gray surfaces in the immediate assessment area should be used to give reliable color rendition. Lighting should be uniform, shadow-free, and controllable.

Booth Design

Booths should be set up so that the assessors can make independent personal judgments with the minimum of distraction from the surrounding environment and from other assessors. The space allocated to booths is determined by the number of tests, people, and space available. The minimum number is 3, and usually 5 to 10 booths are sufficient for most purposes. Temporary booths can be set up if there is little space available. These are similar to booths used at polling stations and are collapsible for easy storage.

Each booth should have enough space to hold the samples, utensils, expectoration cups, palate cleansers, questionnaire, and pencil, and computer-based data input if this is to be used. General dimensions are 70 to 90 cm wide and 45 to 60 cm deep. The height can be either a normal desk sitting height of 70 to 80 cm or a kitchen counter height of 90 to 95 cm. Lateral dividers between booths extend beyond the counters, usually by 30 to 45 cm. Seats should preferably not be fixed or on castors but should be easily movable. Spittoons are sometimes used, but their operations can be noisy, and drain smells can occur.

Correct lighting is important if appearance judgments are to be carried out. Color rendition should be constant for all booths, shadow-free, and comfortable for the assessor. Types of light that can be used include tungsten and fluorescent, including point-of-sale lighting. Diffusion filters should be placed over all lighting to minimize the spotlight effect. Dimmer switches are useful, as are colored green and red light filters if color masking is needed for the product.

Serving hatches to the preparation area are important if there is to be minimum disturbance to the assessors. Hatches can slide horizontally or vertically or be "bread box" style. The detailed design will often depend on the available space. Ready signals are useful to allow the assessors to signal both their arrival and their completion of tasting.

Selection of Suitable Test Procedures

Sensory test methodologies fall into three main classes:

1. *discrimination/difference tests:* "Is there a difference?"
2. *descriptive tests:* "What is the difference?" and "How big is the difference?"
3. *hedonic/affective tests:* "Which is liked?" and "Why is it liked?"

The first two classes are quite different from the third class, as illustrated in Figure 3–5. They are analytical, and their purpose is to use human subjects as a form of instrument to measure properties of the food. Hedonic tests measure the response of consumer populations to the food in terms of likes or dislikes. In general, there is no simple linear relationship between the two types of data. Of great practical importance, the types and numbers of subjects used for the analytical and hedonic tests are quite different.

Difference tests are used to identify relatively small overall differences in sensory characteristics. They find extensive use in product maintenance and product matching. Most new product development requires sensory tests with much higher information content, and descriptive profiling is the most powerful test class available. Hedonic tests require larger numbers of respondents, who act as a subsample of the target population. Although often regarded as the preserve of marketing functions in companies, they should be used at early stages of new product development to give development programs more accurate direction.

Selection and Training of Suitable Test Subjects

The subjects to be used are defined by the objective of the test and by the consequential choice of test. The numbers of subjects to be used depends on the level of expertise and training of the assessors. Recommended minimum numbers are given in British Standard BS 5929, Part 1 (the equivalent of International Standard ISO 6658; British Standards Institution, 1986), which also discriminates among assessors, selected assessors, and experts.

Analytical Tests

Both discriminative and descriptive tests use small panels of assessors chosen for their abilities to carry out the tests. Guidelines for establishing such assessors are given in British Standard BS 7667, Part 1 (the equivalent of International Standard ISO 8586-1; British Standards Institution, 1993). A general scheme for establishing a panel requires the following steps.

Recruitment. Assessors can be recruited from within the company, or dedicated part-time assessors can be recruited from the local population (in no circumstances should company employees be *compelled* to participate).

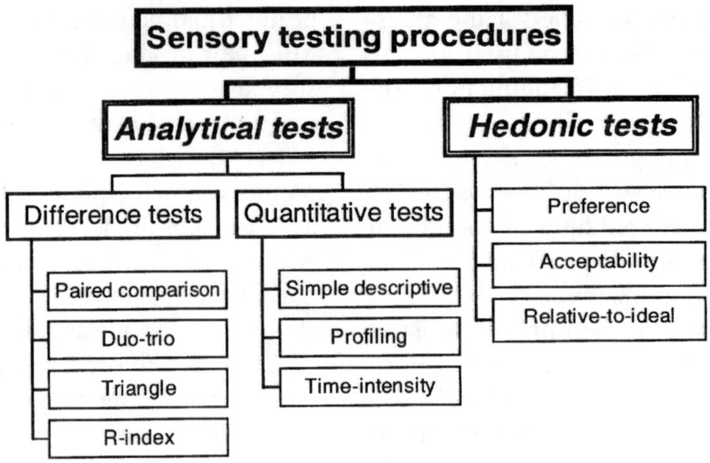

Figure 3–5 Classification of the Main Types of Sensory Testing Procedures

In either case, a suitable advertisement should be devised, and responses should be followed up with a questionnaire that gathers information on availability, likes and dislikes, state of health, and known food allergies. Important aspects that should be established at this stage are interest and motivation, availability, ability to articulate, and literacy.

Screening. These preliminary tests are used to establish that sensory impairment is absent, to establish sensitivity to appropriate stimuli, and to evaluate the ability to verbalize and communicate responses. These tests will depend mainly on the defined objectives of the sensory program but will typically consist of the following:

- The ability to detect and describe the four basic tastes: sweet, sour, salt, and bitter. Identification of tastes may be extended to cover metallic, *umami*, and astringent.
- The ability to detect and recognize common odorants, together with those characteristic of the product range of interest.
- The ability to order correctly increasing intensities of a specific stimulus—for example, increasing sweetness or increasing firmness.
- The ability to describe textural terms characteristic of relevant food types.
- Absence of color vision deficiencies. Approximately 1 in 13 men, but only 1 in 250 women, suffer color vision deficiencies (Hutchings, 1994). Tests can be carried out using Ishihara charts (Tokyo, Kanehara Shuppan).

Selection of suitable assessors is usually made on the basis of a good performance across the entire range of tests rather than excellence in some and poor response to others. If the assessors are to be used for a specific purpose, then the tests relevant to that purpose can be weighted appropriately.

Training. In the initial stages, training is limited to the basic principles and operations, following which further selection can be made. More closely targeted training can then be carried out using the products of interest and aimed toward the specific tests to be used in practice.

Monitoring. Close monitoring of assessors' performance is essential, and any drift that is identified must be corrected by retraining procedures. Care should be taken to ensure adequate motivation, especially when carrying out long-term routine testing.

Hedonic Tests

Subjects (respondents) for hedonic tests are chosen to represent the target consumer population and to reflect any inhomogeneity in that population. Consequently, they need to be used in sufficient numbers to give statistical confidence that they are representative, and they must be given the opportunity to behave as they would in a real consumption environment. In particular, they must not be selected on the basis of sensory ability and must not be given any training. Numbers in excess of 100 respondents are normally used. For the early stages of concept development, qualitative studies using focus groups with small numbers of respondents can be used, but the data generated should be treated carefully, and conclusions must not be generalized.

The same subjects must not be used for both types of test, and, in particular, in-house staff must not be used to generate hedonic data that may be viewed as consumer related.

Data Handling, Analysis, and Presentation

Sensory experiments can generate large amounts of data, and it is rare that reliable conclusions can be drawn without validation using statistical techniques. Details of suitable statistical methods can be found in a number of texts (e.g., Lahiff & Leland, 1994; Meilgaard et al., 1991; O'Mahony, 1986; Smith, 1988). Different types of sensory test procedures generally use specific analysis procedures, but in the case of the more sophisticated profiling techniques, a wide range of options is available, both univariate and multivariate. These procedures will be described in more detail following the descriptions of the individual tests. Many statistical software packages are now available. The most sophisticated require a sound understanding of statistical principles, but more user-friendly packages are available that satisfy most requirements. However, it is usually found that no single package can cover the entire range of basic requirements.

Clear and effective presentation of sensory data, including the results of statistical tests, is essential. Most standard spreadsheets are now able to offer a wide range of presentation possibilities for both univariate and multivariate data.

DISCRIMINATION TESTS

Discrimination tests are perceived as one of the easiest classes of sensory testing to apply in an industrial environment and are consequently heavily used. The tests can be used in two ways: to determine whether there is an overall difference between two samples or to determine whether one sample has more or less of a specific attribute than another. However, there are inherent limitations of such tests—for example, the restricted information content and the difficulty in determining whether the absence of a difference can be interpreted as the samples being the same. Consequently, such tests are often overused in circumstances in which alternative methods such as profiling would be superior. At the time of this writing, difference tests are almost universally used to ascertain whether two samples are different, not to ascertain whether two samples are the same. However, standards of the International Organization for Standards (ISO) are under preparation that will advise sensory analysts on how to use the tests for the latter purpose.

In addition to determining if a sensory difference exists between two samples, the tests can be used as a means of measuring thresholds and sensitivity. Three types of difference tests are used most frequently: paired comparison, duo-trio, and triangular. Two other tests, the two-out-of-five test and the "A," "not A" test, are used less commonly, and many other forms of the test have been described but have found little application. A shortcut signal detection test (*R* index) has been described and may have future applications. In this section, the main types of test with practical value will be described in detail, and other forms will be described only briefly.

Paired-Comparison Test

In the most common form of the test (less commonly referred to as the 2-AFC [alternative forced-choice] test), two coded samples are presented either sequentially or simultaneously in a balanced presentation order (i.e., AB and BA). There are two variations on the test. In the directional difference variant, the assessors are asked to choose the sample with the greater or lesser amount of a specified characteristic. They are usually instructed to make a choice (forced-choice procedure) even if they have to make a guess, but they may be allowed to record a "no-difference" response. The distinction between these two procedures will be expanded on in the discussion on statistical analysis later in this chapter. In the directional form, it is important that the assessors clearly comprehend the nature of the attribute of interest. This might not necessarily be clearly defined to either the test organizer or the assessors, however, especially if interactive effects are operating or if textural differences are sought that might be difficult to describe precisely. It has been pointed out that if time is needed to train assessors to recognize a specific characteristic, then a descriptive test should be selected (Stone & Sidel, 1993). These authors have also pointed out that the test can be adapted to a nondirectional format in which the assessors are asked to judge whether pairs of samples are the same or different (simple difference test). In this case, all possible pairs of samples must be presented (AB, AA, BB, BA). However, little or no use has been made of this format. The most common use for nondirectional paired-comparison testing is consumer preference testing.

An additional form of paired-comparison testing is the "A," "not A" test. The assessors are presented with a single product and asked to evaluate the sample and remember its characteristics. They are then presented with the test product and asked whether the products are the same or different. The test is more frequently used as an identification test, however, in which a series of samples is presented for evaluation.

Duo-Trio Test

In the most common variant of the duo-trio test, the assessors are presented with a sample that is identified as a reference followed by two coded samples, one of which is the same as the reference and the other of which is different. These coded samples are presented in a balanced presentation order:

A (reference)	AB
A (reference)	BA

The assessors are asked to identify which sample is the same as the reference. A more rigorous procedure, the balanced reference procedure, is to use both samples as the reference in a balanced presentation order:

A (reference)	AB
A (reference)	BA
B (reference)	AB
B (reference)	BA

The constant reference variant is usually used, especially in circumstances in which one sample can be used as the reference that is familiar to the assessors. This can be especially useful for quality control procedures. The duo-trio test is particularly useful when testing foods that are difficult to prepare in identical portions. Testing such heterogeneous foods using the triangle test, which relies on identical portions (see next section), can give rise to difficulties, but in the duo-trio test there are no inherent difficulties in asking the question: Which sample is most similar to the reference?

Triangle Test

Three coded samples are presented to the assessors, two of which are identical, using all possible sample permutations:

ABB	AAB
BAB	ABA
BBA	BAA

The assessors are asked to select the odd sample in either fixed-choice or no-difference procedures. The increased number of samples compared with a paired-comparison test can result in problems with flavor carry-over when using strongly flavored samples, making identification of the odd sample more difficult. Difficulties can also be encountered in ensuring presentation of identical samples of some foods.

A less commonly used variant of the triangle test is the 3-AFC procedure. This procedure uses the same sample permutations in a triad format, but either the difference of interest between the samples is revealed to the assessors in advance, or the assessors identify the nature of any difference in advance. In the test itself, the assessors are then asked to identify the sample (or samples) with the specified characteristic. For example, a typical question might be "One of these samples is stickier than the others; please identify this sample." O'Mahony (1995) has identified the reasons that this test can be more sensitive than the triangle test, but the test suffers from the need to positively identify the nature of the difference in advance.

Two-out-of-Five Test

Assessors are presented with five samples: three of one material and two of another. They are asked to pick out the two samples that are the same but different from the remaining three. There are 20 possible orders of presentation:

AAABB	AABAB	ABAAB	BAAAB	AABBA
ABABA	BAABA	ABBAA	BABAA	BBAAA
BBBAA	BBABA	BABBA	ABBBA	BBAAB
BABAB	ABBAB	BAABB	ABABB	AABBB

This test is highly sensitive, but the five-sample presentation can cause severe fatigue when carrying out oral testing and is more commonly used for visual, auditory, and tactile assessments. The test also requires good memory capability.

Difference-from-Control Test

In the difference-from-control test, which is sometimes used when a control is available, the assessors are presented with an identified control and a range of test samples. They are asked to rate the samples on suitable scales anchored by the points *not different from control* and *very different from control.* The test results are usually analyzed as scaled data. This type of test is sometimes classified as similarity/dissimilarity scaling and analyzed using multidimensional scaling methods (see the section "Multidimensional Scaling" later in this chapter).

R-Index Test

This method is a relatively recent development (O'Mahony, 1979, 1986) and is a shortcut signal detection method. The test samples are compared against a previously presented standard and are rated in one of four categories. For difference testing, these categories are "standard," "perhaps standard," "perhaps not standard," and "not standard." The test can also be carried out as a recognition test, in which case the categories are "standard, recognized," "perhaps standard, recognized," "perhaps standard, not recognized," and "standard, not recognized." The results are expressed in terms of *R* indices, which represent probability values of correct discrimination or correct identification. The method claims to give some quantification of magnitude of difference, but its use has not been widely reported in the literature. One important limitation is that a relatively high number of judgments are needed in this form of test, leading to the risk of severe assessor fatigue or increased numbers of sessions.

Analysis of Discrimination Tests

The basic principle underlying the analysis of difference is to test the actual response obtained against the response that would have occurred purely by chance. In the case of the main difference tests described above, the probabilities of guessing the result are shown in Table 3–3. Frijters (1988) has tabulated the probabilities of guessing for a more generalized set of difference tests.

One consequence of the different probabilities is that the statistical power of the tests differs, together with the numbers of responses that are needed to give a meaningful and reliable result. These numbers are related to the levels of risk that are deemed acceptable. These are the Type 1 risk (incorrectly concluding that there is a difference that does not exist) and the Type 2 risk (not identifying a difference that is present). Table 3–4 shows the minimum numbers of assessors recommended in British Standard BS 5929, Part 1 (equivalent to International Standard ISO 6658; British Standards Institution, 1986). It is possible to generate the required number of judgments by replicated tests with a smaller number of assessors. Such a procedure should be used with care (e.g., generating 15 responses by using three assessors in five replicates is not recommended), and each replicate should be set up as a separate test. This table also illustrates the principle that the number of assessors required decreases with increasing expertise. However, these numbers should be used for guidance only, and it is probable that future revisions of ISO standards will recommend the use of higher numbers of assessors.

The test results are usually analyzed using tables of the binomial expansion, although

Table 3–3 Probability of Guessing the Result in Discrimination Tests

Test	*Probability*
Paired comparison	1/2
Duo-trio	1/2
Triangle	1/3
Two-out-of-five	1/10

Table 3–4 Minimum Numbers and Experience of Assessors[a]

Test	*Experts*	*Trained Assessors*	*Assessors*
Paired comparison	7	20	30
Triangular	5	15	25
Duo-trio			20
Two-out-of-five		10	
"A," "not A"		20	30
Ranking	2	5	10 (100[b])
Simple descriptive	5	5	
Profile (quantitative descriptive analysis)	5	5	

a. British Standard BS 5929, Part 1 (the equivalent of International Standard ISO 6658; British Standard Institution, 1986.

b. Consumer tests.

other distributions have been used. The 5% level of significance is frequently used in sensory tests, but an increasingly common procedure is to calculate exact probability levels. The duo-trio, triangle, and two-out-of-five tests are all inherently one-tailed tests, but the paired-comparison test can be either a one- or a two-tailed test. The one-tailed test is used only when a difference in a specific direction is sought.

An important consideration in all difference tests is whether they should be operated as forced-choice tests or whether assessors should be allowed to record a no-difference response. If a strict statistical interpretation is required, then a forced-choice response must be used. Similarly, if relatively inexperienced assessors, or consumers, are being used, then a forced-choice test must be used to prevent "fence sitting." However, if highly experienced assessors are used, then a no-difference response can be highly informative in specific circumstances.

Other useful information can be acquired from discrimination tests, although some authors warn that the tests should be strictly limited to establishing a difference (Meilgaard et al., 1991; Stone & Sidel, 1993). Descriptions of the nature of any difference can provide useful guidance for further testing. A simple scaled assessment of the degree of confidence in the decision (*absolutely sure, fairly sure, not very sure, only guessed*) is very useful, especially when using forced-choice procedures. Assessment of the degree of difference is likely to be of value only if assessors have been trained in scaling procedures. More controversially, assessors can be asked which samples they prefer, but this type of procedure is of value only for crude guidance; preference tests should be set up separately as consumer tests. All such information is supportive only and can be used only from assessors who have given the correct response.

QUANTITATIVE DESCRIPTIVE TESTS

The major advantages of discrimination tests are their relative simplicity to set up and operate and their high sensitivity. However, they have two important limitations. First, only two sample treatments are compared together. If sensory analysts find themselves carrying out multiple discrimination tests, then it is highly probable that other types of sensory tests should be employed. Second, the information content of discrimination tests is limited, even when operated in an extended format, incorporating a range of questions. More informative tests can produce more quantitative data that can be subjected to a wider range of statistical treatments.

Scaling Procedures

Quantification of sensory data is needed in many applications, and the recording of perceived intensity of attributes or liking requires some form of scaling procedure. This can simply involve ordering the samples on the basis of intensity or grading by placing them into categories that can be predefined on a number of attributes. Most uses of scaling, however, are carried out with the intention of generating numeric data that can be manipulated and analyzed statistically. Before this can be carried out, however, thought must be given to how the scales used are seen and interpreted by the assessors and how this may influence the type of analysis that can be safely applied. True scaling techniques rely on the use of three types of scales, where a scale is defined as a continuum divided into spaced successive values, which may be graphic, descriptive, or numerical.

Nominal scales specify only class identification, in which the classes have no quantitative relationship; for example, *apples*, *oranges*, *pears* or *hard*, *sticky*, *tough*.

Ordinal scales specify more or less of an attribute or class without any defined quantitative implication about size. The scale may rank samples, may specify amounts in categories, or may be a line with verbal anchors at the ends. In each case, no assumptions can be made about the size of differences between numbers, and normal parametric statistical tests cannot be used.

Interval scales specify that successive categories or unit intervals are equal and that the origin is not necessarily a real zero, although many sensory scales do have a zero. Many sensory scales are treated as interval scales, but usually without justification, and parametric statistical tests are used for analysis.

Ratio scales specify equal ratios between successive unit intervals and have a true zero—for example, the absolute scale of temperature. Parametric statistical tests can be used for analysis, provided that geometric means and logarithmic transformations are used on the data.

Ranking

Ranking is a method in which three or more samples, presented at the same time, are arranged in order of increasing (or decreasing) intensity or degree of some specified attribute. Ranked data give information on ordering only and carry no information on relative intensity. Ranking tests are easy to carry out, and relatively large numbers of samples can be assessed. Although the information from them is strictly limited, they can give valuable information with little expenditure of effort. Ranking scales are *ordinal scales*.

Grading

Grading is a quality classification of samples on the basis of one or more attributes. The classification is based on experience and is arbitrary, and any numeric gradings have no arithmetic meaning. The simplest grading is the in/out specification but grading can be more complex. Grading scales are usually *nominal scales* but can be *ordinal scales*.

Rating

Rating is classification into categories on an ordered scale in which perceived differences between categories are not defined. The term is often incorrectly used synonymously with *scoring*.

Scoring

Scoring is a form of rating using a numeric scale, where the numbers form an interval or ratio scale and the different scores have a defined and demonstrated mathematical relationship with each other. The use of ratio scales for intensity scoring is usually known as *magnitude estimation*. Quantified sensory data are usually treated as scored data, although checks are rarely made to see if the data are suitable for this treatment.

Practical Types of Scales

Category scales use a defined number of boxes or categories (often 5, 7, or 9, although other numbers are often used). The scale ends are defined by verbal anchors, and intermediate scale points are often given verbal descriptions. More verbal descriptions can often be a help in improving consistency of assessor response, especially for inexperienced assessors, but care should be taken to avoid ambiguous words and descriptions that distort the scale.

Graphic scales (line scales) consist of a horizontal or vertical line with a minimum number of verbal anchors, usually at the ends. Other anchors can be used—for example, to define a central point or to denote the position of a reference sample.

Unipolar scales have a zero at one end and are most commonly used in profiling, especially for flavor attributes.

Bipolar scales have opposite attributes at either end. Definition of the central point can often give rise to logical difficulties, as can ensuring that the extreme anchors are true opposites. This can be a particular problem for textural attributes—for example, when using *soft*-to-*hard*–type scales. Bipolar scales are frequently used for consumer acceptability testing, especially using the *like extremely*–to–*dislike extremely* format.

Hedonic scales are used to measure consumer liking or acceptability. Category scales are usually used.

Relative-to-ideal scales are a type of hedonic scale that measures deviation from a personal ideal point.

Selection of Scales

The type of scale used and its construction depend on a number of factors.

- *purpose of test*. In consumer testing, ranking or category scaling methods are usually used. Ranking and category can also be used with trained panels, but graphic and ratio scaling are also heavily used.
- *expertise of assessors*. Untrained assessors are generally poor discriminators and can discriminate only over a small number of scale points. Consequently, 5- or even 3-point category scales are often used with consumers. Trained panels can start with 5- or 7-point category scales, but as their discrimination ability increases, they can use effectively more scale points or graphic scales. When one is using inexperienced assessors or consumers, scales incorporating a "neutral point," such as the central point in an odd-numbered category scale, are sometimes avoided to minimize the risk of "fence sitting."
- *number of assessors*. Using small assessor numbers with a low number of category scale points will limit analysis to nonparametric tests.

- *data-handling facilities*. Category scaling responses can be entered relatively quickly onto a spreadsheet, whereas data from line scales must be measured, which can be a time-consuming procedure that is open to errors. Computerized data acquisition, either directly from a terminal or indirectly from optical readers, can avoid this problem.

In practice, establishing a trained sensory panel can often proceed from a category scale with a small number of scale points (e.g., 5) through a category scale with more points (e.g., 9) to a line scale. In a comparative study of the use of unstructured line scales, semistructured line scales, and category scales for profiling the texture of noodles by trained profile assessors (Galvez & Resurreccion, 1990), unstructured line scales were found to be more discriminating.

Scale Biases

Sensory analysts should be aware of difficulties that assessors have in using scales, and careful training is needed to ensure that scales are unambiguous and can measure the intended response. Common problems encountered are

- noncontinuous scales, often identified by an illogical central point
- confounded scales, which are attempting to measure more than one attribute
- central tendency, in which assessors are reluctant to use scale extremes
- contrast and convergence effects from the use of a sample very different from the others

The primary means of minimizing these scale biases is by thorough panel training with extensive use of references if possible.

Simple Descriptive Procedures

Scaling may often be needed to quantify a single, well-defined attribute. However, it should be established that there is no ambiguity in the attribute of interest. This is particularly relevant during product development or modification, when the assumption that a process or ingredient modification will change only a single attribute is frequently violated. Such changes are especially common when textural changes are a consequence of process or ingredient modifications. If it is suspected that several attributes might be of interest, then the profiling procedures described in the subsequent sections should be considered.

Quantitative Descriptive Analysis

Quantitative descriptive analysis (QDA) is a comprehensive system covering sample selection, assessor screening, vocabulary development, testing, and data analysis (Stone & Sidel, 1993). Variants of the original QDA procedures are probably used more than any other profiling procedure. The QDA technique uses small numbers of highly trained assessors. Typically, 6 to 12 people are screened for sensory acuity and trained to perform the descriptive task of evaluating the product. Three major steps are required: development of a standardized vocabulary, quantification of selected sensory characteristics, and analysis of the results by parametric statistics.

Vocabulary Development

Development of the vocabulary is a group process for creating a complete list of descriptors for the products under study. The assessors freely describe the flavor, appearance, odor, mouth-feel, texture, and aftertaste characteristics of different samples. No hedonic (*good* or *balanced*), general (*full* or *typical*), or intensity-based (*strong* or *weak*) terms are permitted. Well-defined attributes to describe the products should originate from the panel members themselves, as all members need to understand and feel comfortable with the descriptive terms being used. Published terms can provide starting points for further discussion in the panel orientation. Terminology should be consistent from product to product and tied to reference materials. The references decrease assessor variability, reduce the amount of time necessary to train sensory assessors, counterbalance cultural differences in interlaboratory studies, and allow calibration of the panel in the use of intensity scales. References should be simple, reproducible, and clear to the assessors and should illustrate only a single sensory descriptor; they can be single chemical substances or finished products and can be made available during both the training and the testing phase at various concentrations or intensities.

The attributes are collected and compiled into a master list. This individual preliminary evaluation of the samples may be revised during an open discussion to eliminate any redundant or synonymous descriptors. New terms might be added and physical references proposed. The panel leader condenses and formats the information into a proposal for standardized vocabulary. This vocabulary is then modified and improved in several interactive sessions. Multivariate statistical methods (e.g., factor analysis) are sometimes used to reduce the number of descriptors.

Intensity Measurement

Once assessors feel comfortable with the vocabulary, further training is performed. The number of training sessions is dependent on the subject's performance, product and attribute difficulties, and the time allowed for QDA testing. Replicates of each sample and session allow monitoring of within-sample variation and within-assessor variation of reliability, performance, response in time, and learning effects. Panel training increases assessors' sensitivity and memories and helps them to make valid, reliable judgments independent of personal preferences. A need for further training is indicated by significant differences among assessors and a lack of significant differences among samples.

Selection of the verbal anchors for labeling the scale is another key to valid measurements and depends crucially on the selection of appropriate training samples that illustrate the extreme scale points. Care must be taken not to use one scale for evaluating two different attributes. Terms such as *ideal*, *optimum*, *objectionable*, *appealing*, or *desirable* are hedonic and should not be used as anchors for intensity scales.

Once the training sessions have established satisfactory panel performance and removal of ambiguities and misunderstandings, the test samples can be evaluated. This is usually carried out in replicated (commonly three) sessions, using experimental designs that minimize biases. Analysis of QDA data is described later in this chapter.

The Spectrum Method

This more recent method provides a tool with which to design a descriptive procedure for a given product category (Meilgaard et al., 1991). The method resembles QDA in many respects: for example, the panel must be trained to fully define all product sensory attributes, to rate the intensity of each, and to include other relevant characterizing aspects such as change over time, difference in the order of appearance of attributes, and integrated total aroma and/or flavor impact.

Assessors develop their lists of descriptors by first evaluating a broad array of products that define the product category. The process includes using references to determine the best choice of term and to best define that term so that it is understood in the same way by all assessors. Words such as *vanilla, chocolate,* and *orange* must describe an authentic vanilla, chocolate, and orange character for which clear references are supplied. All terms from all assessors are then compiled into a list that is comprehensive yet not overlapping.

The Spectrum method is based on an extensive use of reference points. The choice of scaling technique may depend on the available facilities for computer manipulation of data and on the need for sophisticated data analysis. Whatever the scale chosen, it must have at least two, or preferably three or five, reference points distributed across the range.

Free-Choice Profiling (FCP)

FCP is a very different concept that removes the need to generate a compromise consensus vocabulary (Williams & Langron, 1983) and that can also be used in consumer research (Jack & Piggott, 1992; Jack, Piggott, & Paterson, 1993). Assessors are allowed to develop their own individual vocabularies to describe sensory perceptions and to use these to score sets of samples. As a consequence of removing the need to agree on vocabularies, FCP requires little training—only instruction in the use of the chosen scale. Assessors merely have to be objective, capable of using line scales, and able to use their developed vocabulary consistently. Thus, assessors can still be regarded as representing naive consumers. Characteristics being judged can be restricted by the panel leader, but the number of descriptors produced is limited only by the perceptual and descriptive skills of the assessors. A range of sensory characteristics such as appearance, flavor, aroma, and texture can be examined.

FCP has been used extensively in sensory analysis but might be expected to have a particularly important role in food texture research in situations where the different use of textural descriptors can cause considerable difficulties in assessors' reaching a consensus. In work with cheese analogues (Marshall, 1990; Marshall & Kirby, 1988), the method was found suitable for investigating the effects of moisture and fat content on textural characteristics.

The Texture Profile Method

Developed during the 1960s, this is the only method designed specifically for the assessment of texture. The starting point for the method is the classification of texture into mechanical characteristics, geometric characteristics, and other characteristics, as described earlier in this chapter. Assessors are selected on the basis of discrimination ability, interest, availability, and other general factors described previously. They are exposed to a wide range of products from the class of interest and are also given training in the underlying textural principles and the concepts of stress and strain. This is claimed to minimize problems with redundant terms.

In carrying out the training procedures, extensive use is made of standard rating scales, with each scale point referenced by a food product. An example of a hardness rating scale is shown in Table 3–5 (from International Standard ISO 11036; ISO, 1994). In practice, flexibility in the use of the standard scales has developed since the introduction of the method. Samples are evaluated individually by each assessor, and various types of

Table 3–5 A Standard Hardness Scale Used for the Texture Profile Method

Popular Term	*Panel Rating*	*Reference Product*	*Type*	*Sample Size*	*Temperature*
Soft	1	Cream cheese		1.25-cm cube	7°C–13°C
	2	Egg white	Hard boiled 5 min	1.25-cm tip	Room
	3	Frankfurter sausages	Large, uncooked, skinless	1.25-cm–thick slice	10°C–18°C
	4	Cheese	American, yellow, processed	1.25-cm cube	10°C–18°C
	5	Olives, green	Giant-sized, stoned	1 olive	10°C–18°C
	6	Peanuts	Cocktail type, under vacuum	1 whole nut	Room
	7	Carrots	Uncooked	1.25-cm slice	Room
	8	Peanut brittle	Candy part		Room
Hard	9	Rock candy			Room

Source: From International Standard ISO 11036:1994 and is reproduced with permission of the International Organization for Standardization, ISO. This standard can be obtained from any ISO member, or from the ISO Central Secretariat, Case postale 56, 1211 Geneva 20, Switzerland. Copyright remains with ISO.

scales have been used, such as category, line, and magnitude estimation scales. Data from these scales can be analyzed statistically, but a group consensus can also be reached by the assessors.

Time-Related Methods

Time-Intensity Profiling

Sensory attributes are not perceived instantaneously and can change in intensity with time in the mouth. Time-intensity methods are used to measure intensity of a specific attribute as a function of time and have been used extensively to investigate the temporal behavior of tastants, such as sweet and bitter molecules, and the release of volatile flavor materials from foods (Overbosch et al., 1991; Shamil, Wyeth, & Kilcast, 1992). Such studies are particularly important in the reformulation of foods that results in structural modifications. These structural modifications are often accompanied by textural changes, and these often result in complex perceptual phenomena that are direct consequences of the changes in texture with time, producing different flavor-release phenomena. Although the use of time intensity for flavor measurement is relatively well established, textural changes can also be monitored using the method (Burger, 1992; Duizer, Gullett, & Findlay, 1993). However, since most textural attributes have a nonzero intensity at the zero time point, there are additional practical difficulties.

Time-intensity data can be recorded simply on scales on paper, but other recording systems based on chart recorders or potentiometers have also been used. Computerized data acquisition has facilitated the recording of time-intensity data, using various data input methods such as moving a hand-held computer mouse or light pen, or touching a touch-sensitive screen. Assessors need training in the technique, but suitable procedures have yet to be established.

Progressive Profiling

A major limitation of the time-intensity method is that only a single attribute can be tracked with time; thus, if a number of important attributes are thought to be time dependent, separate sessions are needed for each attribute. In addition, concerns are being voiced that there is insufficient understanding of the processes occurring when assessors are carrying out a continuous scaling with time. Difficulties encountered in time-intensity profiling prompted the development of a hybrid technique, *progressive profiling* (Jack, Piggott, & Paterson, 1994). In this technique, assessors carried out a profile on a set of texture descriptors at each chew stroke over the mastication period. Such a method has a number of potential advantages: several attributes can be assessed in one session, scaling is reduced to a unidimensional process, and the most important aspects of the shape of a time-intensity curve are retained.

UNIVARIATE ANALYSIS OF SCALED DATA

Univariate procedures are the starting point for the analysis of any sensory data, regardless of their apparent complexity. The procedures can be used at different stages of a sensory program but are particularly useful in assessing the performance of assessors undergoing training for profile panels and for exploratory investigation of scaled data. The techniques used can be found in numerous standard texts on statistics, but a number of useful publications specific to sensory evaluation are available (e.g., Lahiff & Leland, 1994; Lea, Næs, & Rødbotten, 1997; Meilgaard et al., 1991; O'Mahony, 1986; Smith, 1988).

An important consideration in selecting appropriate analysis techniques is the nature and distribution of the data. Prior to the use of any statistical procedure, the form of the data should be examined by visualization techniques, such as the use of scatterplots. Ranked data and data that are not normally distributed are analyzed by nonparametric methods. The most commonly used method for analyzing ranked data is the Friedman test (British Standard BS 5929, Part 6, the equivalent of International Standard ISO 8587; British Standards Institution, 1989b), which determines whether there are significant rank differences among samples. This can be followed by a multiple-comparison test for differences in rank between pairs of samples, or Page's test to see if the samples fall into a specified order. Earlier texts recommended the use of the Kramer test for the analysis of such data, but this test has been heavily criticized as statistically flawed and should be avoided (Joanes, 1985).

It is frequently assumed that sensory data are normally distributed and that parametric tests can be used. The distribution of all sensory data should be examined, however, especially when relatively small numbers of responses are being used. If in doubt, nonparametric tests can be employed.

The procedures most commonly used to examine sensory data are *t* tests, analysis of variance (ANOVA), and multiple-comparison tests. The *t*-test procedure can be used to compare the mean scores from two samples, usually in the paired format if the same assessors have assessed both samples. If more than two samples are to be compared, two-way ANOVA is used with the assessors and samples as factors. Assessor × sample interactions are also usually examined. If significant differences for a given attribute are identified by ANOVA, then multiple-comparison tests can be used to identify which samples differ. The various types of multiple-comparison tests and their use have been described by O'Mahony (1986).

MULTIVARIATE ANALYSIS OF SCALED DATA

In most applications of any form of sensory testing, the intensities of many attributes are being measured, leading to highly complex data sets. Multivariate analysis (MVA) methods are increasingly being seen as essential in interpreting such data sets, and several different uses are evident.

- assistance in panel training, including assessment of assessors' performance and reduction of attribute lists in forming profile vocabularies
- simplifying the complexity of data presentation. Visualization of the relationships between two attributes is easily accomplished, and visualization of three attributes presents few difficulties, but greater numbers of attributes present substantial problems
- identification of redundancy in the use of descriptive attributes
- investigation of the underlying structure among products and among the attributes characterizing them
- construction of "maps" visualizing the similarities and dissimilarities between products

Numerous MVA techniques are available for the analysis of sensory data, and several valuable texts and papers describe their use (e.g., Aishima & Nakai, 1991; Martens & Russwurm, 1983; Meilgaard et al., 1991; Piggott, 1986, 1988). The most common uses of MVA techniques are described briefly below, with some applications to food texture measurement.

Multivariate Analysis of Variance (MANOVA)

The multivariate equivalent of ANOVA, MANOVA, has seen relatively limited use in sensory analysis but can be useful in the assessment of individual assessors' performance (Noronha, Damásio, Pivatto, & Negrillo, 1995; Powers, 1988b; Powers, Shinholser, & Godwin, 1985) or in situations in which ANOVA cannot make distinctions between samples. MANOVA can reduce the evaluation of large numbers of probability levels to a single value but needs relatively large numbers of measured data points.

Cluster Analysis

Cluster analysis methods have not been widely used on sensory data but are a very useful preliminary before progressing to other multivariate methods; a comprehensive description of uses is given in Jacobsen and Gunderson (1986). Cluster analysis methods identify groups of observations on the basis of their degree of similarity and are classified into two types of algorithms: hierarchical and nonhierarchical. The main distinction is that once an observation has been assigned to a cluster by hierarchical methods, it cannot be moved to another cluster, as is possible in nonhierarchical methods.

Hierarchical methods can operate in two different directions, but the most common is initially to assign each observation into a cluster of one (i.e., its own) and then to find the closest observations successively until all observations are in one cluster. The region of practical interest lies in an intermediate region in such a procedure. The results of such an analysis are often displayed in the form of a dendrogram, as shown in Figure 3–6. In this figure, the lower region shows the 11 individual observations, which are then progressively clustered on the basis of closeness. At the similarity level of 58%, for example, four clusters are evident. The first contains Observations 1, 5, 4, 10, 6, and 2; and the second, Observations 7 and 8; the third, Observation 9; and the fourth, Observations 3 and 11. The different cluster analysis algorithms use different ways of measuring the distance between clusters.

Nonhierarchical methods include the *k*-means method and the fuzzy objective functions (FOV) method. For both, the user must indicate the number of clusters that are believed to exist, and iterative mathematical methods are used to assign observations until there is no movement to the center of the cluster.

The methods can be used to look for groupings in assessors, consumers, attributes, or samples. However, there may be some software limitations to the shape of the data matrix that can be analyzed. Godwin, Bergmann, and Powers (1978) used cluster analysis to identify two distinct assessor groupings in a study of processed green beans. In an extensive study of the texture and flavor properties of ice cream, cluster analysis was used in conjunction with principal component analysis (PCA) to study the relationships among ice cream samples (King & Moreau, 1995).

The many different algorithms used can give rise to distinctly different groupings, and it is often recommended that a number of different methods be used to check the stability of the analyses: if very different groupings are a consequence, then alternative methods should be employed.

PCA

Presenting and interpreting the relationships among a small number of variables is relatively easy, using either a number of two-dimensional plots or, with the increasing use of modern computer software and with a little

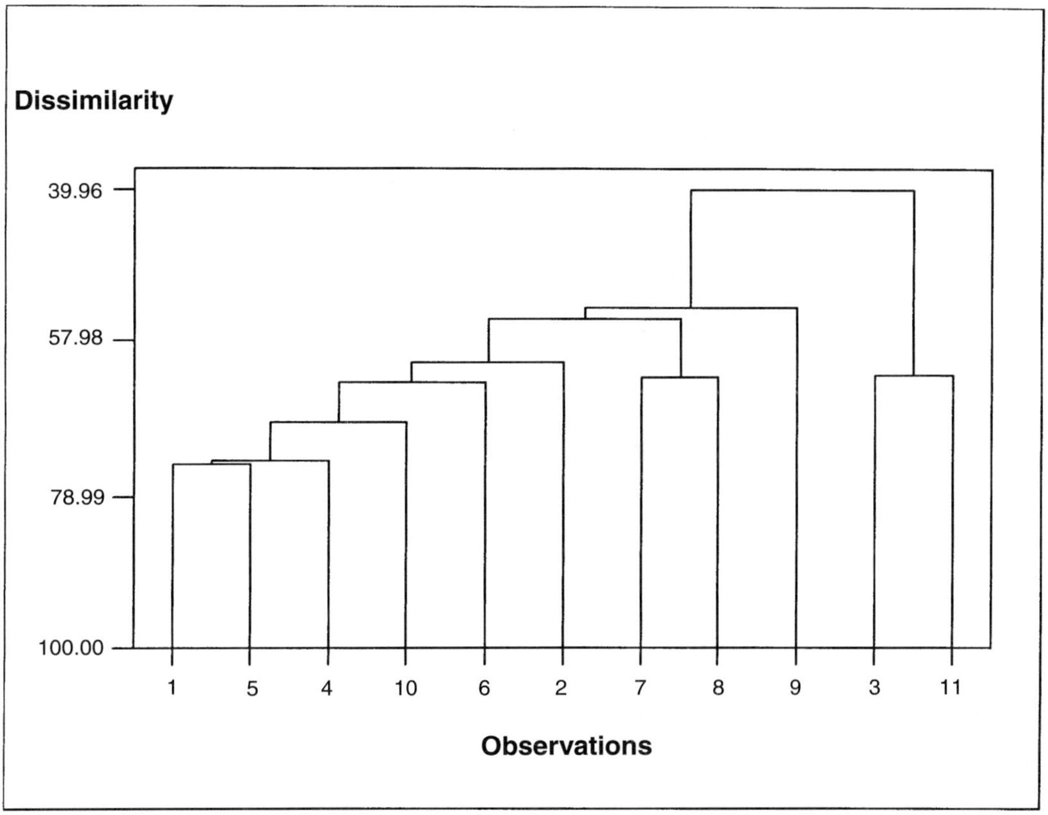

Figure 3–6 Dendrogram Showing Intersample Groupings Derived from a Hierarchical Cluster Analysis

more difficulty, three-dimensional plots. PCA is a data reduction technique that replaces a large number of original variables by a smaller number of linear combinations, while still explaining a substantial proportion of the original variation in the data. Essentially, PCA projects an *n*-dimensional space onto a two-dimensional plot.

PCA analyzes the correlation structure in the data set and identifies the axis along which the maximum variation occurs. A second principal axis is then identified orthogonal to the first axis, corresponding to the second greatest amount of variation, and so on. A geometrical representation of the process is shown in Figure 3–7 for a trivial two-dimensional data set. The first step is equivalent to a translation to center the coordinate system at the center of gravity of the points, followed by a rotation to align the direction of maximum variation with the first principal axis.

The new axes are linear combinations of the original axes, and the coefficients, or loadings, measure the importance of the original variables on each principal component. A useful reduction will often retain 70% to 80% of the variation in the first three dimensions. Plots of the loadings give valuable information on the relationships between variables, especially on the correlation structure between attributes and the possible use of different terms for the same perceptual response. In addition, the locations of the samples can be plotted and groupings identi-

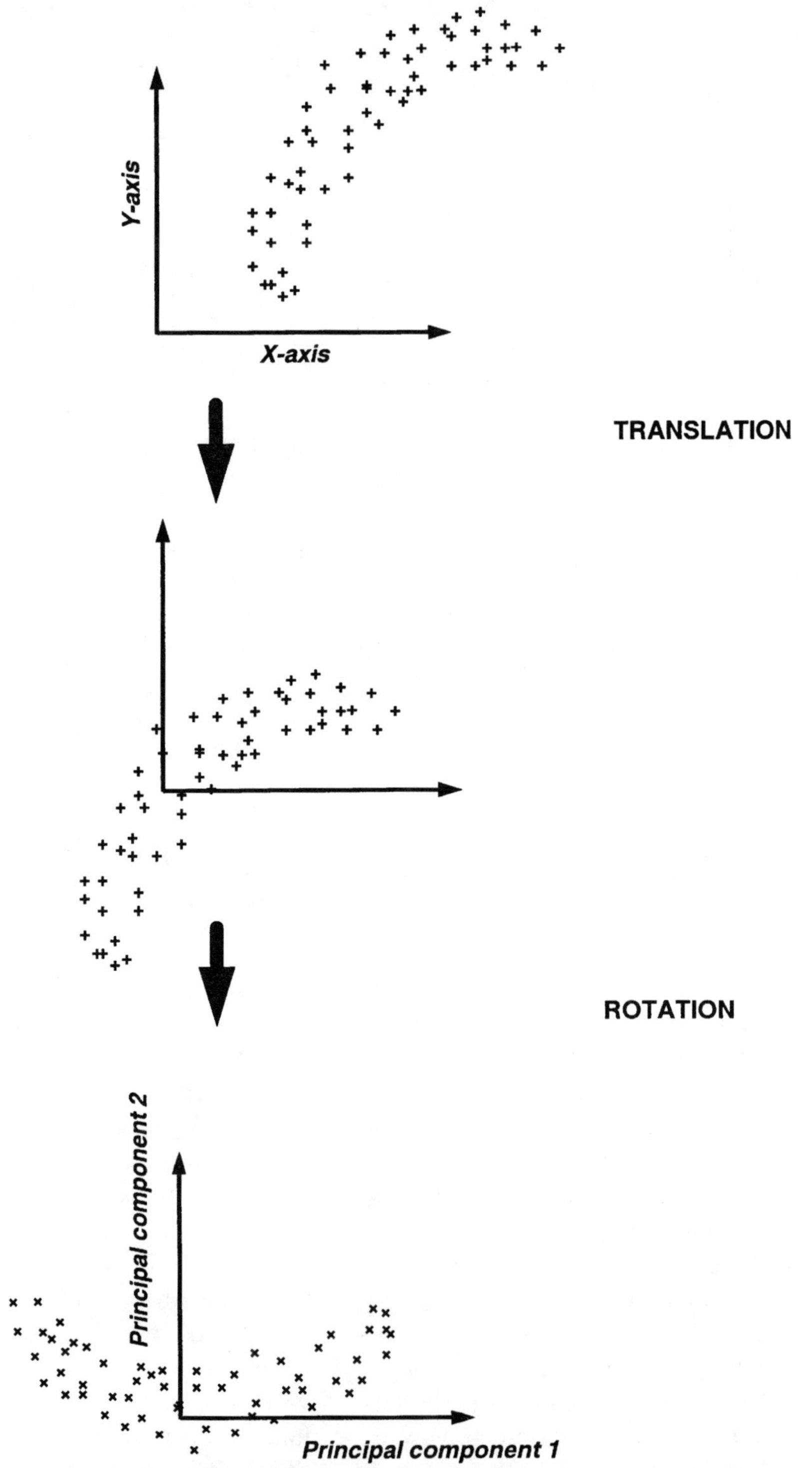

Figure 3–7 Geometric Representation of the Steps Involved in a Principal Component Analysis

fied in such a way that closely positioned samples are perceived as similar and samples positioned far apart are perceived as very different. The samples plot is often superimposed onto the attribute plot to aid interpretation. It should be remembered, however, that information may be contained in minor dimensions that could be important in determining consumer response and that these dimensions, even if they contribute less than 10% to the total variance, should not be rejected out of hand.

PCA is rapidly becoming a routine statistical procedure for analyzing sensory profile data, and extensive examples of its use are given in Piggott and Sharman (1986) and Powers (1988a). In addition to being used to analyze sensory profile data, PCA is increasingly used to analyze combined sets of sensory and instrumental data.

Factor Analysis

A closely related method to PCA, factor analysis (FA) is less frequently used in analyzing sensory data. Whereas PCA separates out the signal from the noise in the data and models the signal only, FA also models the noise. Although FA is less suitable than PCA for nonstatisticians, use has been made of FA in detecting underlying structure in data sets—for example, in the investigation of the texture of chicken meat (Lyon, 1980). Piggott and Sharman (1986) have pointed out that the terms *PCA* and *FA* are sometimes used interchangeably, especially in the United States, and that care should be taken in identifying which technique is being referred to.

Generalized Procrustes Analysis

Generalized Procrustes Analysis (GPA), described in detail by Arnold and Williams (1986), is an essential means of analyzing FCP data but also has value in analyzing conventional profile data. As used to analyze FCP data, GPA carries out an iterative matching of the individual configurations generated by the individual assessors to give a consensus configuration.

In using GPA for the analysis of conventional profile data, three distinct steps are involved:

1. translation of the coordinate system to a common origin. This makes adjustment for assessors using different parts of the scale.
2. rotation/reflection of axes, while maintaining intersample distances. This allows for the different interpretation of attributes.
3. rescaling to adjust for differences in the range of scaling for different assessors

The configurations reached after these processes are called *principal axes* and, like principal components, have decreasing proportions of variance associated with them. One particular value of GPA with conventional profile data lies in the examination of assessor performance, particularly the identification of outliers. Hunter and Muir (1995) carried out a detailed comparison of the use of PCA and GPA on sensory profile data of cheeses and found that each method gave different configurations and that GPA gave a more detailed summary of the data.

The general procedure for analyzing FCP data is similar, but the reflection/rotation stage becomes highly significant, since up to this stage no attempt is made to match the terms used by one assessor to the terms used by any other. In addition, scaling of the sample space of each assessor by stretching or shrinking is carried out to improve agreement between assessors. The interpretation of the new axes is carried out by examining the correlations between the original attributes for each assessor with the transformed attributes of the consensus. Jack et al. (1993) used GPA in a consumer profile of the appearance and texture of cheddar cheese using FCP. The first three dimensions accounted for 56% of the total variance and were readily interpreted.

Although GPA has obvious applications in the analysis of both conventional profile data and FCP data, there are concerns that interpretation of the consensus configurations can be difficult and can rely too heavily on the subjective interpretation of the sensory analyst.

Multidimensional Scaling

Multidimensional scaling (MDS) methods are frequently called *similarity scaling* and make up the analysis of sensory data in the form of a similarity-dissimilarity matrix. Details of the procedure can be found in Schiffman and Beeker (1986) and in MacFie and Thomson (1988). A number of computer packages are available that can carry out different types of analysis. In their usual form, MDS methods collect data on overall similarities from pairwise comparisons of all possible pairings. Since this is carried out by all assessors, a formidable number of comparisons is involved, and this has limited the applications of the method for sensory studies. More recently, however, Bertino and Lawless (1993) have studied alternative methods of data collection and have claimed some advantages. In a study on the sensory characteristics of yogurts, a combination of overall similarity scaling and attribute scaling was used (Poste & Patterson, 1988).

Analysis of Time-Intensity Data

The analysis and interpretation of time-intensity data present two specific problems that have limited the use of the technique. First, the information content is very high, and, especially if computerized acquisition systems are used, a very large number of data points are collected. Computerization can ease the parameterization and subsequent analysis of these curves, but Cliff and Heymann (1993) noted that different terminology and abbreviations have been used by different researchers and presented a consolidated list. Second, although trained assessors can generate highly reproducible curve shapes, the curve shapes from different individuals are unique and can vary substantially. Analysis of these different shapes can be complex, since a simple averaging of intensities at fixed times can mask discrimination between samples if the individual curve shapes are very different. Averaging can be restricted to those curves (assessors) exhibiting similar shapes, but more complex methods have been proposed.

- normalizing the data in the intensity direction and calculating averages in the time direction for the ascending and descending curve portions (Overbosch Enden, & Keur, 1986)
- normalizing in both intensity and time directions (Liu & MacFie, 1990)
- using PCA for the generation of summary principal curves (van Buuren, 1992)

Variants of these methods are under development, but a more meaningful approach may be to carry out research into the underlying factors that can generate such varied shapes from different individuals. This could involve factors such as mastication patterns, salivation patterns, and even motor difficulties in coordinating the movement of the recording device.

CONCLUSION

Meaningful quantification of food texture attributes presents some of the most daunting difficulties that can face the sensory analyst, but also some of the most exciting and rewarding challenges. In the evaluation of flavor characteristics, problems in descriptive terminology can be relatively easily simplified by the use of chemical references that are quite precise in defining the attribute in question. In the case of texture, however, it is much more difficult to devise references that exemplify a single textural attribute, and this necessitates the use of large numbers of complex commercial products in some of the profiling methods. Even such extensive reference use does not guarantee that all assessors are using descriptive vocabularies uniformly, however, and sophisticated statistical analysis methods are of particular value in extracting the underlying information from textural data sets. It is perhaps inevitable that free-description methods such as FCP are being used to circumvent the perceived difficulties in the use of consensus methods, although these methods are not without their own problems.

Generation and analysis of textural data will therefore continue to demand the development and use of the most sophisticated sensory methods, and it is inconceivable that measurement of such an important and complex modality by instruments can ever replace sensory evaluation. However, the future is likely to see much more emphasis being put on the synergistic use of combined sensory and instrumental methods to understand the important relationship between flavor and texture perception and how this is influenced by oral processes. Development of practical combined in-mouth sensory and instrumental techniques will provide important new tools for the food industry.

REFERENCES

Aishima, T., & Nakai, S. (1991). Chemometrics in flavor research. *Food Reviews International, 7*(1), 33–101.

Arnold, G.M., & Williams, A.A. (1986). The use of generalised Procrustes techniques in sensory analysis. In J.R. Piggott (Ed.), *Statistical procedures in food research* (pp. 233–254). London: Elsevier Applied Science.

Bertino, M., & Lawless, H.T. (1993). Understanding mouthfeel attributes: A multidimensional scaling approach. *Journal of Sensory Studies, 8,* 101–114.

Bourne, M.C. (1982). *Food texture and viscosity.* New York: Academic Press.

Brandt, M.A., Skinner, E.Z., & Coleman, J.A. (1963). Texture profile method. *Journal of Texture Studies, 28,* 404–409.

Brennan, J.G. (1988). Texture perception and measurement. In J.R. Piggott (Ed.), *Sensory analysis of foods* (2nd ed., pp. 69–102). London: Elsevier Applied Science.

British Standards Institution. (1986). *British Standard BS 5929, Part 1: Introduction and general guide to methodology.* Milton Keynes: British Standards Institution.

British Standards Institution. (1989a). *British Standard BS 7183: Guide to design of test rooms for sensory analysis of food.* Milton Keynes: British Standards Institution.

British Standards Institution. (1989b). *British Standard BS 5929, Part 6: Methods for sensory analysis of food, Part 6. Ranking.* Milton Keynes: British Standards Institution.

British Standards Institution. (1992). *British Standard BS 5098: Glossary of terms relating to sensory analysis.* Milton Keynes: British Standards Institution.

British Standards Institution. (1993). *British Standard BS 7667, Part 1: Assessors for sensory analysis. Part 1. Guide to the selection, training and monitoring of selected assessors.* Milton Keynes: British Standards Institution.

Burger, J. (1992). Sensory evaluation techniques for chocolate with different types of cocoa butter products. *Manufacturing Confectioner, 72*(10), 56–60.

Cardello, A.V. (1994). Consumer expectations and their role in food acceptance. In H.J.H MacFie & D.M.H. Thomson (Eds.), *Measurement of food preferences* (pp. 253–297). London: Blackie A&P.

Cardello, A.V. (1996). The role of the human senses in food acceptance. In H.L. Meiselman & H.J.H. MacFie (Eds.), *Food choice, acceptance and consumption* (pp. 1–82). London: Blackie A&P.

Cliff, M., & Heymann, H. (1993). Development and use of time-intensity methodology for sensory evaluation: A review. *Food Research International, 26,* 375–385.

Drake, B. (1989). Sensory textural/rheological properties: A polyglot list. *Journal of Texture Studies, 20,* 1–27.

Dubose, C.N., Cardello, A.V., & Maller, O. (1980). Effects of colorants and flavorants on identification, perceived flavor intensity and hedonic quality of fruit flavored beverages and cake. *Journal of Food Science, 45*, 1393–1415.

Duizer, L.M., Gullett, E.A., & Findlay, C.J. (1993). Time-intensity methodology for beef tenderness perception. *Journal of Food Science, 58,* 943–947.

Frijters, J.E.R. (1988). Sensory difference testing and the measurement of sensory discriminability. In J.R. Piggott (Ed.), *Sensory analysis of foods* (2nd ed., pp. 134–154). London: Elsevier Applied Science.

Galvez, F.C.F., & Resurreccion, A.V.A. (1990). Comparison of three descriptive analysis scaling methods for the sensory evaluation of noodles. *Journal of Sensory Studies, 5,* 251–263.

Godwin, D.R., Bergmann, R.E., & Powers, J.J. (1978). Use of cluster analysis to evaluate sensory-objective relations of processed green beans. *Journal of Food Science, 43*, 1229–1334.

Guinard, J.-X., & Mazzucchelli, R. (1996). The sensory perception of texture and mouthfeel. *Trends in Food Science and Technology, 7,* 213–219.

Hunter, E.A., & Muir, D.D. (1995). A comparison of two multivariate methods for the analysis of sensory profile data. *Journal of Sensory Studies, 10,* 89–104

Hutchings, J.B. (1994). *Food colour and appearance.* London: Blackie A&P.

International Organization for Standards. (1994). *ISO 11036: Sensory analysis: Methodology: Texture profile.* Available from ISO Central Secretariat, Case postale 56, 1211 Geneva 20, Switzerland.

Jack, F.R., & Piggott, J.R. (1992). Free choice profiling in consumer research. *Food Quality and Preference, 3*, 129–134.

Jack, F.R., Piggott, J.R., & Paterson, A. (1993). Discrimination of texture and appearance in cheddar cheese using consumer free-choice profiling. *Journal of Sensory Studies, 8*, 167–176.

Jack, F.R., Piggott, J.R., & Paterson, A. (1994). Analysis of textural changes in hard cheese during mastication by progressive profiling. *Journal of Food Science, 59,* 539–543.

Jacobsen, T., & Gunderson, R.W. (1986). Applied cluster analysis. In J.R. Piggott (Ed.), *Statistical procedures in food research* (pp. 361–408). London: Elsevier Applied Science.

Joanes, D.N. (1985). On a rank sum test due to Kramer. *Journal of Food Science, 50,* 1442–1444.

Jowitt, R. (1974). The terminology of food texture. *Journal of Texture Studies, 5,* 351–358.

King, B.M., & Moreau, N. (1995). Sensory evaluation of aroma, texture and taste. In S. Poretta (Ed.), *Consumer preference and sensory analysis* (pp. 98–121). Maarssen: Miller Freeman.

Lahiff, M., & Leland, J.V. (1994). Statistical methods. In G. Reineccius (Ed.), *Source book of flavors* (pp. 743–787). New York: Chapman & Hall.

Lea, P., Næs, T., & Rødbotten, M. (1997). *Analysis of variance for sensory data.* New York: John Wiley.

Liu, Y.-H., & MacFie, H.J.H. (1990). Methods for averaging time-intensity curves. *Chemical Senses, 15,* 471–484.

Lyon, B.G. (1980). Sensory profiling of canned boned chicken: Sensory evaluation procedures and data analysis. *Journal of Food Science, 45,* 1341–1346.

MacFie, H.J.H., & Thomson, D.M.H. (1988). Preference mapping and multidimensional scaling. In J.R. Piggott (Ed.), *Sensory analysis of foods* (2nd ed., pp. 381–410). London: Elsevier Applied Science.

Marshall, R.J. (1990). Combined instrumental and sensory measurements of the role of fat in food texture. *Food Quality and Preference, 2*, 117–124.

Marshall, R.J., & Kirby, S.P.J. (1988). Sensory measurement of food texture by free-choice profiling. *Journal of Sensory Studies, 3*, 63–80.

Martens, H., & Russwurm, H., Jr. (1983). *Food research and data analysis*. London: Applied Science Publishers.

Maruniak, J.A. (1988). The sense of smell. In J.R. Piggott (Ed.), *Sensory analysis of foods* (2nd ed., pp. 25–68). London: Elsevier Applied Science.

Meilgaard, M., Civille, G.V., & Carr, B.T. (1991). *Sensory evaluation techniques* (2nd ed.). Boca Raton, FL: CRC Press.

Moskowitz, H.R., & Krieger, B. (1995). The contributions of sensory liking to overall liking: An analysis of 6 food categories. *Food Quality and Preference, 6,* 83–90.

Muñoz, A.M., Civille, G.V., & Carr, B.T. (1992). *Sensory evaluation in quality control.* New York: Van Nostrand Reinhold.

Noronha, R.L., Damásio, M.H., Pivatto, M.M., & Negrillo, B.G. (1995). Development of the attributes and panel screening for texture descriptive analysis of milk gels aided by multivariate statistical procedures. *Food Quality and Preference, 6,* 49–54.

O'Mahony, M. (1979). Short-cut signal detection measures for sensory analysis. *Journal of Food Science, 44,* 302–303

O'Mahony, M. (1986). *Sensory evaluation of food: Statistical methods and procedures.* New York: Marcel Dekker.

O'Mahony, M. (1995). Who told you the triangle test was simple? *Food Quality and Preference, 6,* 227–238.

Overbosch, P., Afterof, W.G.M., & Haring, P.G.M. (1991). Flavour release in the mouth. *Food Reviews International, 7,* 137–184.

Overbosch, P., Enden, J.C. van den, & Keur, B.M. (1986). An improved method for measuring perceived intensity/time dependent relationships in human taste and smell. *Chemical Senses, 11,* 331–338.

Piggott, J.R. (1986). *Statistical procedures in food research.* London: Elsevier Applied Science.

Piggott, J.R. (1988). *Sensory analysis of food* (2nd ed.). London: Elsevier Applied Science.

Piggott, J.R. (1995). Design questions in sensory and consumer science. *Food Quality and Preference, 6,* 217–220.

Piggott, J.J., & Sharman, K. (1986). Methods to aid interpretation of multivariate data. In J.R. Piggott (Ed.), *Statistical procedures in food research* (pp. 181–232). London: Elsevier Applied Science.

Plattig, K.-H. (1988). The sense of taste. In J.R. Piggott (Ed.), *Sensory analysis of foods* (2nd ed., pp. 1–24). London: Elsevier Applied Science.

Poste, L.M., Mackie, D.A., Butler, G., & Larmond, E. (1991). *Laboratory methods for sensory analysis of food.* (Pub. No. 1863/E). Ottawa: Agriculture Canada.

Poste, L.M., & Patterson, C.F. (1988). Multidimensional scaling: Sensory analysis of yoghurt. *Journal of the Canadian Institute of Food Science and Technology, 21,* 271–278.

Powers, J.J. (1988a). Descriptive methods of analysis. In J.R. Piggott (Ed.), *Sensory analysis of foods* (2nd ed., pp. 187–266). London: Elsevier Applied Science..

Powers, J.J. (1988b, November). Uses of multivariate methods in screening and training sensory panels. *Food Technology, 42*(2), 123–124, 126–127, 136.

Powers, J.J., Shinholser, K., & Godwin, D.R. (1985). Evaluation assessors' performance and panel homogeneity using univariate and multivariate statistical analysis. In J. Adda (Ed.), *Progress in Flavour Research 1984; Proceedings of the 4th Weurman Flavour Research Symposium* (pp. 193–209). Amsterdam: Elsevier Applied Science.

Schiffman, S.S. (1973). *The dietary rehabilitation clinic and a multi-aspect, dietary and behavioural approach to the treatment of obesity. B) taste and smell of foods.* Paper presented at the Meetings of the Association for Advances in Behavioral Therapy, Miami Beach, FL.

Schiffman, S.S., & Beeker, T.G. (1986). Multidimensional scaling and its interpretation. In J.R. Piggott (Ed.), *Statistical procedures in food research* (pp. 255–292). London: Elsevier Applied Science.

Shamil, S.H., Wyeth L.J., & Kilcast, D. (1992). Flavour release and perception in reduced-fat foods. *Food Quality and Preference, 3*(1), 51–60.

Shepherd, R., & Sparks, P. (1994). Modelling food choice. In H.J.H. MacFie & D.M.H. Thomson (Eds.), *Measurement of food preferences* (pp. 202–226). London: Blackie A&P.

Sherman, P. (1969). A texture profile of foodstuffs based on well-defined rheological properties. *Journal of Food Science, 34,* 458–462.

Smith, G.L. (1988). Statistical analysis of sensory data. In J.R. Piggott (Eds). *Sensory analysis of foods* (2nd ed., pp. 335–380). London: Elsevier Applied Science.

Stone, H., & Sidel, J.L. (1993). *Sensory evaluation practices.* Orlando, FL: Academic Press.

Szczesniak, A.S. (1963). Classification of textural characteristics. *Journal of Food Science, 28,* 385–389.

Szczesniak, A.S. (1971). Consumer awareness of texture and other food attributes II. *Journal of Texture Studies, 2,* 196–206.

Szczesniak, A.S., Brandt, M.A., & Friedman, H.H. (1963). Development of standard rating scales for mechanical parameters of texture and correlation between the objective and sensory methods. *Journal of Food Science, 28,* 397–403.

Szczesniak, A.S., & Kahn, E.L. (1971). Consumer attitudes to and awareness of food texture 1: Adults. *Journal of Texture Studies, 2*, 280–295.

Szczesniak, A.S., & Kleyn, D.H. (1963). Consumer awareness of texture and other food attributes. *Food Technology, 17*, 74–77.

van Buuren, S. (1992). Analyzing time-intensity responses in sensory evaluation. *Food Technology, 2*, 101–104.

Vickers, Z.M. (1991). Sound perceptions and food quality. *Journal of Food Quality, 14*(1), 87–96.

Watts, B.M., Ylimaki, G.L., Jeffery, L.E., & Elias, L.G. (1989). *Basic sensory methods for food evaluation*. Ottawa, Canada: International Development Research Centre.

Williams, A.A., & Langron, S.P. (1983). A new approach to sensory profile analysis. In J.R. Piggott (Ed.), *Flavour of distilled beverages: Origin and development* (pp. 219–224). Chichester, UK: Ellis Horwood Ltd.

Yoshikawa, S., Nishimaru, S., Tashiro, T., & Yoshida, M. (1970a). Collection and classification of words for description of food texture. I. Collection of words. *Journal of Texture Studies, 1*, 437–442.

Yoshikawa, S., Nishimaru, S., Tashiro, T., & Yoshida, M. (1970b). Collection and classification of words for description of food texture. II. Texture profiles. *Journal of Texture Studies, 1*, 443–451.

Yoshikawa, S., Nishimaru, S., Tashiro, T., & Yoshida, M. (1970c). Collection and classification of words for description of food texture. III. Classification of multivariate analysis. *Journal of Texture Studies, 1*, 452–463.

CHAPTER 4

Rheological Classification of Foods and Instrumental Techniques for Their Study

Ton van Vliet

The rheological properties of foods vary widely, from thin liquids such as water, wine, and whiskey to hard, solid products such as some biscuits and candies. The latter products have to fracture or melt during consumption. Although, strictly speaking, fracture properties do not belong to the field of rheology, their determination will be incorporated in this chapter.

The choice of an instrumental technique for the determination of the rheological properties of a food will depend to a large extent on these properties. For liquids, different techniques have to be used than for solids. Another aspect is that small-deformation properties and large-deformation and fracture properties of solid and solidlike materials often cannot be determined by the same technique. A frequently neglected factor is that the choice of a technique has to depend on the purpose of the measurement. For quality control, a simple, cost-efficient, empirical method may be appropriate. However, designing an energy-efficient heat exchanger for a liquid product requires precise information on its flow properties under the prevalent conditions, so an accurate but much more expensive test will be appropriate.

Below, first, the most important basic rheological concepts are discussed. There follows a short section on the phenomenological classification of foods according to their rheological behavior. Finally, an overview is given of the main instrumental techniques available at present for the determination of the rheological behavior of food materials. The emphasis is on the fundamentals of the various techniques. For details on specific techniques, the reader is referred to specialized books and papers (e.g., van Vliet & Luyten, 1995; Vincent, 1992; Whorlow, 1992).

BASIC RHEOLOGICAL CONCEPTS

Rheology is the study of the relation between forces exerted on a material and the ensuing deformation as a function of time. To correct for variations in the size and shape of the materials, forces have to be recalculated to force per unit area (*stress*, usually expressed in Pa $\equiv$ N $\cdot$ m^{-2}) and the deformation to be taken relative to the relevant size (length) of the material. Forces are characterized by both a size and a direction; they are vectors. The direction is fixed relative to a (often rectangular) coordinate system and characterized by decomposing the force into three components, one normal and two in the plane of the surface at the point on which the force acts. The first gives rise to the so-called *normal stresses* and the other two to the so-called *shear stresses*. It can be shown (Whorlow, 1992) that nine components are needed to describe all stress components when an arbitrary force is exerted on a material. Stress is a so-called *second-order tensor*.

When a stress is applied to a mechanically nonhomogeneous material, as many foods are, the ensuing relative deformation may vary over the product. Therefore, the change in distance between two points in the material relative to the original distance should be taken as a quantitative measure of the local relative deformation. This ratio is called the *strain*. Only in the case of a homogeneous deforming material, will the macroscopic strain be equal to the local strains. A second complication is that a change in distance always involves a size and a direction; it is a vector. The same applies to the original distance between the points considered. Thus, a strain is a vector divided by a vector, resulting in a second-order tensor, and nine components are needed to describe it completely.

This tensorial character of both stress and strain makes it essential to perform rheological experiments in such a way that most of the stress and strain components can be neglected for the calculation of a specified characteristic parameter of the rheological behavior of a material.

Before defining these basic rheological parameters, it is useful to distinguish two ideal types of rheological behavior: namely, ideal liquid or viscous behavior and ideal solid or elastic behavior. Viscous materials start to flow at a certain rate when a stress is applied to the material, and after removal of the stress, they keep the shape they had at the moment stress was taken away. Elastic materials deform instantaneously to a certain extent when a stress is applied and regain their original shape after the stress has been removed.

In general, three simple types of deformation of a food product can be distinguished: all-sided compression; uniaxial extension, or compression; and shear. In the case of all-sided compression, the test piece is pressurized, and the relative change in volume $\Delta V/V$ is determined. The behavior of the material is characterized by the (all-sided) compression modulus or bulk modulus K, which is calculated from $P/(\Delta V/V)$, where P is the applied pressure. Since all-sided compression is not very discriminating between materials, except between those with a differing air content, it will not be dealt with further.

Shear Deformation

When a shear stress σ is applied to an elastic material, it will deform in shear (Figure 4–1). The shear strain γ is given by $\tan \alpha = \Delta x/\Delta y$. During such a deformation, only the shape of the product will change while the volume stays (in a first approximation) constant. For an elastic material,

Equation 1

$$G \equiv \frac{\sigma}{\gamma}$$

where G is the shear modulus.

In the case of a liquid, the material will start to flow at a certain shear rate $\dot{\gamma} = \mathrm{d}\gamma/\mathrm{d}t$, where t is time. For a liquid confined between two parallel plates of which the lower one is fixed, the upper plate will start to move with a velocity v on application of a shear stress (Figure 4–2). The shear rate $\dot{\gamma}$ is then equal to $\mathrm{d}\gamma/\mathrm{d}t = \mathrm{d}(\Delta x/\Delta y)/\mathrm{d}t = (\mathrm{d}(\Delta x)/\mathrm{d}t)(1/\mathrm{d}\Delta y) = v/h$, so $\dot{\gamma}$ can be calculated easily if the distance and the velocity difference between the two plates are known. The velocity gradient is perpendicular to the direction of the flow. For an ideally viscous material,

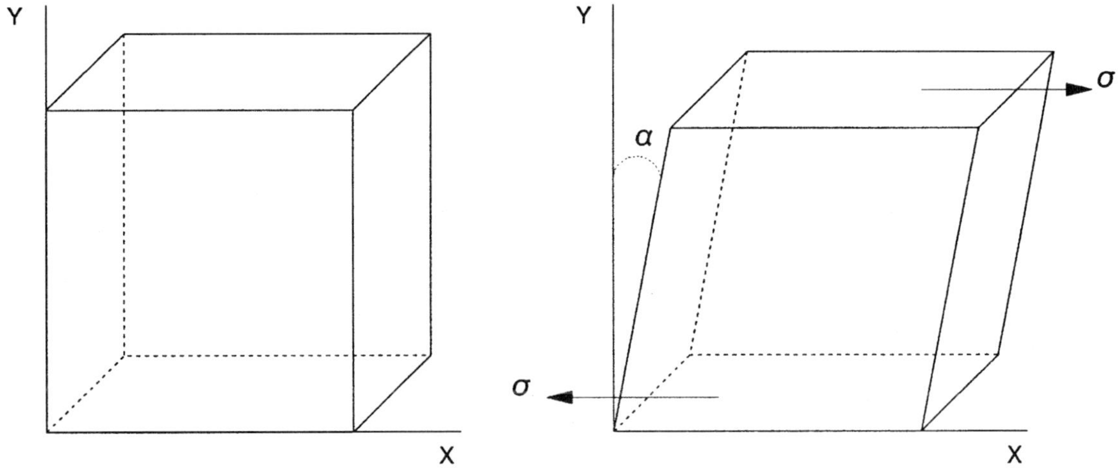

Figure 4–1 Shear Deformation

Equation 2

$$\eta \equiv \frac{\sigma}{\dot{\gamma}}$$

where η is the viscosity. For so-called Newtonian liquids, η is a constant and independent of $\dot{\gamma}$.

Uniaxial Compression or Extension

A material will deform in uniaxial compression or extension if a normal compressive or tensile stress σ_E is applied (Figure 4–3). The tensile or compressive strain ϵ is (for small deformations) given by the change in

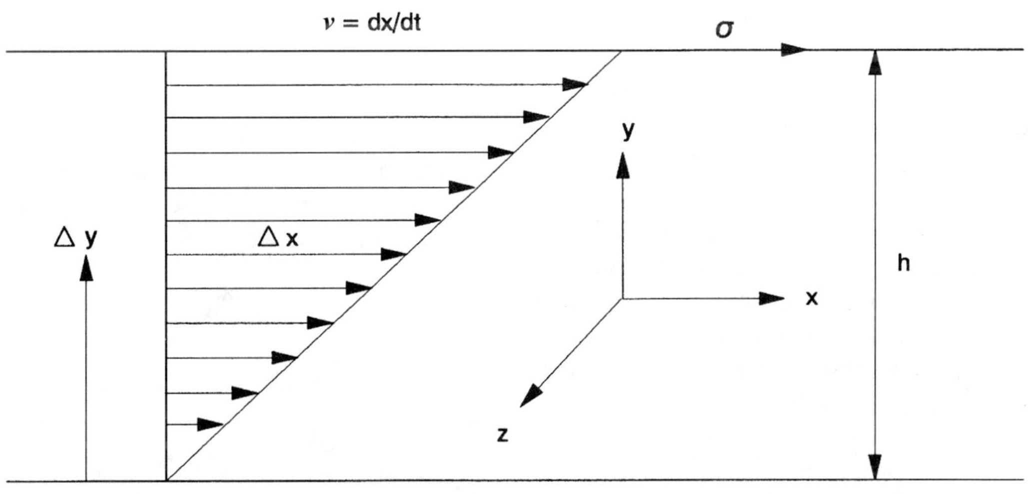

Figure 4–2 Shear Flow between Two Parallel Plates at a Distance *h* from Each Other Due to a Shear Stress σ. Arrows denote flow velocity v_x.

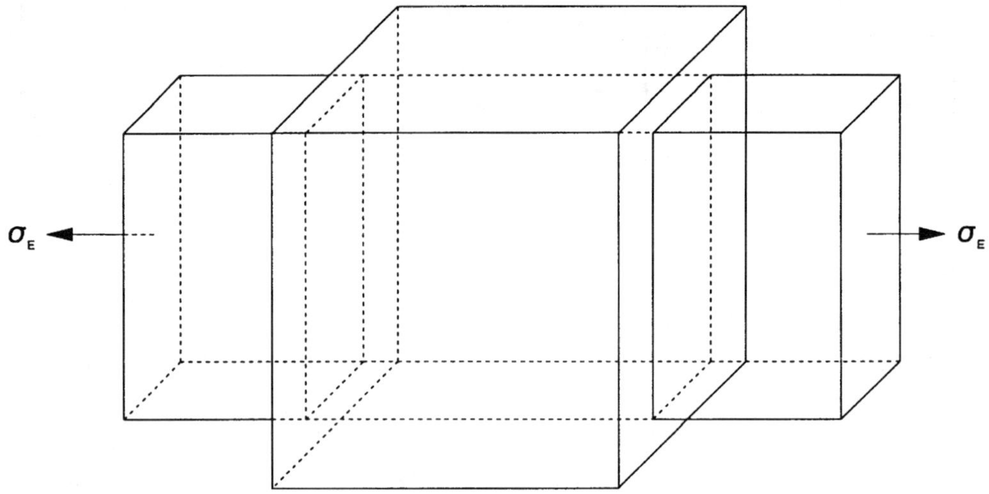

Figure 4–3 Tensile Deformation

length ΔL divided by the original length L_0. For an elastic material,

Equation 3

$$E \equiv \frac{\sigma_E}{\epsilon}$$

where E is Young's modulus. During a compressive or tensile deformation, not only the shape but often also the volume of the test piece will change.

In the case of a fluid, the deformation will increase in time at a certain strain rate, which is given by $\dot{\epsilon} = d\epsilon/dt$. In contrast to shear flow, the velocity gradient is in the direction of the flow. This type of flow is called *elongational* or *extensional flow*; it is always transient. Flow with a clear elongational flow component occurs, for instance, in a constriction or a widening of a flow stream or if a flowing liquid encounters a "solid" wall. The so-called elongational or (uniaxial) extensional viscosity η_E is defined as:

Equation 4

$$\eta_E \equiv \frac{\sigma_E}{\dot{\epsilon}}$$

Relation between Various Rheological Parameters

The relation between the shear and the Young's modulus of a product depends on the extent of the volume change during uniaxial compression or extension, which is expressed in the Poisson ratio μ. This parameter is given by

Equation 5

$$\mu = \frac{1}{2}(1 - \frac{\partial V}{V \partial \epsilon})$$

In the case that $\partial V/\partial \epsilon = 0$, $\mu = 0.5$. When Equation 5 and the expression for the bulk modulus K are used, the next expressions can be deduced (Whorlow, 1992):

Equation 6

$$E = 3K(1 - 2\mu)$$

and

Equation 7

$$E = 2G(1 + \mu)$$

The Poisson ratio will effectively be 0.5 if K is large compared to E and G.

Analogous to the relation between E and G for solid materials, the following relation can be derived for liquids:

Equation 8

$$\eta_E = Tr\eta$$

where Tr is the Trouton ratio. For Newtonian liquids, $Tr = 3$, but for polymer solutions it may be much larger, up to 10^3.

RHEOLOGICAL CLASSIFICATION OF FOODS

As a starting point for a classification of foods according to their rheological properties, the above-described division in elastic and viscous materials will be used. A first complication is that material properties depend often on the strain or on the rate of strain (nonlinear behavior). Moreover, many foods' reaction on a stress or strain consists partly of a viscous contribution and partly of an elastic one: they behave *viscoelastically*. Second, the length of time that a material is stressed or strained may play a part, resulting in nonequilibrium behavior over "short" deformation times. In the following discussion, foods will be classified, first according to their equilibrium behavior and then according to their nonequilibrium behavior.

Equilibrium Behavior

In Figure 4–4, an overview is given of the simple basic relations between stress and strain rate that can be observed for various foods in the event of a shearing deformation. In practice, sometimes a combination of two or more of these simple relations can be seen when the mechanical behavior of the product is studied over a large range of shear rates or strain.

Curve 1 in Figure 4–4A illustrates the relation for a linear viscous (so-called *Newtonian*) liquid. The viscosity η is given by the slope of the line and is independent of the shear rate $\dot{\gamma}$ (Figure 4–4B). Examples of Newtonian liquids are water, sugar solutions, pasteurized milk, alcohol, tea, and dilute dispersions of spherical particles.

For *non-Newtonian* liquids, the relation between σ and $\dot{\gamma}$ is not linear. A shear rate–dependent viscosity is obtained, called *apparent viscosity* η_{app}, with $\eta_{app} = f(\dot{\gamma})$. If η_{app} decreases with increasing $\dot{\gamma}$, one speaks of *shear thinning* (or *pseudoplastic*) behavior (Curve 2 in Figure 4–4A and B) and if an increase is observed, of *shear thickening* behavior (Curve 3). If the latter is accompanied by a volume increase of the material, often the term *dilatant behavior* is used. Some food products showing shear thinning behavior are concentrated milk; solutions of concentrated macromolecules, such as xanthan and guar gum; and several fruit juices. Shear thickening behavior of foods is only rarely observed and then only over $\dot{\gamma}$ ranges normally not observed in practice. Examples of shear thickening liquids are a concentrated suspension of starch granules and concentrated solutions of various macromolecules at rather high $\dot{\gamma}$.

Shear rate thinning and thickening behavior can often be modeled over a limited range of $\dot{\gamma}$ by

Equation 9

$$\sigma = k\dot{\gamma}^n$$

where k (the so-called *consistency index* with the strange unit of Pa • s^n) and n (the power law index) are constants. The big advantage of this power law equation is that it is rela-

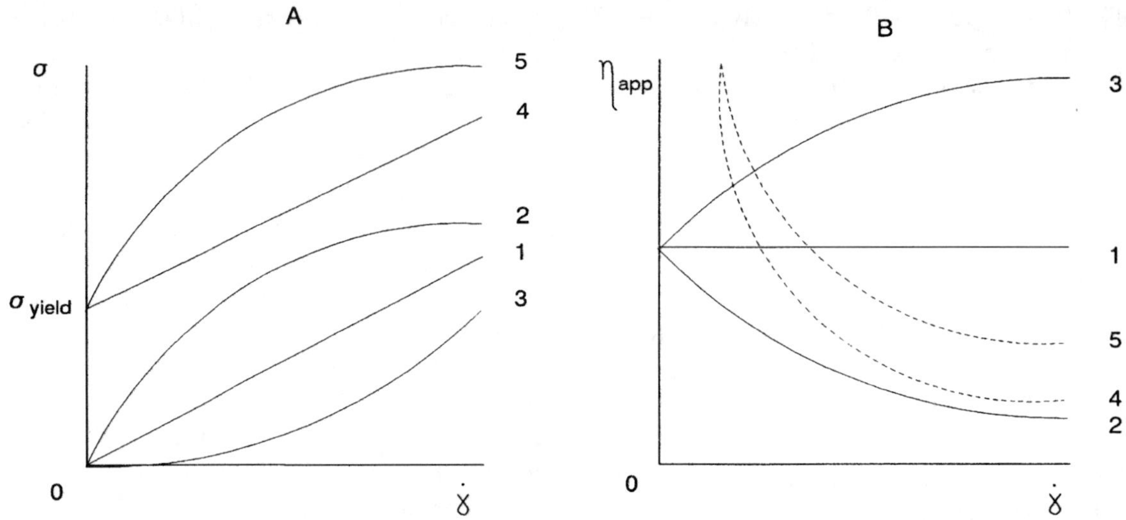

Figure 4–4 The Shear Stress σ (A) and the (Apparent) Viscosity η_{app} (B) as a Function of the Shear Rate $\dot{\gamma}$ for Various Foods with a "Liquid" Character. (1) Newtonian liquid, (2) shear thinning behavior, (3) shear thickening behavior, (4) Bingham liquid, and (5) plastic fluid behavior.

tively easy to use in either analytical or numerical calculations of flow in complex flow situations. One must realize that in practice k and n are frequently only constant over at most a decade of $\dot{\gamma}$. The consistency index k is a measure of η_{app} at $\dot{\gamma} = 1\ s^{-1}$; n is less than 1 for shear thinning and greater than 1 for shear thickening behavior.

Many foods such as margarine, butter, tomato ketchup, applesauce, and buttermilk behave like solids under low stresses and are liquidlike under a large stress. Clear flow is only noticed above a certain threshold value, the so-called *yield stress* σ_y. Foods with a plastic flow character (Curve 5 in Figure 4–4) are very common. Above the yield stress, they exhibit a shear thinning character. Often the flow behavior of plastic fluids is modeled as

Equation 10

$$\sigma = \sigma_y + \eta_B \dot{\gamma}$$

where η_B is a constant, the so-called *Bingham viscosity* (Curve 4 in Figure 4–4). In practice such a behavior is never observed, but for stresses clearly above σ_y, Equation 10 describes real flow behavior often well enough that the Bingham model can be used for engineering calculations. In fact, σ_y and η_B are fitting parameters. It should be emphasized that even for an ideal Bingham fluid, η_{app} decreases with increasing $\dot{\gamma}$ (Figure 4–4B).

The characterization of σ_y may be problematic because its value often depends on time (see below). This implies that depending on the purpose for which one wants to know σ_y, one may have to accept a different value. This does not imply that σ_y is not a useful product characteristic. Moreover, it is a factor that has to be taken into account in engineering calculations.

In Figure 4–5 an overview is given for the various relations between stress and strain that can be observed for predominantly solid materials subjected to a shearing deformation.

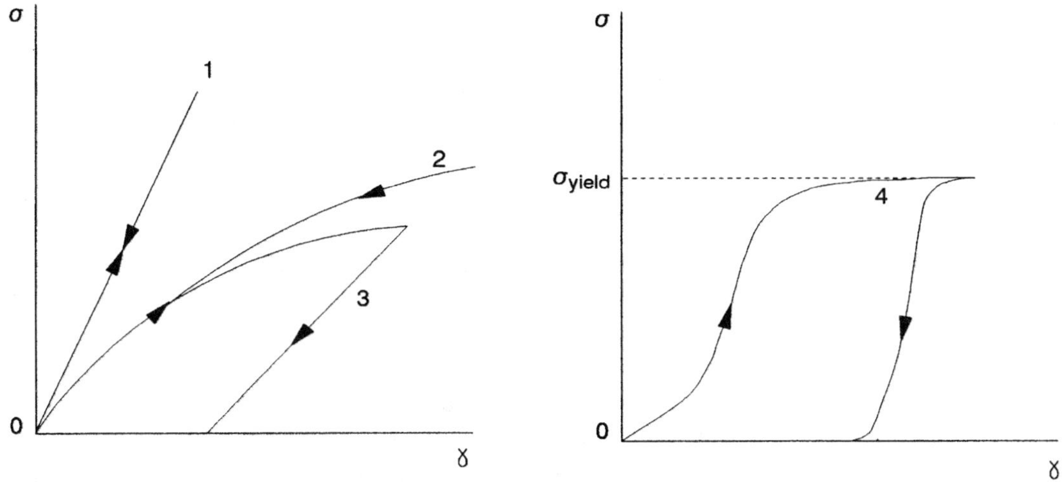

Figure 4–5 The Shear Stress σ as a Function of the Shear Strain γ for Various Foods with a "Solid" Character. (1) Linear elastic behavior, (2) nonlinear elastic behavior, (3) plastic behavior, and (4) plastic fluid behavior.

For an ideally elastic material, the relation between σ and γ is linear (Curve 1). The proportionality constant *G* or *E* for an experiment in shear or uniaxial extension respectively is independent of the strain. Such behavior is observed at small deformations for many elastic materials, such as boiled egg white, gelatin gels, hard biscuits, and margarine. Only a few brittle food products such as potato crisps and several hard biscuits are linearly elastic up to the point where they fracture. For most other elastic products, the behavior becomes nonlinear at larger deformations (e.g., boiled egg white and gelatin gels), and the product may even start to flow at stresses above the yield stress (as do margarine and butter) (Curves 2 and 4, Figure 4–5).

For nonlinear elastic materials, *G* and *E* are a function of the strain, but the material behaves reversibly regarding the deformation. In the case of plastic materials, this reversibility for the shape is lost. During deformation, their behavior is apparently nonlinear elastic, but the material does not come back to the original shape after the stress is taken away. The deformation is partly permanent, indicating a viscous component; the material behaves viscoelastically. Clear examples of viscoelastic materials include most types of cheese, bread doughs, doughs for biscuits and cakes, meat doughs, and liver paté.

Examples of products with a plastic flow character are already dealt with by the predominantly liquid products. Below the yield stress, they may be nonlinear elastic (e.g., margarine) or plastic (e.g., many doughs and spreadable liver patés).

As stated earlier, plastic and many plastic fluid materials behave viscoelastically at low stresses. One speaks of linear viscoelastic behavior if the elastic component can be described by Equations 1 or 3 (linear elastic behavior) and the viscous component by Equations 2 or 4 (linear viscous behavior).

Time-Dependent Flow

For many non-Newtonian fluids and for products with a plastic flow character, the relation between σ and $\dot{\gamma}$ depends on the length of time of the flow and flow history. In general, two main types can be distinguished, *thixotropy* and *antithixotropy* (sometimes referred to as *rheopexy*). In the first case, η_{app} decreases with increasing time of flow, and in the second case it increases (Figure 4–6). Both thixotropy and antithixotropy are reversible phenomena. The apparent viscosity has to come back to its original value during rest or shearing at a lower shear rate, $\dot{\gamma}$. These phenomena must not be confused with situations in which the structure of the product changes irreversibly due to flow, as during the stirring of an acid milk gel for the preparation of stirred yogurt or during the whipping of cream. Thixotropic behavior is common in many foods, but antithixotropy is not often found in foods.

A direct consequence of thixotropic behavior is that η_{app} is affected by all handlings involving flow of the material, such as pouring the material out of the packing or filling the measuring body of a rheometer. This history dependency may strongly affect results of a rheological test.

Thixotropic behavior may be studied by subjecting the product to increasing σ or $\dot{\gamma}$ followed by a decrease at a certain rate. If the length of time that the product is sheared effectively at a certain σ or $\dot{\gamma}$ is shorter than the time required to obtain the equilibrium value of η_{app}, the obtained relation between σ and $\dot{\gamma}$ shows a so-called *hysteresis loop* (Figure 4–7). The size of this hysteresis loop will depend on the rate by which σ or $\dot{\gamma}$ were increased and decreased, and thus on measuring conditions.

Time Scale Effects

As already mentioned, time plays an important role in the mechanical behavior of foods noticed by the consumer and shown during their production. Moreover, it may also strongly affect results obtained in a rheo-

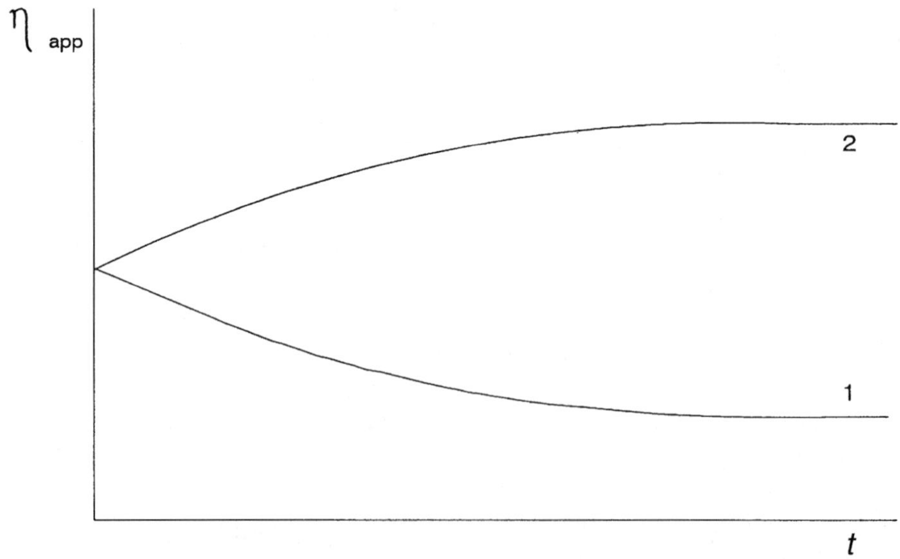

Figure 4–6 Apparent Viscosity η_{app} as a Function of the Length of Time of Flow at Constant Shear Rate $\dot{\gamma}$ (or shear stress σ). (1) Thixotropy and (2) antithixotropy.

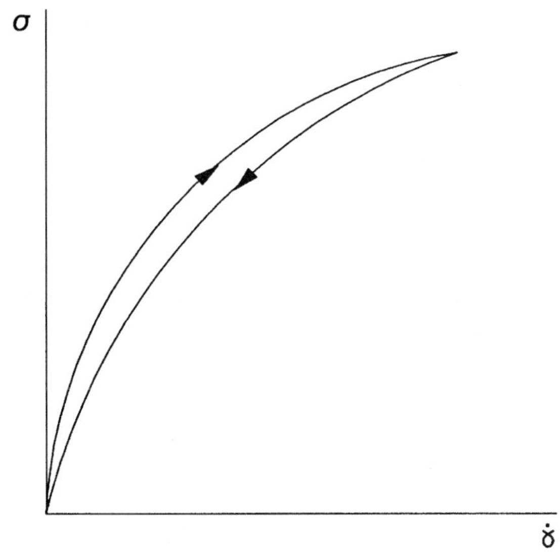

Figure 4–7 Hysteresis Loop in a Shear Stress σ versus Shear Rate $\dot{\gamma}$ Diagram of a Thixotropic Material

logical experiment. One aspect of time—namely, the length of time of shearing at constant $\dot{\gamma}$—has been dealt with in the preceding section. However, another aspect of time is at least as important, *time scale*. The time scale of an experiment is the time that a stress of a certain size and direction acts on a material. For many viscoelastic foods, the reaction due to a stress may be relatively more elastic or viscous depending on the time scale of the stress application. For instance, over short time scales, the inner part of a ripe Camembert cheese behaves elastically—it can be cut with a knife—while over long time scales it flows.

For a steady state experiment in which a certain compressive stress is applied to an elastic material, the time scale and the length of time of the experiment will be the same. However, this is not the case as the material is subjected to varying stresses (e.g., to a sinusoidal oscillating stress). The time scale of the experiment is then roughly equal to 1 over the frequency of the oscillation in rad • s^{-1}.

The origin of time-dependent behavior of materials is situated in their structure. For the sake of simplicity, we will consider homogeneous materials. The structure of (most) materials is not static. Even if the geometric structure is permanent, the bonds between the structural elements are not. In general, these bonds will occasionally break and re-form due to Brownian motion. The rate by which this process proceeds may vary by many orders of magnitude. For an ideal elastic material, the rate will be effectively zero. A characteristic of elastic materials is that they recover their shape after the stress by which they were deformed is taken away. This implies that all the energy supplied during the deformation was stored in the material and could be used to let the material return to its original shape. This can only be the case if during deformation no energy is lost due to disrupting and stress-free re-formation of the bonds between the structural elements in the material (e.g., the gelatin molecules in a gelatin gel). Hence, the bonds between the structural elements must be permanent over the time scale of the experiment. In a liquid, bonds are also present between the structural elements (e.g., the oil molecules), but they will break and re-form much faster than the time scale of the experiment, and all energy supplied will be immediately dissipated as heat.

In other materials on which a stress is applied, the process of disrupting and stress-free re-formation of the bonds may proceed over time scales between seconds and days. This process can be followed easily by first deforming the material to a certain extent and then measuring the stress required to keep this deformation constant. Except in the case of ideally elastic materials, the stress will decay at a certain rate. This process of stress decay is called *relaxation*. The time required for the stress to decrease to $1/e$ (i.e., 36.8%) of its value at the moment the deformation process was stopped is called the *relaxation time,* λ.

The relaxation process can be visualized by considering a so-called Maxwell element (Figure 4–8). Such an element consists of an ideal elastic spring and a dashpot (a piston surrounded by a viscous liquid). If such an element is stressed (e.g., by applying a tensile force) over a short time scale, the resulting deformation will be almost exclusively due to the elastic spring, whereas during the application of a long-lasting stress, the resulting deformation will be due to the displacement of the piston in the dashpot. For intermediate time scales, both the spring and the dashpot contribute to the reaction of the element. The ratio of both contributions depends on the modulus (G_M) of the spring and the effective resistance (which can be modeled as a viscosity η_M) experienced by the piston in the dashpot. It can be shown that the relaxation time λ_M of the Maxwell element is equal to η_M/G_M (Ferry, 1980; Whorlow, 1992) and that for $t \ll \lambda_M$ the reaction of the element is primarily elastic and for $t \gg \lambda_M$ primarily viscous.

For liquids, observed relaxation times are very short (e.g., about 10^{-13} s for water), while for elastic materials, λ is very long. For viscoelastic materials, λ is in between (about 10^{-1} to 10^{6} s) and corresponds roughly with human time scales. This indicates that the ratio between the relaxation time and the characteristic time scale of the observation is important for the observed rheological behavior of a material. The ratio is expressed in the Deborah (De) number (which equals λ divided by observation time). The rheological behavior of materials with a single relaxation time can be classified according to their Deborah number as follows: viscous behavior when De $\ll$ 1, elastic behavior when De $\gg$ 1, and viscoelastic behavior when De is of the order of 1.

In practice, such a classification is less unequivocal than may be suggested by the paragraph above because most foods are composite materials, containing various structural elements connected by various bonds. As a result, these products cannot be characterized by one λ; rather, a whole spectrum of λ's is needed to describe their relaxation behavior. Yet in practice, many viscoelastic materials exhibit stronger viscous behavior over long time scales than over short time scales. However, the picture can be more complicated.

DETERMINATION OF RHEOLOGICAL PROPERTIES

Selecting an instrumental method to determine rheological properties of products must be the end of a fixed procedure that involves the following questions:

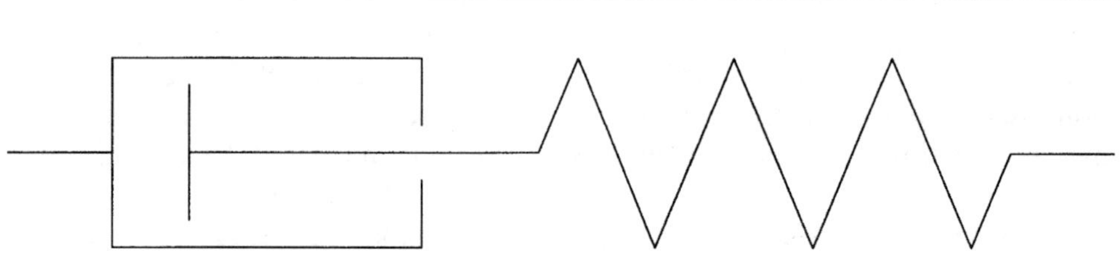

Figure 4–8 Schematic Diagram of a Maxwell Element Consisting of an Ideal Elastic Spring and a Dashpot

1. What is the purpose of the measurements? Is it necessary to determine fundamental rheological parameters, or will a global impression in the context of quality control suffice?
2. Which parameters have to be determined, and under what conditions (e.g., time scale, stress level, temperature)?
3. Which measuring method can be used?
4. Which measuring instrument has to be selected and equipped with which measuring geometry?

A mechanical test always involves a combination of a measuring method with a measuring technique (instrument plus measuring geometries). A complicating factor is that with various instruments a measurement can be taken by differing measuring methods and vice versa.

In the next section, an overview will be given of the various measuring methods. Then an overview of the main measuring techniques will be given in the sections thereupon.

MEASURING METHODS

The determination of the rheological properties of a material implies the measurement of the relation between its deformation and the stresses acting on it as a function of time. A first division of measuring methods can be made depending on whether the test piece is subjected to a stress σ, a strain γ or ϵ, or a strain rate $\dot{\gamma}$ or $\dot{\epsilon}$. Another distinction can be made between methods involving the application of a steady σ, γ (or ϵ), or $\dot{\gamma}$ (or $\dot{\epsilon}$) and methods in which the applied variable varies in time (often sinusoidally)—so-called *dynamic tests*. This latter division is somewhat arbitrary. Experiments in which the reaction of a material on a sudden change in σ, γ (or ϵ), or $\dot{\gamma}$ (or $\dot{\epsilon}$) (so-called *transient tests*) is determined can be considered as either dynamic or steady-state tests.

Stress Relaxation Tests

In these tests, the material is subjected to a certain strain, and the ensuing stress σ is measured as a function of time. Ideally, the material is deformed in a step function, but in practice the deformation always takes some time. For a liquid, σ will be finite during the deformation and will be zero for all positive t (Figure 4–9). For a purely elastic material, σ stays constant with time, whereas in general for most other solid (or solidlike) materials, σ decreases initially quite rapidly and then gradually more slowly. Ultimately, σ may become zero or approach some limited value.

From the stress response, the relaxation spectrum can be calculated (Ferry, 1980). In the analysis of the stress response, one should remember that stress relaxation has already started during the finite time required to attain the set deformation. This relaxation will be more extensive in the case of fast-relaxing bonds and a slow deformation rate. The result is a slower relaxation process after the set deformation has been obtained and a longer apparent relaxation time. Moreover, it follows that the relaxation spectrum can be calculated accurately only for $t \gg$ deformation time.

Tests at a Constant Strain Rate

In many instruments used for the determination of the viscosity of a material, a set $\dot{\gamma}$ is applied, and the resulting σ is measured as a function of time. For products with a yield stress σ_y, σ may display an overshoot (Figure 4–10). In fact, these products are first deformed "elastically" until σ exceeds σ_y and the structure of the material is broken down. With continued deformation, the structure of the material is broken down further, resulting in a decrease of η_{app} with measuring time (thixotropic behavior). "Overshoot" measurements are useful for the determination of

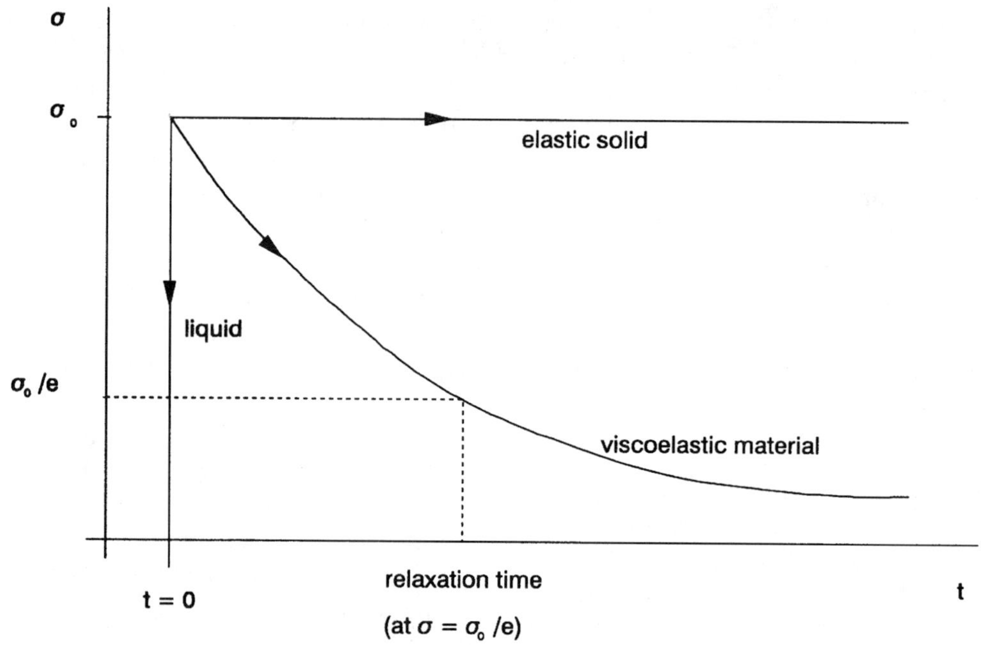

Figure 4–9 The Stress in a Material as a Function of Time after the Sudden Application of a Set Strain

the structure breakdown of plastic fluids due to large deformations.

Fracture properties of solid materials are mostly determined by deforming them at a certain speed in uniaxial compression or extension. However, due to the way the experiments are executed, the strain rate increases (experiments in uniaxial compression) or decreases (experiments in uniaxial extension) to some extent during these experiments. This type of measurement will be further dealt with in the later section "Large-Deformation and Fracture Tests with Tension-Compression Instruments."

Tests at Constant Stress

In many new instruments, a constant stress can be applied instead of a constant strain rate. These instruments can be used to determine η or η_{app} of liquids and liquidlike materials by measuring $\dot{\gamma}$ as a function of σ. A particular type of test that can be done with these instruments is the so-called *creep test.*

Creep tests are preeminently useful for the characterization of viscoelastic materials. A constant σ is applied and the resulting strain determined as a function of time. For liquids, a steady increase in strain with time will be observed, while for an elastic material, the observed strain will be constant in time (Figure 4–11). In the latter case, the material will regain its original shape after removal of the stress.

For a viscoelastic material, there is first an instantaneous increase in the strain followed by a delayed elastic response, and for materials containing a viscous component a steady increase in strain over long times (Figure 4–12). From the slope of the steady increase, a viscosity can be calculated. This viscous component leads to a permanent deformation after removal of the applied stress.

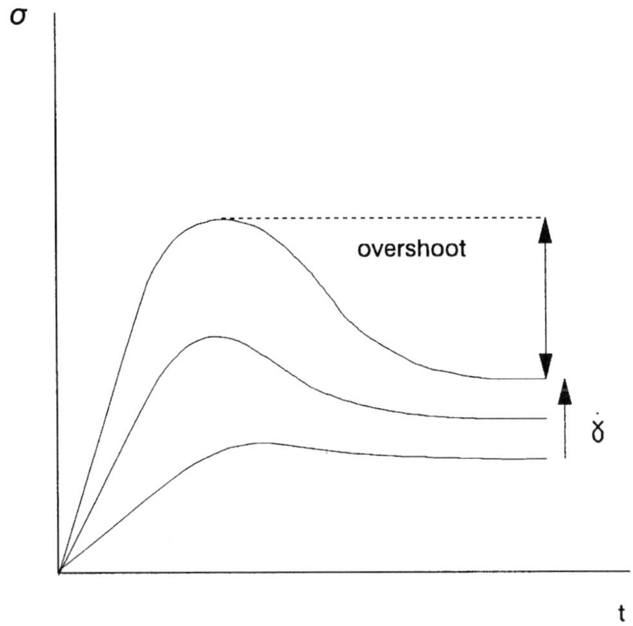

Figure 4–10 The Stress σ as a Function of Time t for Three Different Shear Rates $\dot{\gamma}$ for a Material Exhibiting Stress Overshoot

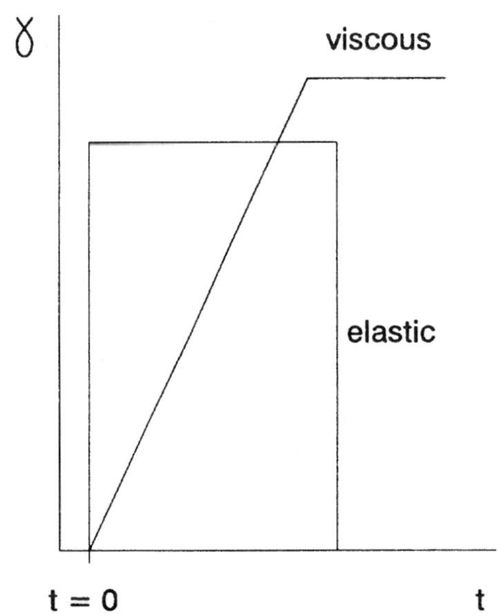

Figure 4–11 The Shear Strain γ as a Function of Time t after Applying a Constant Shear Stress, for a Viscous and an Elastic Material

Normally, results of a creep test are not expressed as moduli but as creep compliance ($J \equiv \gamma/\sigma$). For ideally elastic materials, $J = 1/G$. But for viscoelastic materials, $J(t)$ $1/G(t)$. General relations exist by which $J(t)$ can be recalculated in $G(t)$ and conversely (Ferry, 1980), but for an accurate recalculation, $J(t)$ has to be known over quite a large range of t.

If large enough stresses are applied, the fracture process of viscoelastic materials can be determined (Figure 4–13) and its time dependency studied. This method is not very well suited for studying the fracture behavior of time-independent materials.

Dynamic Tests

In a dynamic test, the applied σ, γ, or $\dot{\gamma}$ oscillates sinusoidally in time with an angular frequency ω (rad • s^{-1}). Dynamic tests are very well suited for the characterization of

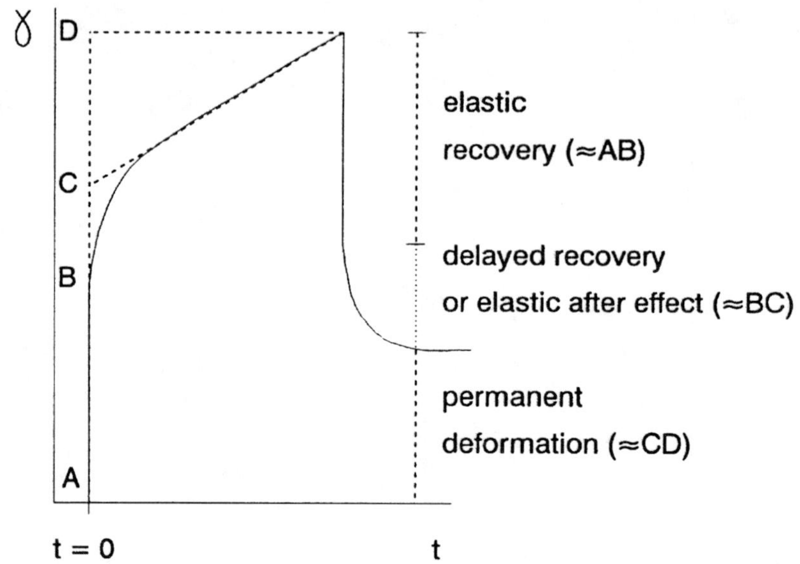

Figure 4–12 The Shear Strain γ as a Function of Time t after Applying a Constant Shear Stress, for a Viscoelastic Material

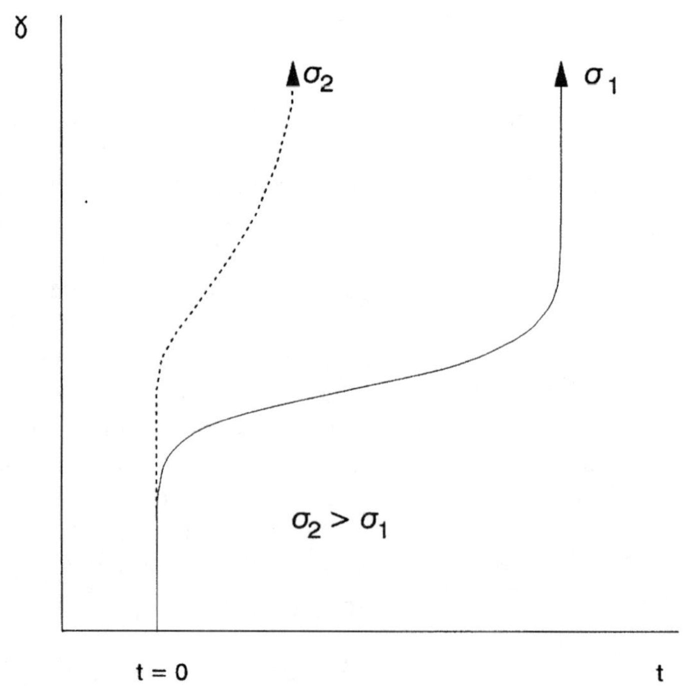

Figure 4–13 The Shear Strain γ as a Function of Time t for a Viscoelastic Material after the Application of Such a Stress σ That after Some Time Fracture Occurs ($\sigma_2 > \sigma_1$) (Schematic Figure)

undisturbed viscoelastic materials as a function of time. Both the elastic and the viscous component can be obtained over a broad range of time scales. The latter can be varied by varying ω (time scale experiment is roughly $1/\omega$).

For an experiment in which a sinusoidal varying shear strain γ is applied with an angular frequency ω, γ is given by

Equation 11

$$\gamma(t) = \gamma_0 \sin(\omega t)$$

where γ_0 is the maximum shear strain. For small maximum strains, the ensuing stress will also vary sinusoidally, but with a phase lag δ compared to the strain. Its magnitude will depend on the type of material. The stress (σ) is given by

Equation 12

$$\sigma(t) = \sigma_0 \sin(\omega t + \delta)$$
$$= \sigma_0 [(\cos\delta \sin\omega t) + (\sin\delta \cos\omega t)]$$

where σ_0 is the maximum stress. For an ideal elastic material, $\sigma_{el} = G\gamma$, and the stress is in phase with the strain (Figure 4–14). For a viscous material, $\sigma_v = \eta\dot{\gamma}$. The stress is 90° or $\frac{1}{2}\pi$ rad out of phase with the strain; σ is at a maximum if $\dot{\gamma}$ is at a maximum. For viscoelastic systems, δ lies in between 0 and $\frac{1}{2}\pi$ rad.

For a linear viscoelastic sample, σ_0 is proportional to γ_0, and Equation 12 can be rewritten as

Equation 13

$$\sigma(t) = \gamma_0 \{[(\sigma_0/\gamma_0)(\cos\delta \sin\omega t)] + [(\sigma_0/\gamma_0)(\sin\delta \cos\omega t)]\}$$

The first term, the elastic part of the stress, is in phase with the strain, and the second term, the part $\frac{1}{2}\pi$ rad out of phase, is the viscous component. It is convenient to define the quantities:

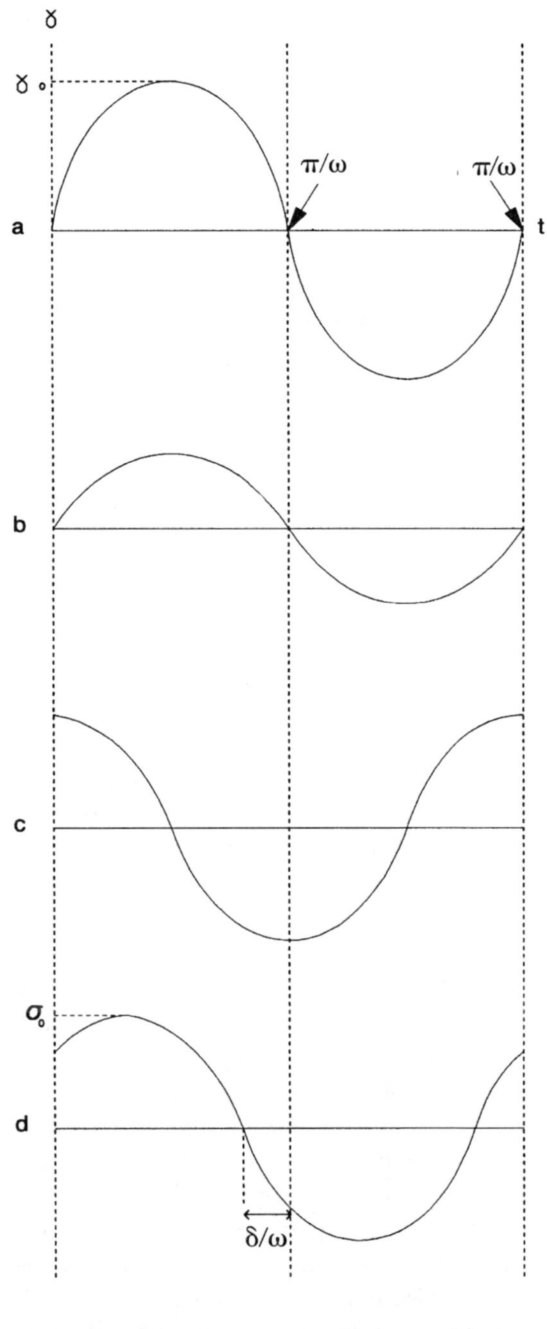

Figure 4–14 The Stress σ as a Function of Time t If a Sinusoidally Varying Shear Strain γ Is Applied at an Angular Frequency ω (rad • s^{-1}). (a) Sinusoidally varying shear strain, (b) reaction in an elastic material, (c) reaction in a viscous material, and (d) reaction in a viscoelastic material.

Equation 14

$$G'(\omega) = \frac{\sigma_0}{\gamma_0} \cos \delta$$

and

Equation 15

$$G'' = \frac{\sigma_0}{\gamma_0} \sin \delta$$

where G' is the storage modulus and G'' the loss modulus. Inserting Equations 14 and 15 in Equation 13 gives

Equation 16

$$\sigma (t) = \gamma_0 [(G' \sin \omega t) + (G'' \cos \omega t)]$$

Moreover:

Equation 17

$$\tan \delta = \frac{G''}{G'}$$

The in-phase component, the storage modulus, is associated with the storage and release of energy during a periodic application of a strain, and the out-of-phase part, the loss modulus, with the dissipation of energy into heat. A higher tan δ means that a material is relatively more viscous and less elastic.

In the above derivation, the moment that the shear strain was at maximum was taken arbitrarily as the origin for the time measurement. Alternatively, the same derivation can be given with the maximum of the stress as the origin resulting in the definition of a storage and a loss compliance (Whorlow, 1992).

Another pair of quantities is obtained if the strain rate is considered, as may be convenient if essentially liquids are considered. Such a derivation results in the definition of the dynamic viscosities η' and η'' (Ferry, 1980), where

Equation 18

$$\eta' = G''/\omega$$

Equation 19

$$\eta'' = G'/\omega$$

For an ideally viscous material, $\eta' = \eta$ and $\eta'' = 0$.

The main advantages of dynamic tests is that the contributions of both the elastic and the viscous component in the reaction of a material on an applied stress or strain can be determined over a range of time scales. Moreover, gel formation and aging can be followed easily if the maximum strains are kept small enough.

The main disadvantage is that the measurements have to be done in the so-called *linear region*—that is, in the strain region where stress is proportional to the strain. Otherwise, an unequivocal interpretation of the data is hard, if not impossible, to perform. Moreover, breakdown of the structure of the test material may occur during the experiment. The moduli obtained have no predictive value for large-deformation properties (van Vliet, 1995; van Vliet & Walstra, 1995).

MEASURING INSTRUMENTS

A large variety of instrumental techniques exist to determine the rheological properties of materials. Regarding the investigation of food products, one can distinguish three main classes of instruments: tube viscometers, rotational instruments, and tension-compression or so-called general material-testing instruments. The first type of instruments can be used to determine the viscosity of liquids and liquidlike materials, and the third type of instruments is primarily used to determine large-deformation and fracture properties of predominantly solid materials. Rotational instruments are well suited to study rheological properties of non-Newtonian liquids and materials that can be solidified in the rheometer (e.g., gels). When equipped with a parallel-plate geometry, they can also be used to study solid materials.

Besides the type of instruments mentioned, one can further distinguish special instruments for determining extensional flow properties and instruments based on wave propagation. However, it is outside the scope of this chapter to treat these instruments, which are used less often in food science. Because they are very popular in food industry, we will discuss some empirical tests in the section "Empirical Test Methods."

TUBE VISCOMETERS

In a tube viscometer, the flow rate of a material due to a known pressure gradient (ΔP) is determined or, conversely, the pressure gradient related to a known flow rate. For thin liquids, flow is mostly due to gravity. Tube viscometers are popular because they are cheap and easy to use and to thermostat. Tube viscometry is a very accurate method to determine the viscosity of Newtonian liquids. For non-Newtonian liquid, it is less suited because the measurement cannot be done at a constant shear rate, since it varies over the diameter of the tube.

Calculations of the shear rate, shear stress, and viscosity from the relation between the flow rate and the pressure gradient are based on the assumptions that the flow in the tube is laminar, the velocity of the fluid at the wall is zero (no slip condition), the fluid is incompressible, and entry and exit effects can be neglected. Then for a capillary of length l and radius R, σ at any point along the radius is given by

Equation 20

$$\sigma = \frac{\Delta P}{2l} r$$

where ΔP is the pressure difference over the length of the capillary and r the distance from the center of the tube. Hence, the shear stress at the wall, σ_w, equals $\Delta PR/2l$. The dependence of σ on r also causes $\dot{\gamma}$ to be dependent on r. For a Newtonian liquid with a viscosity η, $\dot{\gamma}$ is given by

Equation 21

$$\dot{\gamma} = \frac{\Delta P}{2\eta l} \mathrm{r} = \frac{4V}{\pi \mathrm{R}^4 t} r$$

where V is the volume of fluid flowing through the capillary during a time t.

Both σ and $\dot{\gamma}$ are zero in the center of the tube and increase linearly with r. For a Newtonian fluid, this results in the parabolic velocity profile (the so-called *Poiseuille profile*); see Figure 4–15:

Equation 22

$$V = \frac{\Delta P}{4\eta l} (R^2 - r^2)$$

where v is the fluid velocity. The volume flow V through the capillary is given by

Equation 23

$$V = \frac{\pi R^4 \Delta P t}{8\eta l}$$

Equation 23 allows the direct calculation of η from the volume flow per second.

For non-Newtonian fluids, the relation between σ and $\dot{\gamma}$ has to be known to derive the equations equivalent to Equations 21 through 23. The shear rate at the wall (Whorlow, 1992) can be obtained from

Equation 24

$$\dot{\gamma}(R) = \frac{4V}{\pi R^3 t} \left(\frac{3}{4} + \frac{1}{4} \frac{\mathrm{dln}\,(v/t)}{\mathrm{dln}\,(\Delta P/l)}\right)$$

$$\left[= \frac{(3n+1)}{4n} \frac{4V}{\pi R^3 t}\right]$$

where the part between square brackets holds for fluids for which the shear rate dependency can be described by a power law (Equation

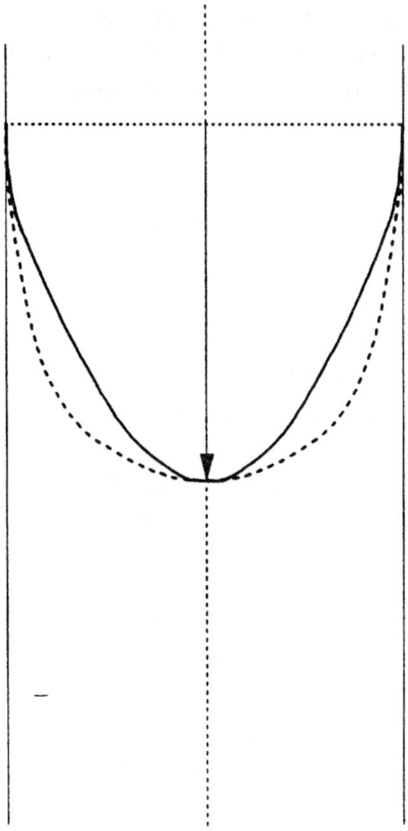

Figure 4–15 Flow Profiles in a Tube Viscometer, Newtonian (Full Curve) and Shear Thinning (Broken Curve) Fluid

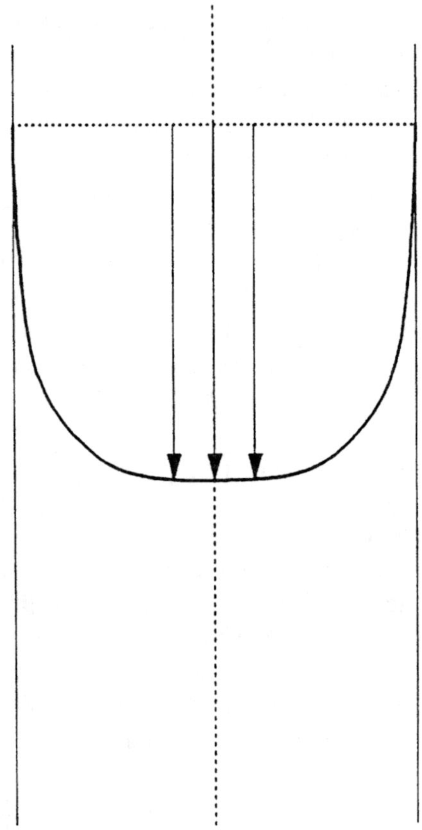

Figure 4–16 Flow Profiles in a Tube Viscometer, Fluid with a Yield Stress

9). When the relation between σ and $\dot{\gamma}$ is not known, the flow curve can be calculated from the relation between the pressure gradient and the volume flow.

To illustrate, two flow profiles are given in Figure 4–15. Compared to the parabolic profile for a Newtonian liquid, the profile for a shear thinning liquid is more blunted. This is even more the case for a liquid with a yield stress. Flow only occurs where σ is greater than σ_y (Figure 4–16). Remember that σ is zero in the center of the tube (Equation 20). For products such as mayonnaise with a "large" yield stress, flow will occur only at the wall, resulting in apparent plug flow. In an extreme case, the flow profile will be nearly the same as that of a material exhibiting wall slip.

Well-known examples of tube viscometers for low-viscosity Newtonian liquids are the Ostwald and the Ubbelohde viscometers (Figure 4–17) which essentially consist of two bulbs connected by a capillary. In these viscometers, an accurately known volume liquid flows due to gravity through the capillary, and the required time is measured. This time is a direct measure of the kinematic viscosity $\nu = \eta/\rho$, where ρ is the density of the liquid

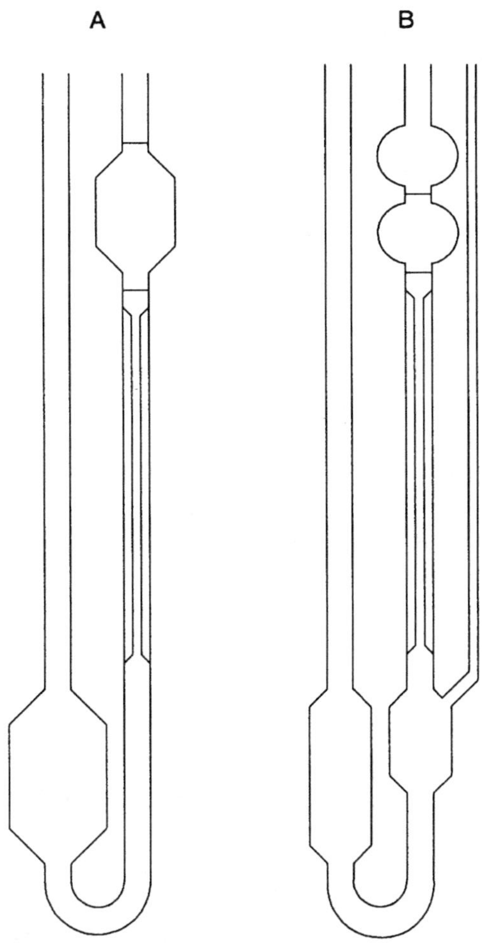

Figure 4–17 (A) Ostwald Viscometer and (B) Ubbelohde Viscometer (Not at Scale)

Equation 25

$$\nu = Ct$$

C is a constant depending on the geometry. The bulbs should have the same diameter to minimize errors due to surface tension. For opaque liquids, reverse-flow hydrostatic head viscometers are available (Whorlow, 1992).

For liquids with a higher viscosity, the driving force can be supplied by gas (or hydrostatic) pressure, and for very thick liquids, it can be supplied by a piston or a constant flow pump. In the latter configuration, the pressure difference over the capillary is measured.

Capillary viscometers are popular for several reasons.

- The viscosity of Newtonian liquids can be determined very accurately. Therefore, capillary viscometers are very suitable for calibrating standard liquids to be used to calibrate various rheometers.
- Thermostating can be done relatively simply and accurately.
- Capillary viscometers are simple and relatively cheap.

There are a number of sources of errors in capillary viscometry.

- *Kinetic energy correction.* Depending on the construction of the viscometer and the properties of the liquid, part of the driving pressure is converted in kinetic energy of the liquid as it leaves the end of the capillary. To correct for this, ΔP in Equation 23 should be corrected by the pressure P' required for providing the kinetic energy. P' is proportional to $\rho\bar{v}^2$, where $\bar{v}$ is the average flow velocity. This results in the next expression for the kinematic viscosity:

  Equation 26

  $$\nu = C_1 t - \frac{C_2}{t}$$

  where C_1 and C_2 are constants for a particular viscometer, which can be determined by calibration. For long flow times, the correction will be small.
- *Entrance effect.* The velocity profiles in the container and the capillary will be different; some length of the capillary is required to obtain the equilibrium velocity profile. Moreover, at the entrance of

the capillary, flow is strongly convergent; it contains a large elongational flow component. This is incorporated into the calibration of the viscometers for Newtonian liquids, but for non-Newtonian liquids with a large Trouton value, this may lead to an extra large energy dissipation at the entrance, which is not accounted for in the calibration.

- *Turbulence.* The flow velocity must be low enough to avoid turbulence that starts at a Reynolds number ($2\rho R\bar{v}/\eta$) of about 2,100.
- *Particle migration.* Particles in a dispersion flowing through a tube exhibit the tendency to move to the center of the tube, leaving a layer with a lower particle concentration and reduced viscosity along the wall.
- *Wall slip.* This is due to the formation of a lubricating layer along the wall.
- *Viscous heating.* For high-viscosity liquids and high shear rates, viscous heating may occur.

ROTATIONAL VISCOMETERS

In a rotational viscometer, the test liquid is sheared between the two parts of the measuring body, which mostly consist of rotating cylinders, plates, or cones. A great advantage of this type of rheometer is that for a well-designed measuring geometry, the shear strain and the shear rate are almost uniform over the test piece. Another advantage is that the sample can be sheared for as long as desired, allowing the study of time-dependent behavior. In general, one can distinguish between apparatuses in which a constant stress is applied and those in which a constant shear (rate) is applied (see the earlier sections "Tests at a Constant Strain Rate" and "Tests at Constant Stress").

Most frequently used are measuring geometries in which the sample is sheared between two coaxial cylinders, a cone and a plate, or two parallel plates (Figure 4–18).

Concentric Cylinder Geometry

In these types of measuring geometries, the test piece is confined between two cylinder surfaces, of which one can rotate. In a *Couette system* the outer cylinder rotates and in a *Searle system* the inner cylinder rotates. For both systems, the equations relating rotation to torsion T are the same.

Neglecting end and inertial effects, the relation between the shear stress σ and the torsion T is given for a cylinder with a radius R_i and length L by

Equation 27

$$\sigma = \frac{T}{2\pi R_i^2 L}$$

This equation implies that σ is not constant over the gap between the two concentric cylinders but decreases from the inner cylinder to the outer one. Only if there is a very small gap between the cylinders compared to their radii may σ be considered as constant over the gap. Then, in the absence of wall slip, the shear strain γ in the gap may be approximated by

Equation 28

$$\gamma = \frac{R_a}{R_2 - R_1}\,\alpha$$

and for laminar flow, the shear rate, $\dot{\gamma}$, by

Equation 29

$$\dot{\gamma} = \frac{R_a}{R_2 - R_1}\,\Omega$$

where α is the angular displacement of the rotating cylinder (rad), Ω the angular velocity ($rad \cdot s^{-1}$), and R_a an average of R_1 and R_2, the radius of the inner and outer cylinders respec-

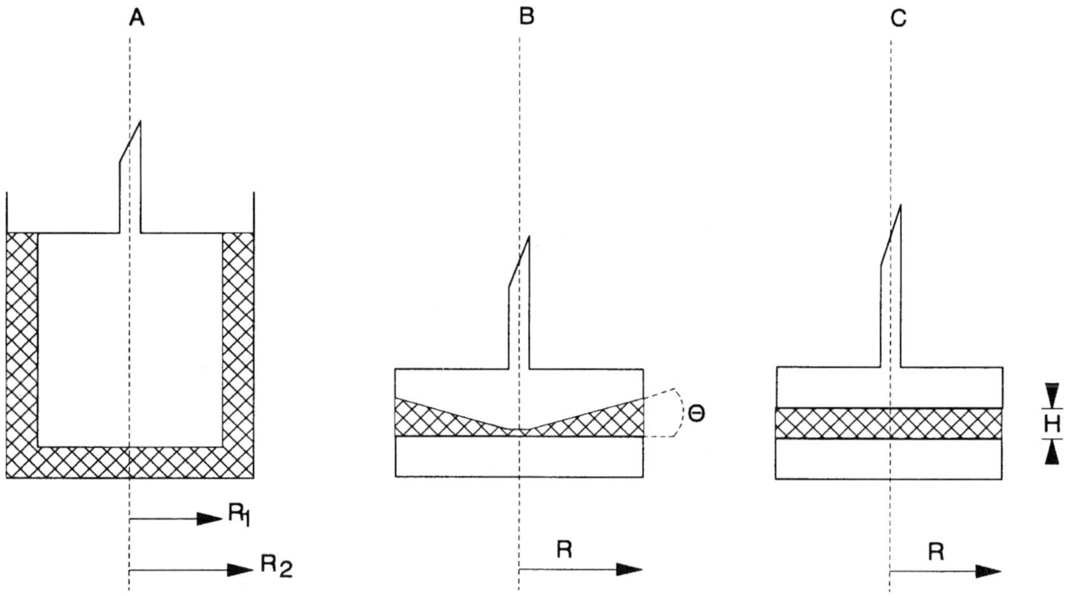

Figure 4–18 (A) Concentric Cylinder, (B) Cone and Plate, and (C) Plate-Plate Geometry

tively. As a first approximation, $R_a = (R_1 + R_2)/2$.

However, if the gap between the cylinders is not very small, σ, γ, and $\dot{\gamma}$ may not be considered constant over the gap. For Newtonian liquids, $\sigma = \eta\dot{\gamma}$, and one gets $\dot{\gamma}$ from

Equation 30

$$\dot{\gamma} = \frac{2\Omega}{R_i^2(R_1^{-2} - R_2^{-2})} = \frac{2R_1^2R_2^2\Omega}{R_i^2(R_2^2 - R_1^2)}$$

Similarly, γ is given by

Equation 31

$$\gamma = \frac{2\alpha}{R_i^2(R_1^{-2} - R_2^{-2})} = \frac{2R_1^2R_2^2\alpha}{R_i^2(R_2^2 - R_1^2)}$$

From Equations 30 and 31, it follows that both γ and $\dot{\gamma}$ decrease over the gap with R_i^2 as σ does. For non-Newtonian liquids, the decrease can be otherwise as the viscosity varies with σ. For liquids that behave according to the power law model

Equation 32

$$\dot{\gamma} = \frac{2\Omega}{nR_i^{2/n}(R_1^{-2/n} - R_2^{-2/n})}$$

Equation 32 implies that for a shear thinning liquid ($n < 1$) the decrease of $\dot{\gamma}$ with R is stronger than for Newtonian liquids (Figure 4–19). This is because η_{app} is lowest where σ is highest, which is near the inner cylinder.

Another consequence of the decrease in σ with R_i is that for liquids with a yield stress σ_y, σ may stay below σ_y near the outer cylinder, causing only a part of the liquid to be sheared. In extreme cases, it may even result in a kind of slip layer near the inner cylinder.

Sources of errors in using a concentric cylinder geometry are

- *Departures from circular stream lines.* If the inner cylinder rotates rapidly, centrifugal forces cause the fast-moving liquid near the inner cylinder to move outwards. Since this is not possible for the liquid as a whole, localized secondary

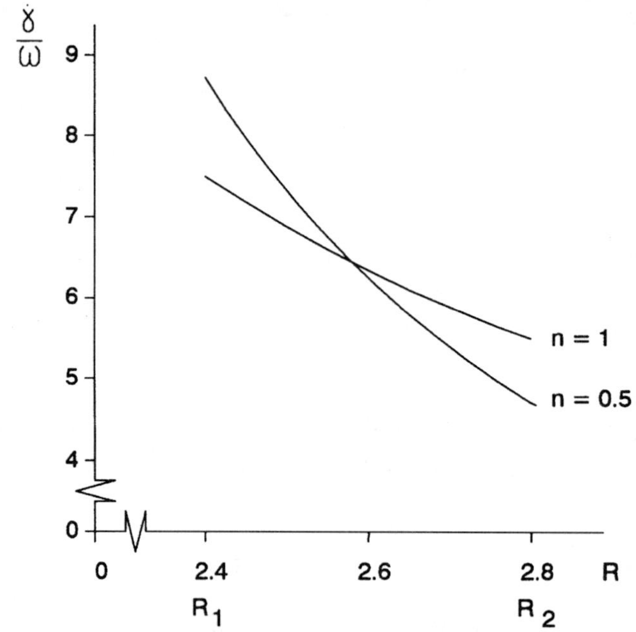

Figure 4–19 The Shear Rate $\dot{\gamma}$ in the Gap of a Concentric Cylinder Geometry as a Function of the Radius for $R_2 = 2.8$ cm and $R_1 = 2.4$ cm. n is the power in the power law model $\sigma = k\dot{\gamma}^{n}$, ω is the angular velocity of the rotating cylinder.

flow patterns develop, the so-called *Taylor vortices*. They occur when

Equation 33

$$Re \frac{(R_2 - R_1)^{1/2}}{R_2} = \frac{v\,(R_2 - R_1)^{3/2}}{\nu R_2} > 41.3$$

where v is the fluid velocity. For this flow situation, the characteristic length in the Reynolds number (Re) is $R_2 - R_1$. Flow is still ordered. Turbulence will occur at still higher Reynolds numbers (Re). Taylor vortices cause extra energy dissipation, resulting in an increase of η determined. They will not occur if the outer cylinder is rotating (Couette system). Then v is zero near the inner cylinder and high near the outer one.

- *End effects*. The resistance against flow of the liquid below (and above) the inner cylinder is often neglected compared to the contribution of the flow between the two cylinders. A correction can be obtained by determining the torsion as a function of the immersed height of the inner cylinder. Be aware that usually the end effect is different for Newtonian and non-Newtonian liquids. Several specially designed measuring geometries have been proposed to eliminate or at least to reduce substantially the end effect: for a few examples, see Figure 4–20. For the configuration shown in Figure 4–20A, the angle of the cone has to be chosen so that $\dot{\gamma}$ below the cone is the same as in the gap between the concentric cylinders. In a double concentric cylinder, the end surface is very small compared to the total surface.
- *Wall slip due to a slip layer*. Its occurrence has to be checked by using various

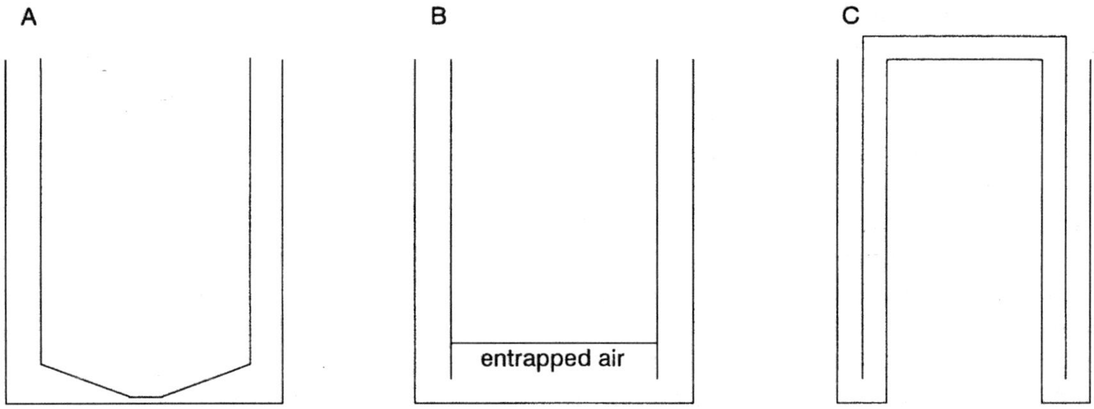

Figure 4–20 Examples of Concentric Cylinder Measuring Geometries with a Reduced End Effect. (A) Cone angle drawn too sharp for clarity; (B) entrapped air bubble under inner cylinder; and (C) double concentric cylinder.

measuring geometries and, if possible, visual inspection. In a dispersion of particles, wall slip can be caused by the excluded volume effect. The center of a particle cannot come closer to the wall than its radius, leading to a layer with a thickness equal to about the particle diameter d. This layer has a lower particle concentration and thus a lower η. This effect may be serious if $(R_2 - R_1)/d$ 20 (Cheng, 1984).

- *Viscous heating*. This will play a role only for materials with a high η sheared at high $\dot{\gamma}$.
- *Incorrect setting of the measuring bodies* with respect to each other.

Cone and Plate Geometry

This measuring geometry consists of a circular plate and a cone with a radius R with its axis perpendicular to the plate and its vertex in the plane of the surface of the plate (Figure 4–18B). The point of the cone is normally ground flat to avoid direct contact between the cone and plate. If the angle θ between the cone and plate is small (<5°), then σ, γ, and $\dot{\gamma}$ are uniform over the material between the cone and plate. This is the main advantage of this measuring geometry compared to a concentric cylinder or plate-plate geometry and it makes the cone and plate geometry, well suited for the study of strongly shear thinning liquids and of materials with a yield stress.

The equations by which σ, γ, and $\dot{\gamma}$ can be calculated are, respectively,

Equation 34

$$\sigma = \frac{3T}{2\pi R^3}$$

Equation 35

$$\gamma = \frac{\alpha}{\theta}$$

Equation 36

$$\dot{\gamma} = \frac{\Omega}{\theta}$$

where α is the angular displacement (rad) of the rotating part and Ω the angular velocity (rad • s^{-1}).

Cone and plate measuring geometries are well suited for testing small samples. End effects do not play a role. On the other hand, thermostating is more difficult for cone and plate geometries than for concentric cylinders. Moreover, drying out of the samples at the edges will have a big effect. The gap setting is harder and has to be done very carefully. Other considerations include secondary flow due to inertial effects at high $\dot{\gamma}$ and viscous heating for liquids with a high η.

Materials containing large particles cannot be studied with a cone and plate geometry. Due to the small distances between the cone and plate, the effect of a layer near the walls with a lower particle concentration, giving apparent slip, will be relatively more important. Moreover, a particle may become trapped between the cone and plate. In these cases a plate-plate geometry may be a useful alternative.

A more extended discussion of sources of measuring errors is given by Whorlow (1992). In that book, several variants of the cone and plate measuring geometry are discussed.

Plate-Plate Geometry

This measuring geometry consists of two parallel plates at a distance H from each other (Figure 4–18C), of which one can rotate. This measuring geometry is specially suited for the study of materials whose structure must be disturbed as little as possible or not at all while one is setting up the rheometer—for example, thixotropic materials or products with a yield stress, which may take a long time to recover. Care must be taken that the sample has two parallel flat surfaces.

The main disadvantage compared to a cone and plate geometry is that σ, γ, and $\dot{\gamma}$ are not constant in the test piece during deformation but vary with the distance from the center, r. They may be calculated by

Equation 37

$$\sigma = \frac{2T}{\pi R^3}$$

Equation 38

$$\gamma = \frac{\alpha r}{H}$$

Equation 39

$$\dot{\gamma} = \frac{\Omega r}{H}$$

Be aware that Equation 37 for σ applies in this form only for Newtonian liquids. For a further discussion of the advantages and disadvantages of using this type of measuring geometry compared to using concentric cylinders, see the earlier section on cone and plate geometries.

LARGE-DEFORMATION AND FRACTURE TESTS WITH TENSION-COMPRESSION INSTRUMENTS

Large-deformation and fracture properties of solid and solidlike food products are very important for handling properties during storage and/or further processing, usage properties, and eating quality. Fracture is considered to occur when all bonds between the structural elements in a certain macroscopic plane break, resulting in a breakdown of the structure of the material over distances much larger than the size of the structural elements and ultimately leading to a falling apart of the material. A short introduction to relevant parts of the theory of fracture mechanics in relation to the structure of food materials is given in Chapter 5. The reader is also directed to van Vliet and colleagues (van Vliet & Luyten, 1995; van Vliet, Luyten, & Walstra, 1993; van Vliet & Walstra, 1995).

A complication in the determination of large-deformation and fracture properties is that more parameters are required to give a full characterization of the material properties than is the case with small-deformation properties. Moreover, these properties are often rate dependent, even as small-deformation properties are not (van Vliet et al., 1993; van Vliet & Walstra, 1995). Fortunately, for many purposes, a few parameters are sufficient to give an adequate characterization of the fracture properties of a material; the specific parameters required will depend on the purpose of the experiment. An extra complication in the determination of fracture properties is that they depend to a larger extent than small-deformation properties on the size of defects in the material. For large-scale inhomogeneous materials (such as cookies and vegetables), this implies the need to use large test pieces or to accept big variations in results. Large variations in measured parameters can as such be regarded as a product characteristic.

There are a large variety of methods to evaluate fracture properties. In many tests, deformation is not one of the simple types dealt with in the earlier section "Basic Rheological Concepts" but a combination, and this makes an exact description of the deformation difficult, if possible at all. There is no test by which the complete large-deformation and fracture behavior of a material can be characterized. In general, the methods that are easier to perform provide less information. In most popular tests, a parameter is determined related to fracture stress. Besides, in many tests, some information is also obtained on the modulus (in literature on fracture mechanics, often called *stiffness*) as a function of the deformation of the material and of the deformation at fracture. In other tests, like the so-called *wedge experiments* and *cutting experiments* (with a knife or wire), the fracture energy (in $J \cdot m^{-2}$) can be determined.

Fracture tests are mostly performed using *tension compression* or so-called *general material–testing instruments*. Essentially, these apparatuses consist of a fixed bottom plate and a moving bar, each of which has a part of the measuring geometry fixed to it. A load cell is placed in the moving bar or bottom plate, by which the forces required to deform the test piece can be registered. The moving bar can be operated at varying speeds. By varying the measuring geometry, one can perform different types of tests.

Recently, a few rather extensive reviews have been published of methods to evaluate fracture properties (Vincent, 1992; van Vliet & Luyten, 1995). Only the most important methods for foods will be dealt with here.

Uniaxial Compression Tests

Uniaxial compression between two flat parallel plates is the most simple and popular technique for determining large-deformation and fracture properties of foods. Normally, a constant compression rate is applied, and the required force is measured. For the calculation of the stress as a function of strain, a correction has to be made for the change in the area of the test piece. For an incompressible material, this correction is easily made by taking for the actual surface area $A(t) = A(t = 0)\ (L_0/(L_0 - \Delta L))$, where L_0 is the test piece height at $t = 0$ and ΔL the change in height afterwards.

The calculation of the strain from the deformation is less straightforward than is often thought (Peleg, 1984). The most simple measure of strain is the Cauchy or engineering strain, which is defined as $\Delta L/L_0$. This measure of strain is adequate for small strains (0.1), but for larger ones it is not (Peleg, 1984; van Vliet & Luyten, 1995). Then the Hencky or natural strain is to be preferred, which is defined as

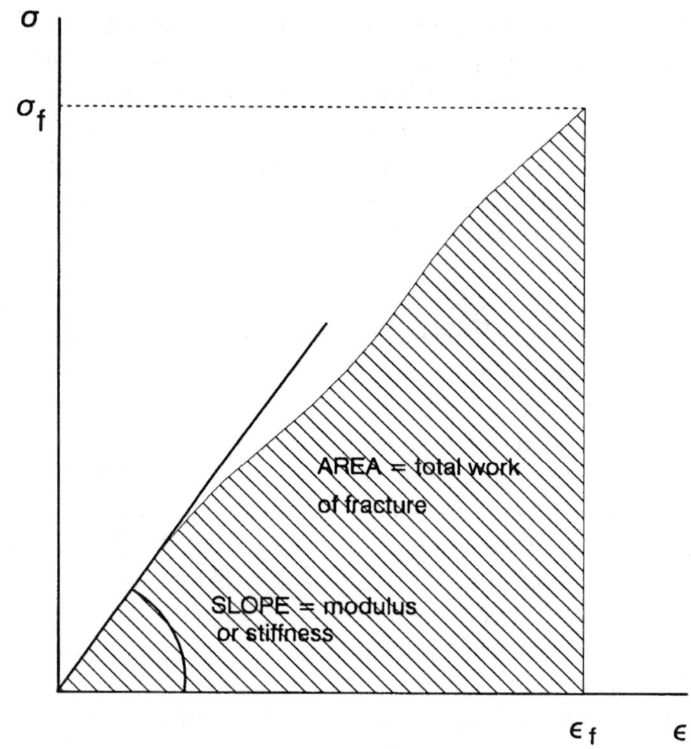

Figure 4–21 Example of a Stress σ versus Strain ϵ Curve for a Nonlinear Deforming Material. Maximum stress and the accompanying strain are often taken as being the fracture stress σ_f and fracture strain ϵ_f.

Equation 40

$$\epsilon_H = \int_{L_0}^{L} \left(\frac{dL}{L(t)} \right) = \ln \left(\frac{L(t)}{L_0} \right)$$

At small deformations, both measures of strain are nearly the same.

The main advantage of uniaxial compression tests is the simple execution of the test, including the relatively simple preparation of the test pieces. If the test piece has flat parallel ends, Young's modulus E may be calculated from the stress-strain curve (Figure 4–21). However, determination of the modulus in the linear region is often not possible. For instance, a linear region of 1% and a test piece height of 2 cm means that the unevenness of the surface must be clearly less than 0.2 mm, a condition that is hardly possible to obtain.

In principle, from the stress-strain curve, the fracture stress σ_f and fracture strain ϵ_f, and from the area below the curve, the total energy of fracture can be obtained (Figure 4–21). However, for many deformable materials, there is no clear fracture point. Often, fracture starts before the maximum in the stress-strain curve, and only after the maximum does the test piece clearly fall apart. Despite this, for most purposes the maximum in the σ-versus-ϵ curve is a workable measure of σ_f and ϵ_f. Other parameters characterizing fracture, like specific fracture energy and notch sensitivity, cannot be determined in uniaxial compression (van Vliet & Luyten, 1995).

During compression, the shape of the test piece may be affected due to friction between the test piece and the plates. Friction may lead to bulging of the middle part of the test piece and thus to a varying cross section of the test piece, which makes the calculation of σ and ϵ difficult. In general, it will be more important for a low height-to-diameter ratio. On the other hand, a too-high ratio (e.g., >1.5) may lead to buckling. Friction can often be diminished by applying a lubricant between the plates and the test piece.

In most modern instruments, the compression speed can easily be varied. Be aware that a constant compression speed implies that the strain rate $\dot{\epsilon}_H$ increases during the experiment according to

Equation 41

$$\dot{\epsilon}_H = \frac{dL}{dt\, L(t)} = \frac{v}{L(t)}$$

where v is the compression speed. Equation 41 also implies that v has to be adjusted if one wants to compare measurements on test pieces with various heights at the same $\dot{\epsilon}_H$.

For materials that fracture at low strains, ϵ_f cannot be determined accurately, especially if their surface is irregular. It becomes impossible if the test piece fractures when the load is still acting only on surface irregularities.

Uniaxial Extension Tests

This type of test largely parallels that of uniaxial compression. From the data obtained, E, σ_f, ϵ_f, and the total work of fracture can be calculated. Since the growth of the crack can be observed, parameters such as crack speed and the difference between fracture initiation and propagation can be measured. Moreover, by applying artificial notches, the notch sensitivity can be determined (van Vliet & Luyten, 1995). Another advantage of tension tests compared to compression tests is that the problem of friction is avoided.

The big disadvantage of tension tests is that they are more difficult to execute. In general, test pieces have to be rather large, so large homogeneous products are required from which they can be cut. A second problem is fixing test pieces to the apparatus without damage. Furthermore, fracture may occur at the grips instead of in the middle of the test piece. To avoid this, specially shaped test pieces that are thin in the middle may be used. Gluing of the test piece to the measuring body often works well.

Three-Point Bending Tests

For three-point bending tests, homogeneous test pieces are required, which are long compared to their thickness and width. These tests are well suited for products such as biscuits and chocolate bars, but for many other products such test pieces are difficult to obtain. The tests are easy to perform. During bending, the upper part of the test piece is compressed and the lower part elongated. In between is a neutral axis (Figure 4–22). For larger strains and smaller span widths, the deformation contains a relatively important shear component.

Fracture stress and strain can be calculated accurately only from data obtained at small strains and when a homogeneous linear elastic deformation of the material occurs on both sides of the neutral axis (van Vliet & Luyten, 1995). This requires that the test material be homogeneous, which is a problem for many foods such as biscuits. Moreover, substantial bending of the test piece due to its own weight must be avoided, so its modulus must be rather high.

Biaxial Extension Tests

If a uniaxial compression test is done in such a way that there is no friction between the

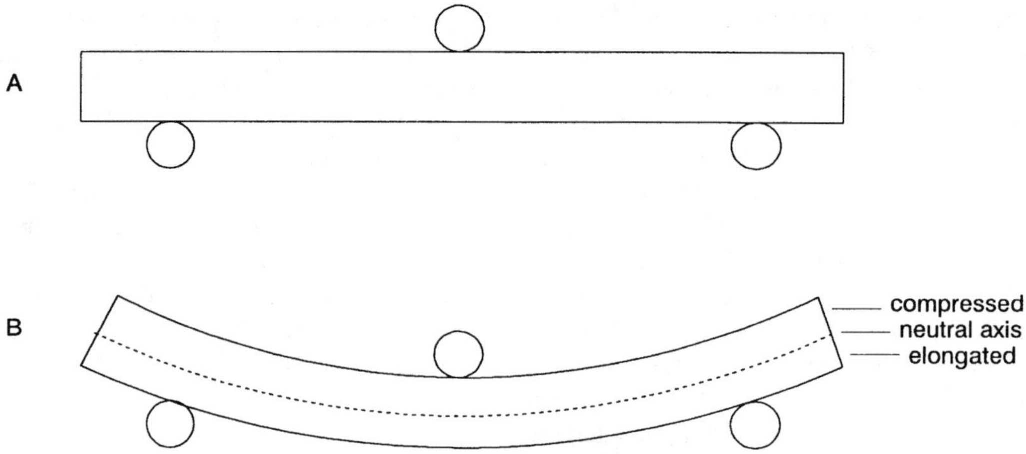

Figure 4–22 Deformation during a Three-Point Bending Test

test piece and the plates, the material investigated is not only uniaxially compressed but also biaxially extended (Figure 4–23) (Chatraei, Makosko, & Winter, 1981). The experiments may be considered as a biaxial extension test. Biaxial flow occurs in various practical situations, such as around growing gas bubbles in bread doughs and cheese (van Vliet, Janssen, Bloksma, & Walstra, 1992) and in the mouth (de Bruijne, Hendrickx, Alderliesten, & Looff, 1993). This type of measurement may also be advantageous for products that exhibit slip during shear measurements, such as mayonnaise. Slip is an essential requirement for good execution of the test. To avoid friction, plungers are used with the same diameter as the test piece diameter. In this situation, the biaxial stress σ is given by

Equation 42

$$\sigma(t) = \frac{F(t)}{\pi R^2}$$

where $F(t)$ is the force recorded at time t and R the radius of the plate. For a material with a Poisson ratio of 0.5, the biaxial Hencky strain ϵ_b can be calculated from

Equation 43

$$\epsilon_b = -\frac{1}{2}\ln\frac{L(t)}{L_0}$$

The biaxial extensional strain rate $\dot{\epsilon}_b$ is given by

Equation 44

$$\dot{\epsilon}_b = \frac{d\epsilon_b}{dt} = \frac{1}{2}\frac{dL}{L(t)\,dt} = \frac{v}{2L(t)}$$

From the test results, an (apparent) biaxial extensional viscosity, $\eta_{b,app}$ ($\equiv\sigma/\dot{\epsilon}_b$) can be calculated, which for Newtonian liquids would be 6η. Be aware that products like dough and mayonnaise will exhibit stress overshoot (see the earlier section "Tests at a Constant Strain Rate"). A real viscosity can be calculated only for strains larger than the strain corresponding with the maximum stress. Before this maximum, the stress response is mainly governed by elastic effects.

EMPIRICAL TEST METHODS

In this type of test, the stresses and strains involved are not well defined, which makes

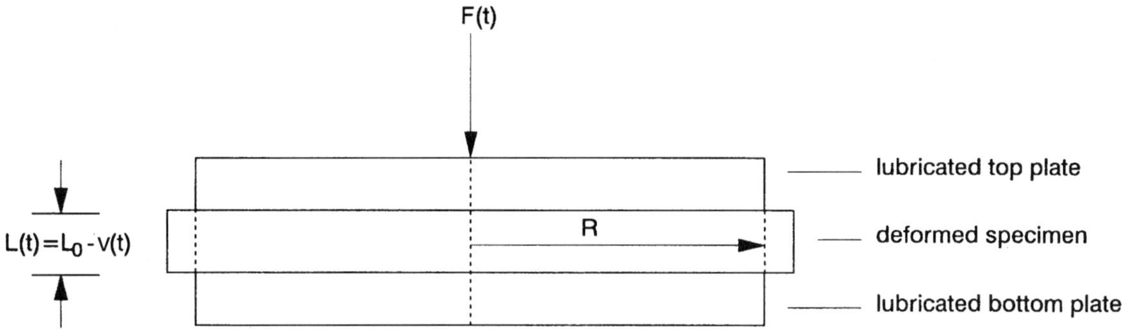

Figure 4–23 Biaxial Extension Configuration

the calculation of fundamental rheological parameters impossible. In the food industry and at control laboratories, a huge range of instruments is in use for empirical tests. Empirical tests can be of great value as long as the following principles are followed:

1. A relevant mechanical property or one that is unequivocally related to it is determined. Measurements based on an accidental correlation may prove to be dangerous if the product, the raw material's properties, or the process conditions are changed.
2. The experiment is done over a time scale that is relevant for the problem studied.
3. The measurements are reproducible regarding test piece preparation and test execution.

Many empirical test methods are variations on fundamental tests—for example, the determination of the torsion required to let a cylinder, a vane, or a helical path rotate in a "liquid" in a large container. For the interpretation of the results, it may be useful to realize which (combination of) fundamental parameter(s) determine(s) the reading obtained. For instance, in a Bostwick consistometer, the distance is measured over which a certain amount of material, initially contained behind a sliding door, flows in a certain time after the sliding door is opened. This distance will be determined by a combination of yield stress, apparent viscosity, and apparent elongational viscosity, at decreasing strain rates (stresses).

In the next section, only a few typical examples of empirical test methods will be discussed.

Compression Tests

For these tests, in principle, the apparatuses used are similar to those discussed in the section "Large-Deformation and Fracture Tests with Tension-Compression Instruments." An object is pressed in or on the food product with a certain force or speed. Many relatively cheap apparatuses are for sale in which the compression speed is fixed or may be varied over a small range. In addition, a large variety of compression geometries can be bought; examples are flat cylindrical plungers of various diameters, cylinders with a rounded tip, needles, perforated plates, and wedges. Depending on how the experiment is done, the data obtained are related to the modulus (stiffness), fracture stress or strain, work of fracture, or a combination of these parameters.

In several tests, the test piece is compressed over a certain set distance. This implies that for test pieces of different height the actual relative deformation will be different. If the force required to obtain a constant relative deformation is determined, some products may have fractured during the test and others not. This makes a comparison of the data obtained difficult.

Penetrometer Test

This is also a kind of compression test. In this test, a conically shaped measuring body falls from a predetermined height onto the product. The penetration depth is measured. This is a simple and relatively cheap apparatus that has been used for applications such as the determination of the spreadability of butter (Walstra, 1980).

The penetration depth depends on measuring geometry properties such as weight and cone angle, on falling height, and on a combination of test material properties. The falling cone will first deform the material and at large deformations the material may yield (e.g., in the case of butter or margarine) or fracture (e.g., in the case of pectin gels). Materials that yield will subsequently flow through the gap between the cone and the unyielding material. This flow contains both a shear and an elongational flow component (Figure 4–24). Thus, penetration depth is determined by a combination of the following fundamental rheological properties, even though their relative contribution may vary: elasticity modulus in shear and compression, yield or fracture stress, and shear and elongational viscosity.

As long as the relative contribution of the various components contributing to the penetration depth does not vary and the experiments are done under standardized conditions, the penetrometer gives an unequivocal characterization of the mechanical properties of the product.

Flow Funnel

A very simple instrument to get an impression of the flow behavior of a liquid is the flow funnel (Figure 4–25). This type of in-

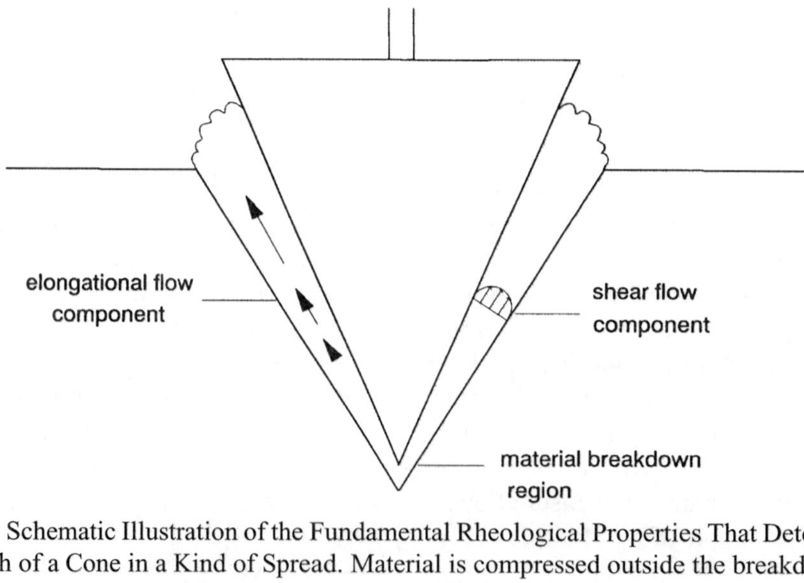

Figure 4–24 Schematic Illustration of the Fundamental Rheological Properties That Determine the Penetration Depth of a Cone in a Kind of Spread. Material is compressed outside the breakdown region.

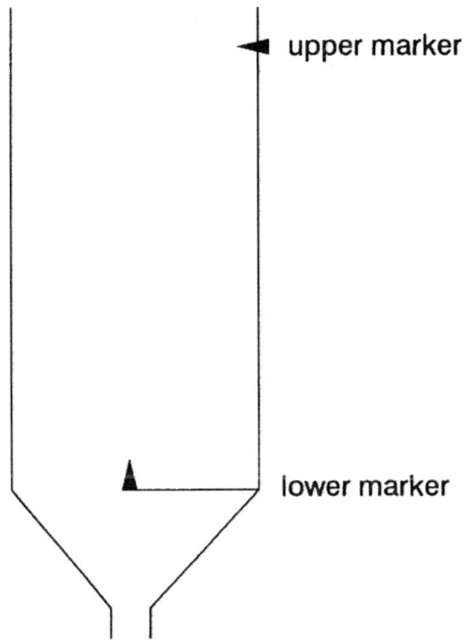

Figure 4–25 Illustration of a Flow Funnel

strument is denoted by various names such as *Ford cup, Marsh funnel, consistency cup*, and *Posthumus funnel*. The instrument consists of a circular container with a conical end in which there is a small orifice leading to a short exit tube (length/diameter less than 1.5). The driving pressure for flow is gravity. One simply determines the time required for a certain amount of liquid to flow out of the funnel. Sometimes pins are set into the inside of the container to allow a better determination of when the required amount of liquid has flowed out.

Because the small exit tube is short or even absent, flow is mainly determined by the convergent flow through the cone, causing a strong contribution of the elongational viscosity to the resistance against flow. The shear component will be zero in the middle of the funnel and at a maximum at the walls. Flow rates will decrease as the funnel empties. Thus, the flow pattern is very complicated. For Newtonian liquids and "power law" fluids, the (apparent) viscosity may be estimated from the flow time (Barnes, 1980). Liquids showing marked thixotropic behavior over long shearing times will give flow times that are much longer than anticipated from steady-state shear viscometry because flow times involved in a flow funnel are mostly relatively short. In addition, the large Trouton value (η_E/η) of polymer solutions implies that their flow time will be long compared to Newtonian liquids with the same viscosity in shear flow.

The presence of a strong elongational flow component is an advantage if one wants to predict the perception in the mouth by consumers. Flow in the mouth also contains a very strong elongational component (de Bruijne et al., 1993). Therefore, a flow funnel may be better for predicting consumers' perception than an expensive rotational viscometer.

CONCLUSION

At present, people working in the food industry, quality control laboratories, and research institutes have a large number of different instrumental tests available for determining process parameters; checking and controlling the quality of foods, intermediate products, or raw materials; and predicting consumer acceptance. In recent years, various relatively cheap and versatile instruments have come onto the market. From this perspective, the situation regarding the determination of the rheological properties of foods, intermediate products, and raw materials is promising. However, often the largest problem is to establish which property must be determined in a certain situation. From this perspective, the situation can still greatly improve, certainly in the case of predicting consumer acceptance. Too often, one simply takes an instrument that is available and determines a parameter that other people have measured before without realizing which fundamental rheological properties determine the property one is interested in. Only after the latter question is answered can one choose an instrumental test that may be fundamental or empirical. The bottleneck is frequently how to use the available instruments to measure the appropriate relevant parameters. To this end, some background knowledge regarding rheology and fracture mechanics of foods must be known, and that is often missing. Moreover, during tests, possible sources of errors are often not checked systematically.

Much more research needs to be done on the topic of fundamental rheological (including fracture) parameters that determine the relevant properties of food. If this area can be addressed, the problem of how to choose a relevant instrumental test can be answered. That problem has been the main topic of this chapter.

REFERENCES

Barnes, H.A. (1980). Detergents. In K. Walters (Ed.), *Rheometry: Industrial applications* (pp. 31–111). New York: Research Studies Press.

Chatraei, S.H., Makosko, C.W., & Winter, H.H. (1981). Lubricated squeezing flow: A new biaxial extensional rheometer. *Journal of Rheology, 25*, 433–443.

Cheng, D.C.-H. (1984). Further observations on the rheological behaviour of dense suspensions. *Powder Technology, 37*, 255–273.

de Bruijne, D.W., Hendrickx, H.A.C.M., Alderliesten, L., & Looff, J. de. (1993). Mouthfeel of foods. In E. Dickinson and P. Walstra (Eds.), *Food colloids and polymers: Stability and mechanical properties* (Special Pub. No. 113, pp. 204–213). Cambridge, UK: Royal Society of Chemistry.

Ferry, J.D. (1980). *Viscoelastic properties of polymers* (3rd ed.). New York: John Wiley.

Peleg, M. (1984). A note on the various strain measures at large compressive deformations. *Journal of Texture Studies, 15*, 317–326.

van Vliet, T. (1995). Mechanical properties of concentrated food gels. In E. Dickinson and D. Lorient, (Eds.), *Food macromolecules and colloids* (Special Pub. No. 156, pp. 447–455). Cambridge, UK: Royal Society of Chemistry.

van Vliet, T., Janssen, A.M., Bloksma, A.H., & Walstra, P. (1992). Strain hardening of dough as a requirement for gas retention. *Journal of Texture Studies, 23*, 439–460.

van Vliet, T., & Luyten, H. (1995). Fracture mechanics of solid foods. In E. Dickinson (Ed.), *New physicochemical techniques for the characterization of complex food systems* (pp. 157–176). Glasgow, Scotland: Blackie.

van Vliet, T., Luyten, H., & Walstra, P. (1993). Time dependent fracture behaviour of food. In E. Dickinson and P. Walstra (Eds.), *Food colloids and polymers: Stability and mechanical properties* (Special Pub. No. 113, pp. 175–190). Cambridge, UK: Royal Society of Chemistry.

van Vliet, T., & Walstra, P. (1995). Large deformation and fracture behaviour of gels. *Faraday Discussions, 101*, 359–370.

Vincent, J.F.V. (1992). Fracture. In J.F.V. Vincent (Ed.), *Biomechanics materials: A practical approach* (pp. 193–217). Oxford, UK: IRL Press.

Walstra, P. (Ed.) (1980). Evaluation of the firmness of butter. *International Dairy Federation Document [Brussels] 135*, 4–11.

Whorlow, R.W. (1992). *Rheological techniques* (2nd ed.). Chichester, UK: Ellis Horwood.

Appendix 4–A

List of Main Symbols

Symbol	*Meaning*	*Unit*
A	surface area	m^2
E	Young's modulus	$N \cdot m^{-2}$
d	diameter	m
F	force	N
G	shear modulus	$N \cdot m^2$
G'	storage modulus	$N \cdot m^{-2}$
G''	loss modulus	$N \cdot m^{-2}$
H	distance between plates	m
J	compliance	$N^{-1} \cdot m^{-2}$
k	consistency index	$N \cdot m^{-2} s^n$
K	bulk or (all-sided) compression modulus	$N \cdot m^{-2}$
l	capillary length	m
L	length/height of test piece	m
n	power law index	
P	pressure	$N \cdot m^{-2}$
r	distance from center of plate	m
R	radius	m
R_1	radius of inner cylinder	m
R_2	radius of outer cylinder	m
R_i	radius at point *i* between inner and outer cylinder	m
t	time	s
T	torsion	$N \cdot m$
v	velocity	$m \cdot s^{-1}$
$\overline{v}$	average velocity	$m \cdot s^{-1}$

Symbol	*Meaning*	*Unit*
V	volume	m^3
α	angular displacement	rad
γ	shear strain	
γ_0	maximum shear strain in dynamic tests	
$\dot{\gamma}$	shear rate	s^{-1}
δ	phase lag between stress and strain in dynamic tests	rad
ϵ	tensile/compressive strain	
ϵ_f	fracture strain	
ϵ_H	Hencky strain	
$\dot{\epsilon}$	elongational strain rate	s^{-1}
$\dot{\epsilon}_H$	Hencky strain rate	s^{-1}
η	viscosity	$N \bullet m^{-2} \bullet s$
η_{app}	apparent viscosity	$N \bullet m^{-2} \bullet s$
η_B	Bingham viscosity	$N \bullet m^{-2} \bullet s$
η_E	elongational or (uniaxial) extensional viscosity	$N \bullet m^{-2} \bullet s$
θ	cone angle	rad or degree
λ	relaxation time	s
μ	Poisson ratio	
ν	kinematic viscosity	$m^2 \bullet s^{-1}$
ρ	density	$kg \bullet m^{-3}$
σ	shear stress	$N \bullet m^{-2}$
σ_E	tensile/compressive stress	$N \bullet m^{-2}$
σ_f	fracture stress	$N \bullet m^{-2}$
σ_w	shear stress at capillary wall	$N \bullet m^{-2}$
σ_y	yield stress	$N \bullet m^{-2}$
σ_0	maximum shear stress in dynamic tests	$N \bullet m^{-2}$
ω	angular frequency	$rad \bullet s^{-1}$
Ω	angular velocity	$rad \bullet s^{-1}$

CHAPTER 5

Measurement of Mechanical Properties of Food Materials in Relation to Texture: The Materials Approach

Bogdan J. Dobraszczyk and Julian F. V. Vincent

Foods are materials and therefore have the mechanical properties of materials. Large parts of the preparation of food involve comminution, which can be related to fracture mechanics. "Texture" is governed by a combination of mechanical and fracture properties and their modification and expression within the mouth during chewing. Difficulties arise due to a number of factors:

- Most foods are mechanically very complex.
- Mechanical processing and chewing combine many processes that materials science needs to separate to quantify and understand them.
- Food in the mouth is continually changing its properties as temperature, water content, pH, and so on change.
- Most food scientists are not trained in materials science.
- Most materials scientists do not admit that food is a material!

It is therefore probably impossible to measure "texture" in a machine, although it *is* probably possible to identify the main factors that govern the texture of a food material and to measure them, thus accounting for a large part of "texture." There are two main types of approach. The first, more common and intrinsically easier, is to apply any type of mechanical deformation to the sample, measure the response, and try to correlate the result with the results from a sensory panel. Food science is replete with tests of this variety, which can be little more than quality control tests since the lack of an initial validated model means that the results cannot be generalized or with any ease ascribed to a particular process within the material. We adopt here the second approach of trying to divine from the way in which the food material responds what sort of mechanical deformation best describes that to which the food is being subjected.

MEASURING FOOD TEXTURE: THE MATERIALS APPROACH

At the heart of the materials approach lies the idea that the mechanical properties of a food and their relationships with the internal structure of the food material and the processing conditions within the mouth should be directly related to the perceived texture of that food during eating. This is based on the structural engineering analysis of materials, where small-scale laboratory measurements of mechanical properties have successfully been extrapolated to the behavior of large engineered structures such as bridges, buildings, and pressure vessels, and also to the behavior of biological materials such as skin, muscle, tendons, gut, and arteries (Vincent, 1990a, 1990b, 1992; Wainwright, Biggs, Currey, & Gosline, 1965). However, the fact that few simple mechanical tests have ever been shown to correlate with sensory perception of texture in foods reinforces the case that sensory assessments of texture involve not only complex interactions between the material properties of the food and its structure but also complex stress states and interactions within the mouth. It is therefore hardly surprising that attempted correlations between sensory assessments and mechanical tests are often inconclusive, especially if approached simplistically.

This chapter introduces you to a range of materials tests and their interaction with structural and process variables so that you can select the tests most applicable to the particular textural attribute being considered. To do this successfully, a thorough knowledge of material properties and structures is necessary, as well as of the nature of the deformations and the conditions the food experiences in the mouth during the assessment of the textural attribute being investigated. We do not build nuclear reactors without first finding out exactly the sort of conditions they are likely to experience in use and whether the materials used can withstand such conditions. Structural engineers want to design a structure such as a building, bridge, or nuclear reactor without its collapsing in everyday use. They do this by using their knowledge of materials' properties (measured on a small scale in the laboratory) and structural principles to predict the behavior of these structures under the sorts of forces expected under real conditions of use. It is not surprising, then, to hear that people take these materials tests very seriously and that a great amount of effort goes into their measuring processes; after all, our lives depend on them! Engineers need to have complete faith in their knowledge of materials' properties; otherwise, their buildings will collapse, and they will be very unpopular. Incidentally, buildings used to collapse quite often in the past when structural engineering principles were not quite as well developed as they are today; bridges collapsed regularly, as did the towers and roofs of the great Gothic cathedrals of Europe. For example, the tower of Beauvais cathedral in France collapsed once, and the roof fell twice. While many were quick to attribute these disasters to acts of God, the contemporary builders knew there was something wrong in their designs; they had a poor understanding of structural engineering principles (Gordon, 1978).

Similarly with foods, if food technologists wish to design a particular textural attribute in a food material rather than just use what is provided by nature, then it is necessary to have a thorough understanding of both the material properties of the food and the conditions it produces in the mouth. If they naively assume that they can find some single material parameter of a food to correlate with some ill-defined sensory attribute, then failure will be almost assured, and they would be well advised not to waste their time by attempting the exercise in the first place. Happily, with foods, this is not a life-or-death situation, and the worst that can happen is that the food will not sell, but this also probably accounts for the lack of rigor in food testing in general. If food technologists were to know precisely which textural attribute they wished to create and measure, how and when that attribute was perceived within the mouth, and what were the relevant conditions within the mouth (deformation, rate, state of the food), then the project would stand more chance of success. Given that the commercial rewards of such an exercise can be considerable, it is worth investing a significant amount of effort into a proper analysis of the processing conditions that occur during the sensory assessment of a desirable textural attribute before any mechanical tests are attempted. For example, if the sensory attribute being sought in a particular product is "chewiness," then it is sensible first to define where and when this chewiness is detected and under which conditions in the mouth. Is it in the first bite (unlikely), or is it after a considerable amount of chewing? What is the state of the food when it is being defined as chewy? What are the deformation boundary conditions of shear, extension, rate, temperature, moisture, and lubrication, and which interactions are the most important within the mouth—surface or bulk interactions between the food and teeth, lips, tongue, or palate? Not until these parameters are well characterized can any meaningful mechanical tests be constructed and performed. Examples of this kind of approach are rare in the food industry; most people prefer the empirical approach. Notable exceptions are detailed in work by Szczesniak (1963), Sherman (1970), Shama & Sherman (1973), Heath (1991), Kilcast & Eves (1991), Brown, Eves, Ellison, & Braxton (in press), and Lucas & Luke (1983).

Most foods are formed from biological materials in which many components are joined into complex structures. For example, bread, cakes, and extruded snacks are foam structures. We usually manipulate these in some way to make the food easier to consume or more attractive to eat, adding another layer of structure. For example, a loaf of bread contains several components arranged in a structure: starch cells and the protein gluten in wheat endosperm are ground into a fine powder (flour) and mixed with various ingredients to form a viscous paste and are then expanded into an open foam by baking.

The perceived crispness of vegetables depends mainly on their cellular arrangement, intercellular adhesion, and turgor pressure within the cells. Fruits and vegetables are complex cellular structures. X-ray microscopy and scanning electron microscopy show that apple flesh is composed of columns of cells radiating from the core. There are air spaces between the columns. The cells in the outer part of the apple, beneath the skin, are arranged in groups around isolated air pockets so that there is no "grain" apparent. These arrangements of the cells are easily seen if you put a slice of the apple into 5% mannitol (which is at about the same osmotic potential as the cell sap) and subject it to slight vacuum (about half an atmosphere, or 50 kPa) for about 5 min. Release the vacuum and allow the mannitol to invade the areas from which the intercellular air has been extracted, then hold the slice up to the light and see the areas where the air has been removed. They appear translucent and occur in the middle area of the flesh. Radial bands and the outer edge of the flesh will appear dark, showing that the air pockets do not communicate in these areas (Vincent, 1989). These differences in arrangement of the cells affect the mechanical properties. Potato tissue is much more dense, with little or no air space between the cells. So while the radial air spaces affect the way in which an apple fractures, a potato is more isotropic (Khan, 1989).

The characteristic texture of meat is due to the fibrous structure of muscle spindles and the way in which the fibers separate during chewing. Quorn, the trade name of a product (mycoprotein) of continuous fermentation of the fungus *Fusarium graminearum*, has been produced as an attempt to reproduce the fibrous structure, and hence texture, of meat by using only vegetable matter (Figure 5–1).

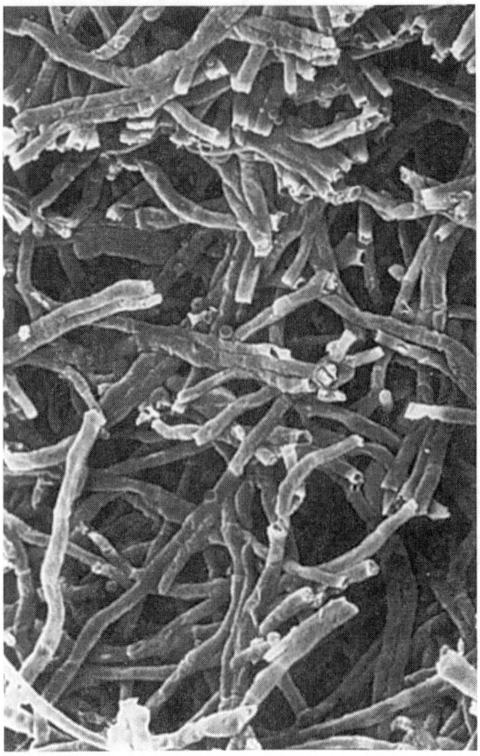

Figure 5–1 Micrograph of Mycoprotein Showing Its Fibrous Structure with Diameter of Fibers ~3–5 μm. Courtesy of Marlow Foods.

The hyphae of the fungus, although not as stiff and well orientated as in meat, produce a soft material with a fracture energy parallel to the hyphal fibers which is about a third that across the fibers. The use of fungal hyphae (usually as mushrooms) as a meat substitute has a history over 2,000 years. The fibrous texture has long been recognized as a factor in common between the two foods. The mechanical analysis most appropriate is that developed for fibrous composite materials.

Both material properties and the structural arrangement of a food material play a part in the perceived texture of a food. When a foamed material such as bread or expanded snacks is deformed in the mouth, the resulting stresses are just as much a product of the structure as is the actual material. The structural variables regarded as important to the architecture of the material include porosity, density, pore size distribution, and cell wall thickness distribution, and the material properties could include stiffness, fracture strength and toughness, yield stress, and Poisson's ratio.

RELATIONSHIP BETWEEN SENSORY TEXTURE AND MECHANICAL TESTS

Sensory assessments such as crispness, chewiness, and toughness are complex processes occurring within the mouth involving interactions among the teeth, tongue, cheeks, auditory signals, saliva, and the food material under complex states of deformation, and it is therefore hardly surprising that attempted correlations between sensory assessments and mechanical tests are often inconclusive. No mechanical test can, or should, attempt to reproduce the complex combination of sensations that make up our perception of texture. The brain is remarkably efficient in integrating these sensations into one overall definition such as "crisp" or "chewy." What we can do as materials scientists is to look at the overall processes occurring while a particular food is being eaten and try to reduce them to a series of individual, definable material properties independent of confounding factors such as shape or geometry, rate, or temperature. This reductionist approach has served humanity well in many areas where complex processes and interactions are reduced into single events, processes, or phenomena that are then amenable to experiment and measurement.

Take, for example, the definition, causes, and quantification of "crisp" and associated sensations such as "crunchy" and "crumbly." We must first translate "crispness" into a form that can be described by materials science and measure single parameters at the material and structural levels that can be related to, and can account for, the defined concept of crispness. Ultimately, it is necessary to decide what is meant, in materials science terms, by the descriptors used by sensory panels to quantify crispness and associated sensations. A starting definition could be related to the rate at which the force applied by the jaw-closing muscles drops when the material fractures, together with the sensing of a reduction in pressure on individual teeth. Crispness may be associated with a rapid drop in force, associated with rapid propagation of fracture, which in turn necessitates that the material be brittle.

The model can then be expressed as follows. As the force on an item of crisp food increases, so does the deformation. This rising part of the curve is a function of the (flexural) stiffness of the object, being affected both by material effects (e.g., the stiffness of the material) and structural effects (e.g., the thickness of the object, its density or degree of aeration, and its shape). The maximum force reached would be equated with hardness; the higher the force, the harder the material. Since a more massive piece of material can support a higher force than a less massive one, it can register as harder simply because of its structure and not necessarily because the material is different. The object then starts to break. If the material is brittle (i.e., requires little work to fracture), the fracture will travel quickly, resulting in sudden unloading of the muscles; this is seen as a sudden (vertical) drop in load on a force-deflection curve. The fracture is then somehow inhibited and stops, only to start again as the food is deformed further (Figure 5–2). In many materials (e.g., potato crisps, many fried or otherwise dehydrated starch products, and some sweets and confections), these drops result in a texture related to crispness. The magnitude of this effect is probably important. For instance, a series of a few large events is typical of a "quality" potato crisp; a longer series of smaller events is typical of an average, light potato crisp. As the events become larger and fewer, there is a transition from "crisp" to "hard" texture; conversely, as the events become

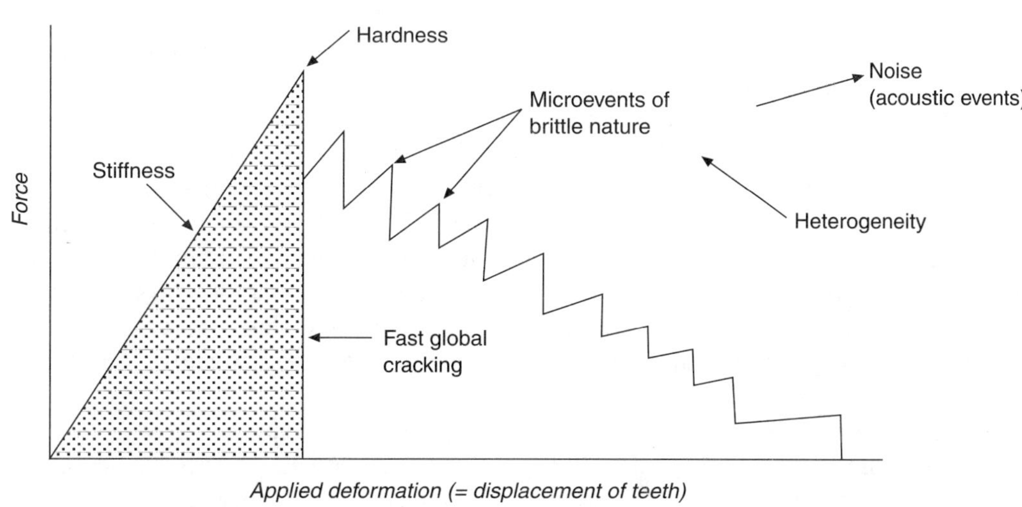

Figure 5–2 Generalized Force–Deflection Trace for a Crisp Material

smaller and more frequent, the texture becomes "crumbly" (Vincent, in press).

Materials' properties should be independent of size, shape, and how they are measured: in other words, universal, rather like the speed of light or density of water, which do not depend on how much light or water is being measured or how it is being measured. It is comforting to know that the modulus of bread measured in my laboratory in Reading will be the same modulus as that measured in a colleague's lab in Hong Kong even if they are measured using different tests, sample sizes, or shapes. The whole point of the materials approach is that the properties that are measured are reproducible and can be compared among different samples, test sizes and shapes, and test methods. Empirical measurements frequently measure a single point number under a particular set of test conditions for one particular material. While this may give the illusion of a "scientific" test by being performed on a machine (frequently with a computer attached) and may give satisfactory correlations with a textural or processing parameter, once you start to apply that test under different conditions or to other materials, the correlations break down or become irrelevant.

For example, one of the most commonly used tests in the food industry is the penetrometer or puncture-type test, in which a probe is driven into the food material and the depth or force of penetration is measured. This has the perceived advantages of being quick, portable, easy to use, and inexpensive (a major requirement for industry!) and of requiring a limited amount of technical expertise or data processing. Belying this comforting sense of simplicity, however, is a very complex and ill-defined deformation process.

In the apple industry, "instrumental texture" is defined as the force required to push a metal probe or penetrometer 8 mm in diameter into an apple. The flesh of the apple is composed of largish (100 μm in diameter) cells stuck together in a more or less columnar fashion radiating from the center of the fruit. The columns are defined by long, radially arranged air spaces that can contribute up to half the volume of the apple in an early variety that grows very quickly. The longer the apple has to grow, the more dense it is, so later varieties are much denser, containing only 5% air spaces. Consider what happens as a probe is pushed into this structure. The cells ahead of it can be squashed, broken, and compressed. But if the cells can be separated from each other without being broken, they will not present sufficient resistance to be broken and compressed and will simply move aside to allow the probe to pass. Therefore, the force needed to push the probe in will be less. This is the state of affairs in a mealy apple, where the cells cannot transfer stress because they are not stuck together well enough. A mealy apple tends also to have more air space because the cells, having lost their mutual adhesion, tend to round up and push against the skin of the apple, splitting it at extreme stages of mealiness. However, the degree of compression of the displaced cells also depends on the density of the apple. The closer the cells are arranged, the greater the force needed to push the probe further. If there is more air space within the apple, the cellular debris will compact into those air spaces, and the force required to push the probe will be less. The apple cultivar Spartan illustrates this. It is an early crisp apple of low density. A penetrometer reading classes it as having poor texture and classes it with mealy apples. A simple fracture test shows that it is brittle and crisp.

Strangely, "crispness" in vegetables is not always dependent on turgor. Some pickled and cooked materials break in a brittle manner even if their stiffness is low. It seems that they do this because the cell walls are still stuck firmly together. The seminal example is the Chinese water chestnut, the corm of a water plant, which retains its crispness after cooking. This is due to the diferulic acid (a phenolic derivative related to lignin), which sticks the cells together. It is hydrophobic and covalently bonded to the cell walls, so it is not dissolved by heat or acid treatment (e.g., pickling) (Waldron, Smith, Ng, Parker, & Parr, 1997). Both crispness and firmness disappear if the cell-cell adhesion is lost. In apples, this is called mealy texture. The cells can still be quite turgid under these conditions: if you run your finger over the flesh of a mealy apple, you will pick up lots of cells that feel like sandpaper between the fingers. These cells are hard and round but are too small, singly, for you to bite between your teeth. A mealy apple is therefore very dry.

Other empirical tests have been developed to characterize a single processing variable such as "handling ability" or "mixing tolerance" while processing foods in a mechanized plant. For example, empirical tests have been developed to attempt to characterize adhesive behavior of bread doughs during processing. Many of these are "single-point" tests that are appropriate only to a single set of conditions under which that test was performed and are generally not applicable to any other conditions. Because the rheological properties of dough are dependent on both time and strain, there is often a discrepancy between such single-point–type tests and actual performance in the plant, where conditions of strain and strain rate may be poorly defined and very different from those in the laboratory test. Most food materials flow over time and are therefore termed *viscoelastic*; their properties depend on how quickly the test is performed (the strain rate or frequency) and also on how quickly the measuring instrument responds to that change. This is important in many aspects of food texture: if a food is deformed quickly, as in the mouth by chewing, then the mechanical properties of a viscoelastic material will be very different if measured at the typical rates of deformation found in testing machines. Similarly, if a food is perceived as sticky in the mouth or during processing, it may not be under the conditions that prevail during a laboratory test. Thus, a test under only one particular set of conditions of rate, temperature, and strain will almost certainly not be applicable to another set of conditions. What is necessary is to define the set of conditions under which the food operates as an adhesive. For example, if a particular dough is sticky at the low strain rates commonly used in instrumental testing, will that same dough be sticky at the higher rates usually experienced in a commercial bakery? Equally, tests made under the small strains typical of dynamic rheological testing may be entirely inappropriate to the much higher strains that occur during processing. Therefore, it is necessary to investigate adhesive behavior over a range of strain rates, strains, and temperatures, since any one of these can influence rheological behavior and hence adhesion (Dobraszczyk, 1997b).

Many of the tests reported in the food literature are totally inappropriate because they do not measure the system under the appropriate conditions. If a test shows no differences in mechanical behavior among materials, despite knowledge of differences in practice, then it is the test that is wrong and not the material. Small-deformation dynamic rheological tests are frequently quoted in the literature and applied almost indiscriminately to any food system regardless of whether they are valid for the system. Dynamic rheology works well when there are chemical changes to a material, such as cross-linking, covalent bonding, or hydrogen bonding, but it is not appropriate if the changes are physical or structural, such as molecular chain entanglements or changes to fiber or cell structure. Further, it is not widely recognized that measurements in one deformation field, such as shear, are not always directly applicable to a different deformation, such as extension. Polymer melt fluid dynamics shows that the results in shear can be very different from those in extension due to the different physical effects these deformations can have on networks of molecules (Cogswell, 1981; Padmanabhan, 1995).

The sensory assessment of food texture depends on the specific material properties of the food material, their structural arrangement into cells, fibers, and pores, and their interaction in the mastication process. Eating involves deformation within the mouth—biting, chewing, and swallowing—and interactions among tongue, cheeks, lips, and teeth, as well as taste, smell, sound, and sight. The deformations comprise complex and intricate movements and stress states. As a result of deformation within the mouth, a food material will respond differently to the stresses induced depending on its size and shape, its inherent material properties, the rates of biting and chewing, and orientation. Materials tests attempt to measure the forces required to produce given controlled deformations, such as squashing (compression) or bending or pulling apart (tension), and to present them in such a way as to be independent of sample size, geometry, and mode of testing. They measure a well-defined property, such as stress and strain or stiffness. A small test piece of the material is usually deformed in a controlled way, normally on a motor-driven machine, and the force is measured as well as the distance moved or displacement of the object. The force is then usually plotted against the displacement to give a force-displacement curve. Normally, we would divide the force and displacement by the original sample dimensions to obtain stress (force/area) and strain (displacement/original dimension); this allows us to remove the sample size as a variable. Thus, a materials stress-strain analysis of a small piece of steel should theoretically scale up to a girder and then to a whole bridge, given that we know all the structural parameters. There is absolutely no reason why food materials should be any different—they do not obey a different set of physical laws than other materials.

It should be noted, however, that these tests were originally developed by engineers for materials normally used in construction, such as steel and concrete, and that food materials often behave very differently. Engineering materials are normally brittle and elastic; on deforming such a material, we can always predict its stresses and strains very accurately. Also, such materials move only a very small amount under applied load—they are very stiff. This is very important for engineering materials such as bridges and buildings; because of the way they are designed, major changes in dimensions would be disastrous! Hence, we do not generally use food materials for buildings (except in fairy tales). The outlook of the structural engineer is very different from that of the food technologist.

Most foods, being biological materials, do not behave like engineering materials. Instead, they do things not allowed by engineers: they flow and change shape dramatically under load, and on releasing that load they stay deformed—they change shape permanently. My bookshelves at home are a good example: whereas the ones made out of glass have stayed flat, the proper wood ones have moved a bit, and the cheap and nasty chipboard ones have adopted a gentle curve! So it is inappropriate to apply the sort of tests performed by the structural engineer directly to measure the mechanical properties of food materials. The types of test and analysis needed for food materials have to take this into account, and quite often the sort of tests used in measuring foods have more in common with tests for polymers and biological materials than with tests for engineering materials.

DEFINITIONS OF MECHANICAL PROPERTIES

Strain

Strain is a pure number, since it compares shape before and after deformation. It is expressed either as the ratio of the dimensional change divided by the starting dimension—the increase in length (δl) or extension divided by the original length (l_o), known as the "engineering" or *Cauchy strain* (e)—or as the natural logarithm of the extended length divided by the original length, $\ln(l/l_o)$, known as the *Hencky strain* (ε_H) (Figure 5–3). Up to strains of 10% (0.1), these two numbers are not very different, but beyond this, they deviate considerably. The engineering strain (e) suffices for most load-bearing materials, which generally operate at small strains, but strains in extensible materials cannot adequately be described by engineering strain. The Cauchy strain definition is satisfactory for small elastic strains where the total elongation is small, but for materials that experience large strains, the original length changes considerably, and the normal engineering definition is inadequate because it is based entirely on the original dimensions. Many food materials can deform to very large strains in excess of 100% and Hencky strain (ε_H) is a more suitable definition of strain for these materials, because it is based on incremental changes in dimension (dl) referred to the instantaneous length during stretching (l):

$$e = \delta l / l_o \quad \text{(Cauchy or engineering strain)}$$

$$\varepsilon_H = \int_{L_o}^{L} dl/l = \ln(l/l_o) \quad \text{(Hencky strain)}$$

The relationship between Hencky and engineering strain is given by

$$\varepsilon_H = \ln(l/l_o) = \ln(1 + e)$$

Shear strain (γ) is defined as the angle moved in response to a shear stress (τ), as shown in Figure 5–4.

$$\gamma = w/l = \tan\theta$$

The *ultimate strain* is as far as anything can be stretched before breaking. It is an important characteristic since if a material can stretch a long way before breaking, it may well survive the load that was stretching it. Biological materials, even crystalline fibers such as cellulose, are commonly able to extend more than 5%, and many materials can extend to strains greater than 100%. Measurement of strain in soft or highly extensible materials is difficult. This is mainly because strain in such materials tends to be nonuniform. This will lead to considerable underestimates of strain if strain is calculated simply

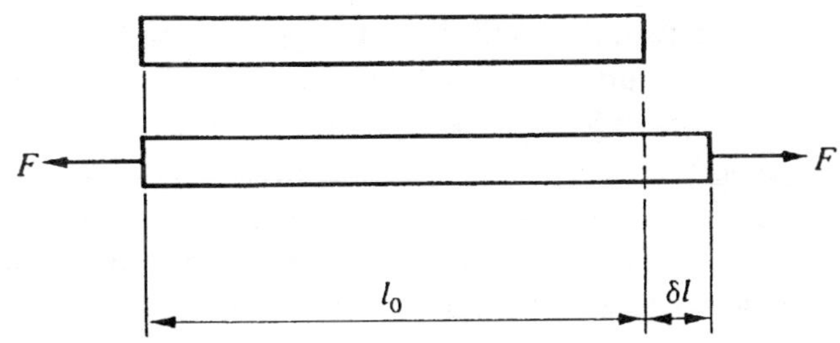

Figure 5–3 A Bar Subjected to Axial Tensile Deformation

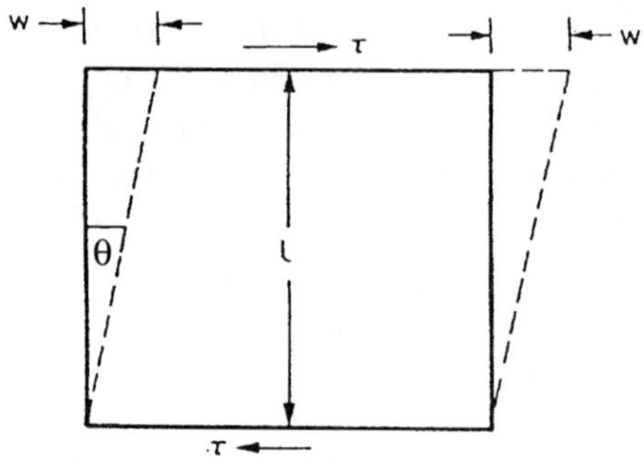

Figure 5–4 A Body Subjected to Deformation in Shear

from the movement of the grips, which hold the sample. In such a case, strain should be measured locally and independently by using contact or noncontact extensometers or by marking the sample in the region of interest and measuring local extension using video or photographic techniques (Shadwick, 1992; Shrive, 1997).

Stress

Stress is the outward expression of the restoring force developed by stretching the interatomic bonds that hold the material together. It is defined as the force exerted (F) divided by the cross-sectional area over which the force is acting (A):

$$\sigma = F/A$$

Normal practice is to divide by the original area (A_o) of the sample, but for large deformations, the instantaneous area (A) at any point should be used. Units of stress are newtons per square meter or ($N \cdot m^{-2}$) or pascals (Pa). Various stress states are defined depending on how the force is applied (Figure 5–5). For example, if the force is applied so as to squash the material, it is defined as compressive force; if the force stretches the material in one direction, it is called uniaxial tension. Other common stress states are shear, biaxial extension, as in an inflated balloon or expanding bubble in a food foam such as bread dough or cells under turgor pressure, and hydrostatic pressure or triaxial stress, in which a material is subjected to force from all sides, as in the case of an object in the deep oceans.

In a sense, these are all abstract idealized states, since in practice most materials rarely achieve uniform stress states and are subject to a complex mixture of stresses. It is general practice to try to resolve any applied forces into the pure components, of stress described above. For example, if an applied force acts at an angle to an interface, the force can be resolved into two components, one perpendicular to the interface (F_c compressive force) and the other parallel to the interface (F_s shear force).

If the material extends more than a few percent, Poisson ratio effects often cause it to become narrower in section so that the real stress is higher than the stress calculated from the original (relaxed) cross-sectional area (the normal way of doing things). This becomes even more important if the specimen becomes narrow in a small zone ("necking"), in which case the actual cross-sectional area should be used to obtain the "true" stress.

Stiffness and Hooke's Law

Either one puts a known load on the material and measures the resulting deformation (e.g., by hanging a weight on the end of a piece of thread), or one deforms the material by a known amount and measures the force generated as a result. Whether the material is being deformed by constant load or stretched by a known amount, one is finding out how difficult it is to deform the material; this is the stiffness and is expressed as the ratio between stress and strain (it therefore has the units of stress), or the slope of the linear portion of the stress-strain plot (Figure 5–6). Hooke's law states that for many materials under low to

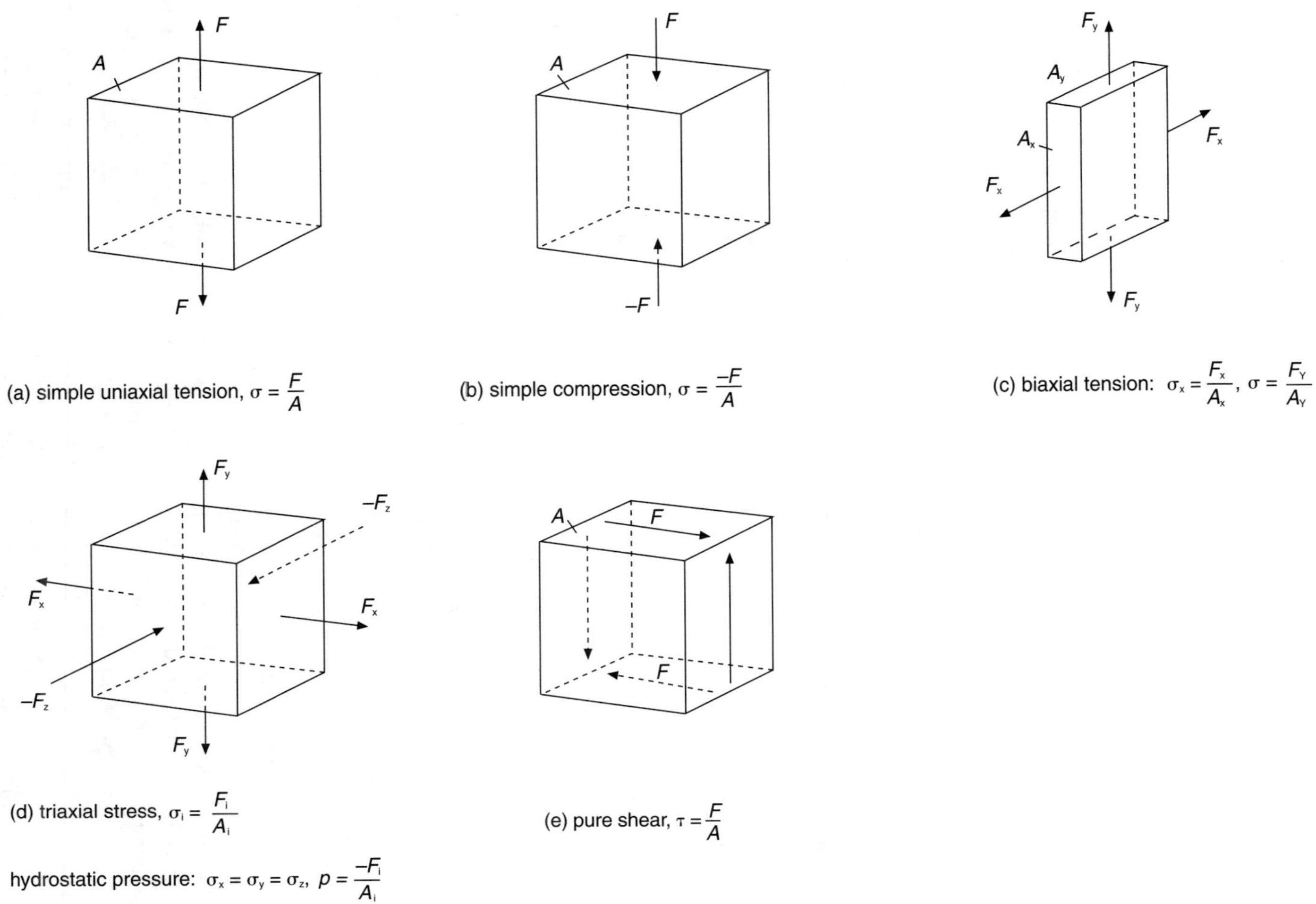

Figure 5–5 Common States of Stress. (a) Simple tension, (b) simple compression, (c) biaxial tension, (d) triaxial stress, (e) pure shear.

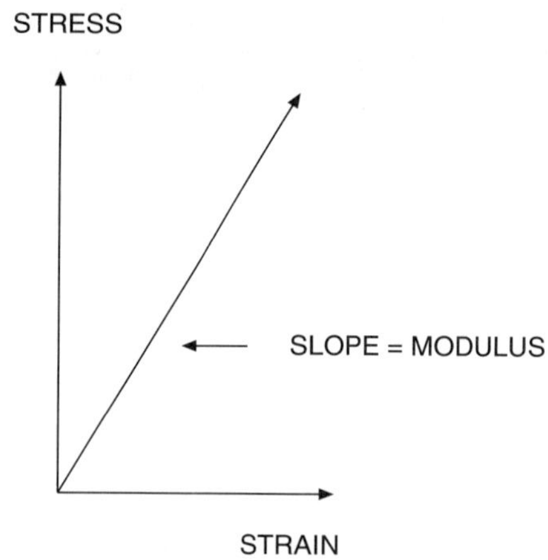

Figure 5–6 Stress-Strain Plot Showing Slope = Modulus

moderate strains, stress is proportional to strain, which is expressed as a linear relationship between stress and strain. Hooke (1635–1703) first published his formula of relationship between force and deformation in 1676: *ut tensio sic vis*, or "As is the extension, so is the force." What Hooke did not realize was that this proportionality is also dependent on the sample geometry. It was not until about 1800 that Thomas Young showed that describing the relationship in terms of stress and strain removed the geometric dependence and gave a material property—named *Young's modulus* in his honor.

Yield, Ductility, and Necking

Most materials' stress-strain relationship remains linear only at small strains, beyond which it becomes nonlinear (Figure 5–7). *Linearity* generally means that the material is behaving elastically, or that it will return to its original dimensions if the force is removed, as in an elastic band. The stiffness or Young's modulus is therefore sometimes referred to as *elastic modulus* or *Hookean modulus*. At some point, the stress-strain curve deviates from linearity, defined as the yield point and yield stress (σ_Y). Beyond the yield point, the material deforms plastically, and on removal of the force, the material does not return to its original dimensions but stays stretched or deformed permanently (Figure 5–7). This can be seen in typically plastic materials such as modeling putty, dough, "blue-tack," and many adhesives. Ductility is the ability of the material to flow in plastic deformation (from the Latin *ductilis*, from *ducere*, to lead). If a material fractures in the ductile region, the fracture surfaces are generally very rough and highly deformed, the surfaces cannot be fitted back together, and fracture occurs over a long period of deformation. Materials that fracture in the elastic region are generally brittle. Where the fracture surfaces are smooth and shiny, they can be fitted back together, and fracture occurs rapidly (Figure 5–8). Some of these differences are apparent in their force-distance relationships. Brittle, elastic fracture shows a linear relationship and a sudden drop in force on fracture. Ductile plastic fracture gives a nonlinear curve, and fracture occurs over a period of deformation.

Necking is a localized nonuniform deformation that generally begins at maximum load during tensile deformation in the plastic region, where the cross-sectional area decreases in an unstable and nonuniform manner.

Viscosity

Stress and strain may be measured without reference to time. Viscosity is defined as the resistance of a fluid (gas or liquid) to flow. It is normally measured in response to a shear stress (τ) and relates the stress to the strain rate $\varepsilon_{(t)}$: that is, the slope of the stress-versus-strain rate plot (Figure 5–9). Units are $N \cdot m^{-2} \cdot s$ or $Pa \cdot s$.

$$\tau = \eta\varepsilon_{(t)}$$

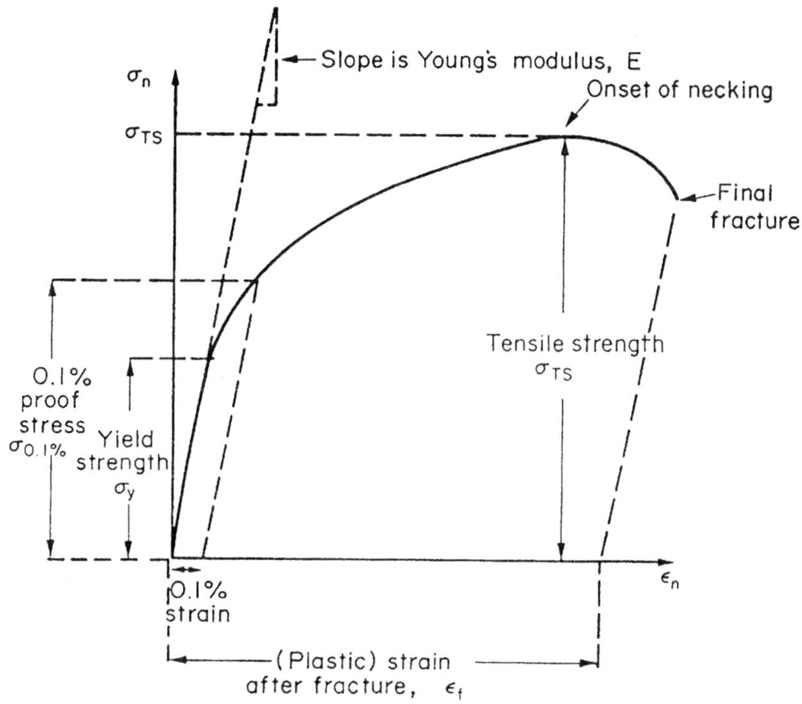

Figure 5–7 Stress-Strain Plot Showing Features of Ductile (Plastic) Flow in a Tension Test. *Source:* M.F. Ashby and D.R.H. Jones, Engineering Materials, *An Introduction to Their Properties and Applications*, p. 78. Reprinted by permission of Butterworth Heinemann Publishers, a division of Reed Educational & Professional Publishing Ltd.

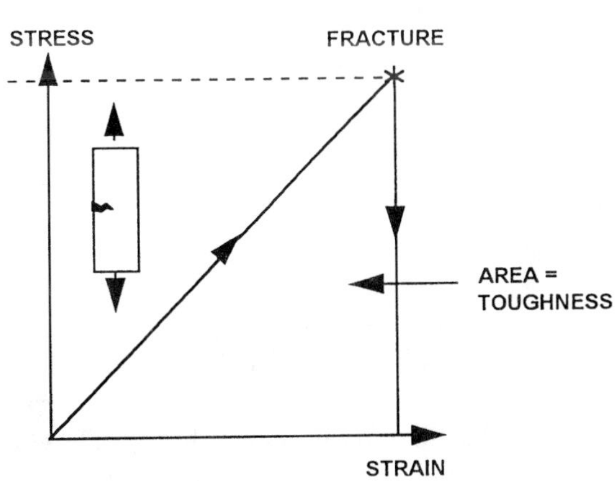

Figure 5–8 Stress-Strain Plot Showing Brittle Fracture in Tension

With most food materials, the mechanical response to force or displacement reduces with time. The simple way to test this is either (a) to stretch or strain the material and hold it at constant strain, the "stress relaxation" test; or (b) to apply a constant force and monitor the strain, the "creep" test. In a creep experiment, the sample is rapidly loaded to give a constant stress—for example, by hanging a weight at the end of a strip of the material (Figure 5–10)—and the elongation is monitored with time, as shown in Figure 5–11. In a stress relaxation experiment, a strain is rapidly imposed and held constant—for example, by rapidly squashing or stretching the material by a fixed amount and allowing the

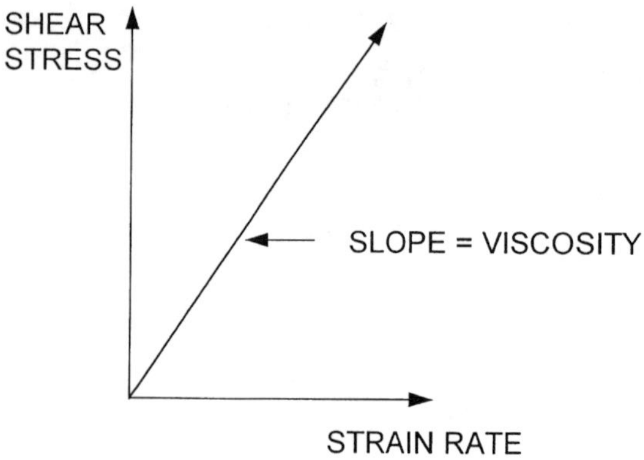

Figure 5–9 Stress versus Strain Rate Plot Showing Slope = Viscosity (Newtonian)

developed force to die away with time, as in Figure 5–12. In both these experiments, either the load gets less as time passes or the strain increases with time, and the stiffness must be falling because the strain or stress is being held constant. This third, time-dependent parameter is viscosity, or the change in stiffness with time.

If, in stress-relaxation, the load reaches a constant level below which it does not fall, then this "relaxed modulus" indicates that the material has some permanent cross-links, and

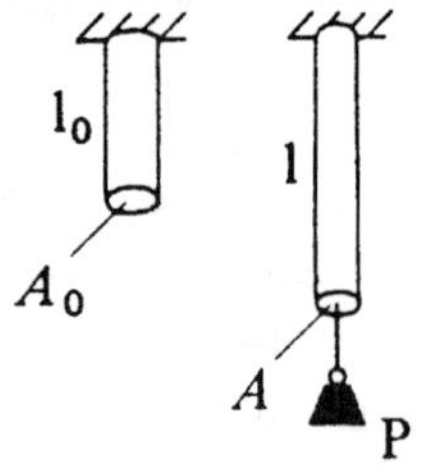

Figure 5–10 Method of Creep Testing by Imposing a Constant Force (P) and Measuring Length (l) to Obtain Strain (ε). *Source:* Reprinted with permission from K.L. Dorrington, The Theory of Viscoelasticity in Biomaterials, in *The Mechanical Properties of Biological Materials*, J.F.V. Vincent and J.D. Currey, eds., pp. 289–314, © 1980, Cambridge University Press.

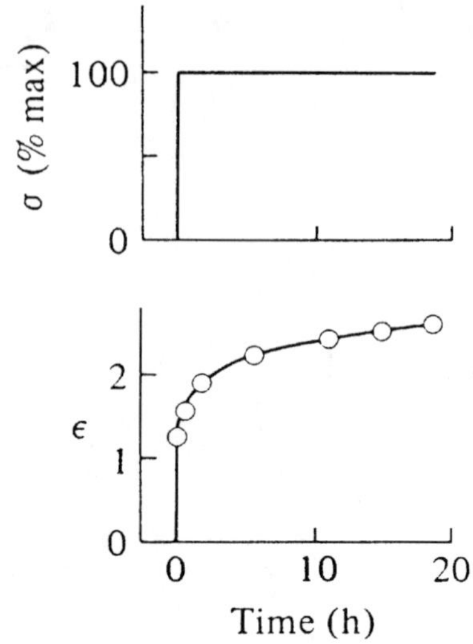

Figure 5–11 Creep: Change in Strain (ε) after Imposition of Constant Stress (σ). *Source:* Reprinted with permission from K.L. Dorrington, The Theory of Viscoelasticity in Biomaterials, in *The Mechanical Properties of Biological Materials*, J.F.V. Vincent and J.D. Currey, eds., pp. 289–314, © 1980, Cambridge University Press.

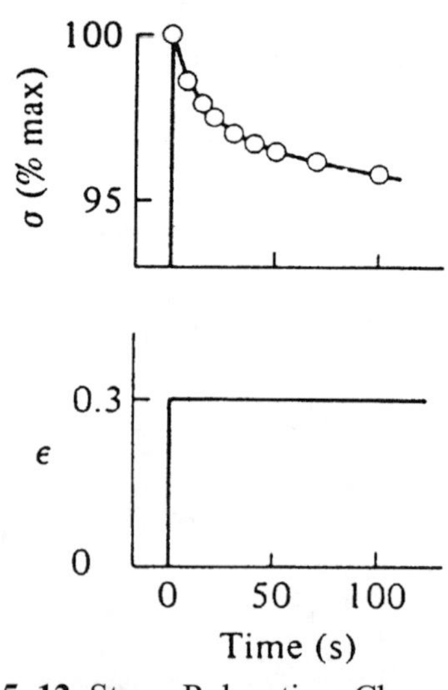

Figure 5–12 Stress Relaxation: Change in Stress (σ) after Imposition of Constant Strain (ε). *Source:* Reprinted with permission from K.L. Dorrington, The Theory of Viscoelasticity in Biomaterials, in *The Mechanical Properties of Biological Materials*, J.F.V. Vincent and J.D. Currey, eds., pp. 289–314, © 1980, Cambridge University Press.

the material is said to be viscoelastic. The viscous component of the mechanical response can be separated from the elastic one by creep or stress-relaxation tests or, more usefully, by a dynamic test in which the specimen is subjected to a small oscillating strain over a range of frequencies. If, however, the relaxed modulus eventually falls to zero, the material is not cross-linked, and the mechanical response is entirely viscous. However, this is not quite as simple as it sounds, since the polymers whose movement give the viscous response can be very large and very entangled. This results in very high viscosity and consequently a material that can take an extremely long time to relax completely, such as glass.

Several useful things can be done with this information. The simplest is to construct "isochronal" stress-strain curves—for example, by taking the stresses (σ_1, σ_2) at various strains (ε_1, ε_2) at fixed times from a stress-relaxation experiment on the same specimen and plotting them on the same axes (Figure 5–13). This is obviously different from the stress-strain curve obtained in a standard tensile test machine, where the strain rate and time vary throughout the test. In this treatment, the strain rate for each isochronal is

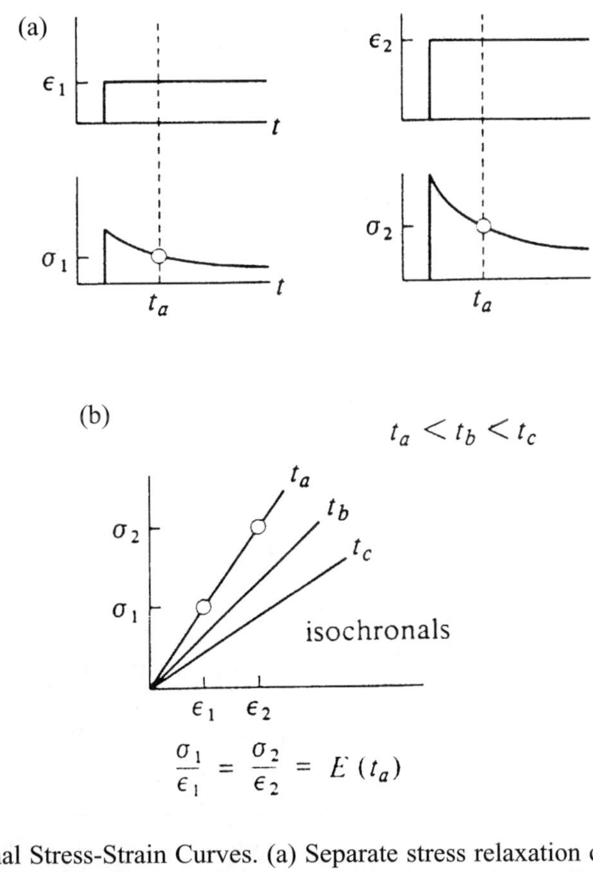

Figure 5–13 Isochronal Stress-Strain Curves. (a) Separate stress relaxation curves on the same specimen measured at different strains (ε_1, ε_2); (b) linear isochronals (σ_1/ε_1), (σ_2/ε_2), etc., measured at fixed times ($t_{a,\,b,\,c}$). *Source:* Reprinted with permission from K.L. Dorrington, The Theory of Viscoelasticity in Biomaterials, in *The Mechanical Properties of Biological Materials*, J.F.V. Vincent and J.D. Currey, eds., pp. 289–314, © 1980, Cambridge University Press.

constant, giving a true relaxation modulus ($E_t = \sigma_1/\varepsilon_1$).

Strain Rate

The speed at which the material is loaded or deformed becomes important, especially if it is a polymer. If this speed is high (on the order of a fraction of a second), then the polymer molecule hardly has time to flow or adjust its shape and registers a high stiffness. But if the speed is low (in some experiments the speed of deformation has been reckoned in mm per day, per week, or even per year), then the polymer flows with the external deformation and registers a lower stiffness. Temperature affects this time by changing the speed at which the polymer responds. Low temperatures slow it down, so the high-stiffness zone is extended into the lower speed zone. A liquid that can act as a solvent for the polymer—with biological materials, this is nearly always water—will plasticize the polymer and have the same effect as increasing the temperature. This effect works because the plasticizer sits between the polymer molecules, breaking any low-energy cross-links and allowing the polymers more room to move. Water is a crucially important plasticizer for biological polymers. Biological polymers are also affected by salts, and these also affect the characteristic times by causing the stiffness of the polymer molecule to change or by changing cross-linking and other interactions between adjacent chains. Some of these effects will be enlarged upon later; for the moment, be aware that all things are mutable. Stiffness as a parameter has to be hedged with provisos, especially in food science.

Poisson Ratio

When a material is stretched, it normally not only gets longer but also contracts in the perpendicular direction. The Poisson ratio is the ratio of strain in one direction to the strain in a direction at a right angle to the first. Since there are three dimensions to any solid, it follows that there are six Poisson ratios. Since also it is common for a piece of material stretched in one direction to contract in one or both of the other two directions, one of the strains is given a negative sign, so the Poisson ratio is usually positive (the normal convention is that an increase in dimension is positive and a decrease is negative). Most materials texts state that the Poisson ratio can vary only between 0 (or very small) and 1. This is not true, especially for biological materials, which are usually much more subtle than the average materials scientist realizes. The problem arises because materials scientists are used to dealing with solid materials. Once you allow holes (and have cellular materials, like sponge), all sorts of new ways of deforming become available. Materials that have a negative Poisson ratio expand sideways when stretched (auxetic behavior). Imagine auxetic foods that swell as they are stretched; as you suck your spaghetti into your mouth, it expands and absorbs all the tomato sauce!

Strength

Strength is the stress at which the material breaks and must not be confused with stiffness. Like stiffness, strength is affected by time, temperature, and other factors, and in the same direction. There are several ways of measuring strength. The simplest and most reliable is direct tension, leading to tensile strength. When the material is tested by bending (perhaps because it is difficult to grip at the ends, or to make a suitably shaped specimen—common enough with biological materials), the result is the modulus of rupture, which is the stress in the tensile surface (on the outside of the curve) of the beam at failure. Strength is measured in pascals (Pa).

MATERIALS TESTING METHODS

Uniaxial Tensile Testing

One of the oldest and most widely used test methods to measure materials properties is the uniaxial tensile test. A strip of material is clamped at both ends and pulled apart at a fixed rate in a suitable testing machine (Figure 5–14), and the force is measured at the same time as the displacement of the object. The force is generally plotted against the displacement (extension) to give a force-extension curve (Figure 5–7). Tensile tests may produce an approximately uniform extension of a sample, provided that necking does not occur. Normally, the force and extension are divided by the original sample dimensions to obtain stress and strain, and this allows removal of the sample geometry as a variable. The slope of the stress-strain curve then gives the elastic modulus or stiffness. Some typical tension test geometries are shown in Figure 5–15.

Figure 5–14 A Typical Mechanical Testing Machine Showing Various Test Geometries Used in Texture Analysis: Stable Micro Systems TA.XT2i. *Source:* Courtesy of Stable Micro Systems Ltd LTD., 1998, Godalming, Surrey, United Kingdom.

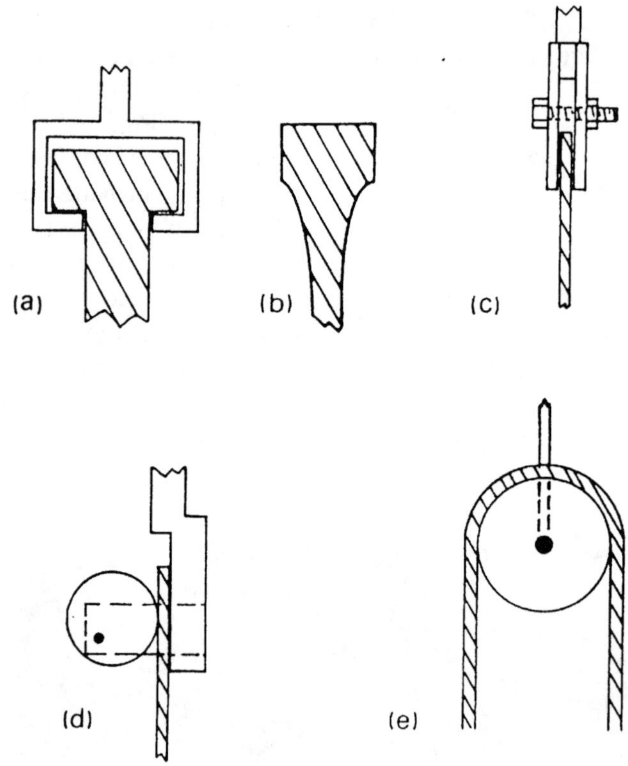

Figure 5–15 Typical Tension Test Geometries. (a) Dumbbell with right-angle shoulders, (b) dumbbell with tapered shoulders, (c) locking clamps for strips or dumbbells, (d) self-tightening grips, (e) pulley arrangement for circular test piece. *Source:* Reprinted with permission from R.W. Whorlow, *Rheological Techniques*, 2nd ed., © 1992, Ellis Horwood Ltd.

In many ways, the tensile test is one of the most difficult to perform properly and the least suited to highly deformable materials. With many food materials, the act of gripping the sample can lead to damage and can distort the sample in the area where it is being gripped. Such end effects are common in tensile testing and many modifications have been developed to mitigate these effects (Figure 5–15). Narrowing the test sample in the central region of the test sample concentrates the stress in that region and decreases the stress on the gripped region, creating the typical dumbbell-shaped sample used in many engineering tests. This leads to nonuniform strain along the length of the sample, which leads to considerable underestimates of strain if strain is calculated simply from the movement of the grips. In such a case, strain should be measured independently by using contact or noncontact extensometers or by marking the sample in the central region and measuring local extension by video or photographic techniques or a traveling microscope. Narrowing of the sample may also lead to necking beyond the elastic limit because if the cross section at any point is slightly less than elsewhere or if there are any inhomogeneities in the sample, the stress will be increased locally and will give rise to a cumulative local reduction in area (necking) and eventual failure. Only if stiffness increases with strain to the extent that it more than compensates for the local reduction in area (strain hardening) will uniform extension occur. Other solutions to end effects include spreading the force at the grips by using protective plates or reinforcing pins or by cutting a ring or loop of material and stretching between two freely rotating pulleys. This method has been used for uniaxial tensile testing of wheat flour doughs suspended in a liquid (Tschoegl, Rinde, & Smith, 1970). This may still lead to end effects because the strain around the pulleys may be different from that along the length of the sample.

Highly extensible materials often come loose from simple clamps during testing because their cross sections decrease substantially during stretching. One possible solution is to glue the ends of the sample to metal plates and grip these. Very soft solids are not easily tested in uniaxial tension, and if a material flows easily and hardly exhibits any elastic properties, then a tensile test may be inappropriate. For these kind of materials, tests such as creep or relaxation give a measure of time-dependent (viscous) behavior, and lubricated compression or biaxial extension will be more appropriate for strain-dependent behavior. These will be described later.

Strain Hardening and the Stability of Deformation during Plastic Flow

During stretching of materials beyond their yield stress, the plastic strain is uniform throughout the sample up to the point of maximum force $dF = 0$ (Figure 5–16). Beyond this point, force begins to decrease, and at this point localized and nonuniform plastic deformation begins to occur. The cross-sectional area begins to change in a nonuniform way, and a neck or localized constriction forms, which can either stabilize or propagate in an unstable manner to failure. If the cross section at any point is slightly less than elsewhere or if there are any irregularities in the sample when the force is increasing in plastic flow, the stress will increase locally. While the force is increasing, the deformation will be stable; that is, any local constrictions will be self-arresting. In contrast, under a decreasing force, the deformation is no longer stable, leading to the formation and cumulative increase of necking and eventual failure. Hence, the force maximum ($dF = 0$) defines a point of instability in tension beyond which fracture is inevitable. The occurrence of strain hardening (sometimes called *work hardening*) in a material stabilizes any regions of incipient localized thinning that could lead to unstable necking and eventual fracture during high extensions and can allow much larger extensions before rupture than would otherwise be possible.

Strain hardening is shown as an increase in the slope of the true stress–Hencky strain curve with increasing extension, giving rise to the typical J-shaped stress-strain curve observed for highly extensible materials (Figure 5–17), and has been shown to be necessary for stability in any operation that requires large extensions, such as cold-drawing of polymer fibers, inflation of polymer films and blow molding of plastic bottles, metal forming, the extension of biological tissues, and the expansion of gas cells within expanding bread doughs (Dobraszczyk & Roberts, 1994; Kokelaar, van Vliet, & Prins, 1996; Wikström, 1997). Strain hardening allows the material to resist further thinning by locally increasing resistance to further deformation.

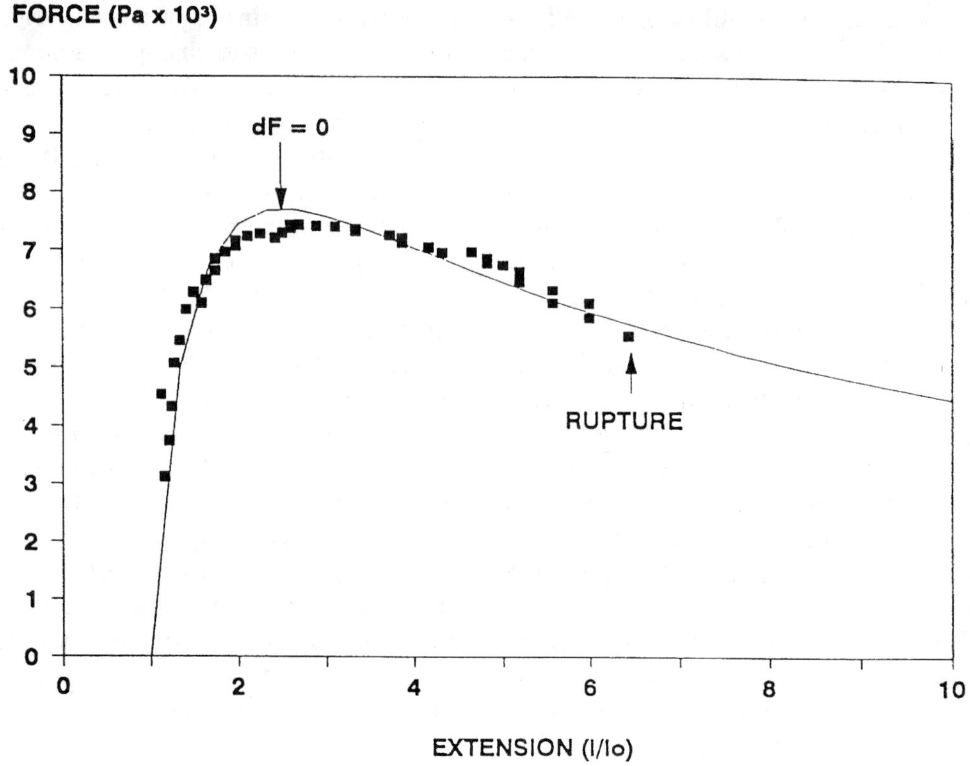

Figure 5–16 Force-Extension Plot in Biaxial Extension Showing Yield and Point of Load Instability at Maximum Force ($dF = 0$).

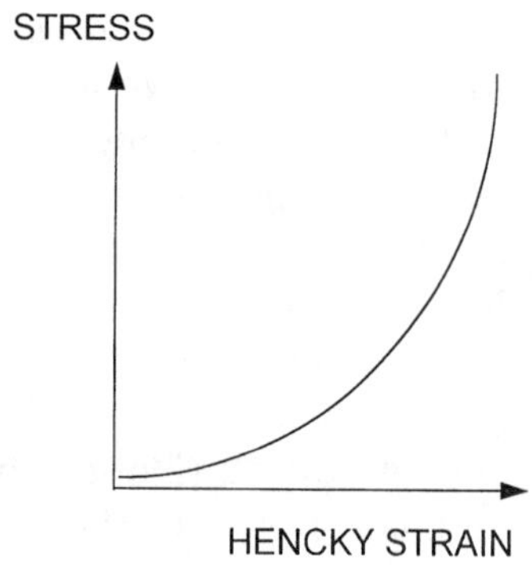

Figure 5–17 Typical J-Shaped Stress-Strain Curve in Highly Extensible Materials Showing Strain Hardening

The Considere Criterion for Instability in Tension

The condition of necking instability in tension is defined by $dF = 0$ (the maximum in the force-extension plot), and since $F = \sigma A$, where F = applied force and A = cross-section area at the point of necking, during a small increment of deformation (dF), the change in force is given by

$$dF = \sigma dA + A d\sigma = 0$$

At the point of instability in tension $dF = 0$,

$$A d\sigma = -\sigma dA$$

Rearranging this gives:

$$d\sigma = \frac{-\sigma dA}{A}$$

During plastic flow, volume remains constant ($dV = 0$); therefore,

$$dV = A dl + l dA = 0$$

and

$$\frac{dl}{l} = \frac{-dA}{A}$$

and since strain $d\varepsilon = dl/l$, then

$$\frac{-dA}{A} = d\varepsilon$$

substituting into $d\sigma = -\sigma dA/A$,

$$d\sigma = \sigma d\varepsilon$$

which defines the point of tensile instability (Considere criterion) as

$$\frac{d\sigma}{d\varepsilon} = \sigma \text{ (at } dF = 0)$$

The true stress σ at the point of necking instability is equal to the slope of the true stress-strain curve at that point. For a material whose true stress-strain curve follows a power law $\sigma = K\varepsilon^n$, the Considere instability criterion can be substituted into this equation to give a relationship between the strain-hardening index (n) and the critical strain (ε_{crit}) at which instability occurs:

$$n = \varepsilon_{crit}$$

This shows that the strain at instability is defined by the strain-hardening index (n) of the material. The higher the value of n, the greater the strain hardening and the greater the strain at which instability occurs. In terms of extension, this means that a material will stretch to a larger extension before failure occurs. Thus, the strain-hardening properties of a material will be critical in determining the extent of its deformation. This has been shown to be the case for wheat flour doughs, where the strain-hardening index (n) showed a high degree of correlation with failure strain (Dobraszczyk & Roberts, 1994; Kokelaar et al., 1996). (See Figure 5–18.)

Compression

This is one of the most common tests for measuring the deformation properties of many food materials. Such tests may be preferable to tensile tests because no gripping problems occur, the tests are easy to perform, and the sample geometry is simple. Normally, a short cylinder (approximately 10 mm in height by 10 mm in diameter) is compressed between parallel plates on a suitable testing machine, and force is measured as a function of constant deformation rate. Strain (ε) is calculated by the relative change in height (dh/h), and stress (σ) from the force divided by the adjusted area:

$$\varepsilon = \int_{H_0}^{H} dh/h = \ln(h/h_0)$$

$$\sigma = Fh/\pi r_0^2 h_0$$

Initial and final radii and heights are related if volume remains constant and no friction occurs:

$$\pi r_0^2 h_0 = \pi r^2 h$$

Problems can arise with friction of the sample against the loading plates. If there is friction between the plate and sample, then the sample will bulge or "barrel" in the central region, making calculation of stress and strain difficult (Figure 5–19). Normally, a lubricant is applied that allows the sample to deform uniformly as it is compressed, although it is by no means clear whether such lubrication works effectively if the sample is highly deformable. Casiraghi, Bagley, and Christianson (1985) measured the deformation and fracture properties of cheese in com-

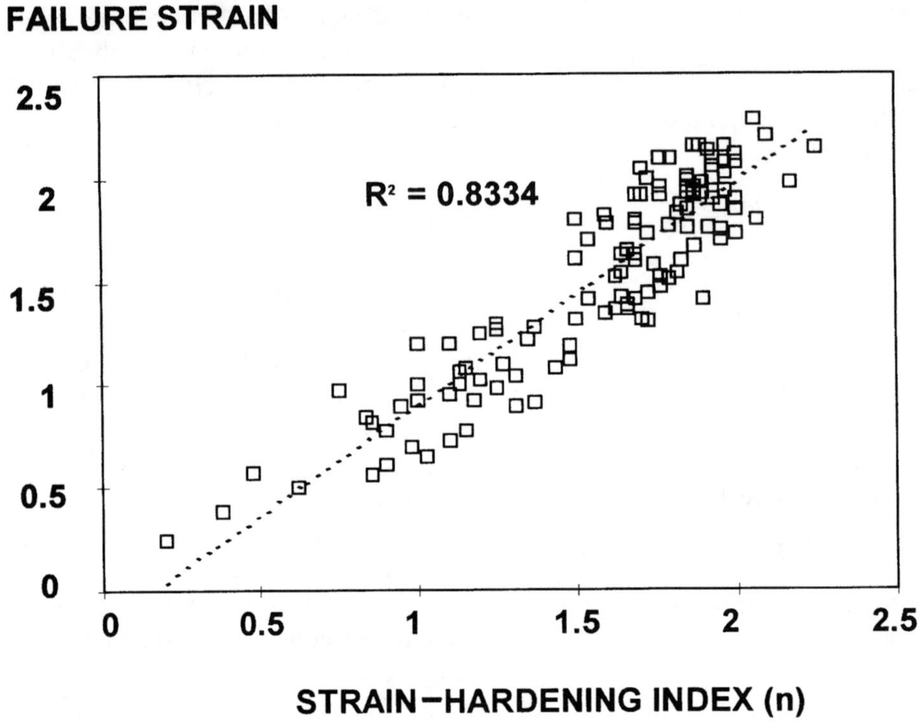

Figure 5–18 Relationship between Strain-Hardening Index and Failure Strain for a Number of Wheat Flour Doughs Inflated in Biaxial Extension.

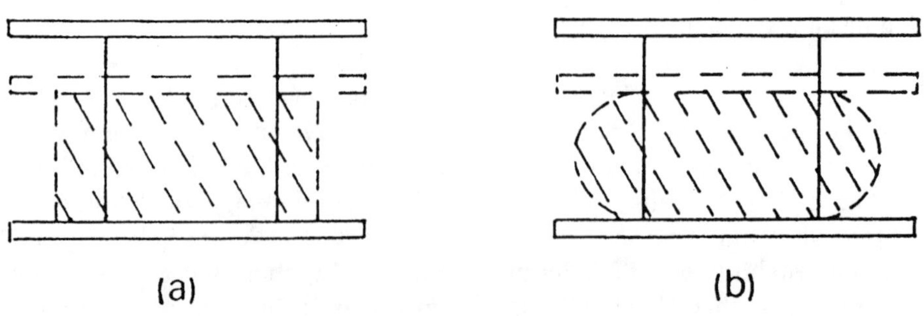

Figure 5–19 Compression of Cylindrical Samples with (a) Perfectly Lubricated Surfaces; (b) Friction at the Surfaces. *Source:* Reprinted with permission from R.W. Whorlow, *Rheological Techniques*, 2nd ed., © 1992, Ellis Horwood Ltd.

pression using lubricated and bonded samples; Bagley and Christianson (1987), Huang and Kokini (1993), and Kokelaar et al. (1996) measured extensional properties of wheat flour doughs using lubricated compression.

Bending

Bending or flexure is another common test geometry and in many ways is preferable to tension testing because it gets away from gripping and clamping problems. If samples are available in the form of a reasonably rigid rod of circular or rectangular cross section, then deflections for a given force are generally much larger in bending than in extension or compression, making deflection measurements easier.

The major disadvantage of bending is that it does not produce a uniform stress state; rather, different parts of the test piece undergo different strains. The inner curved surface will be in compression, and the outer surface will be in tension (Figure 5–20). Tensile stress is at a maximum at the outer curved surfacc, and comprcssivc strcss is maximum on the inner surface. This leads to sensitivity to any surface flaws that may have been produced during sample preparation.

Rods or strips are bent using one of the methods shown in Figure 5–21. Three-point bending is the most commonly used method, but this can give rise to significant shear stresses within the sample if an inappropriate sample geometry is used. Ideally, samples should have a length-to-thickness ratio of greater than 10. Three-point bending has loading at a single central point for a beam supported at either end. The deflection (ρ) at the center of the beam (width = w, thickness = t) produced by a given force (F) is given by

$$\rho = \frac{FL^3}{48EI}$$

where L = length between supports, E = Young's modulus, and I = second moment of area, a function that varies according to the cross-section area (values given in Figure 5–22). For a rectangular-section beam, $I = wt^3/12$, and

$$\rho = \frac{FL^3}{4Ewt^3}$$

or

$$E = \frac{FL^3}{4wt^3\rho}$$

and the maximum stress (σ_{max}) is given by

$$\sigma_{max} = \frac{3FL}{2wt^2}$$

Four-point bending (Figure 5–20) has the advantage of a constant radius of curvature in the portion between the supports so that shear deformation is excluded.

Shear and Torsion

It is not easy to produce a pure shear stress field using conventional materials testing equipment. It is possible to apply a shear stress by inserting a sample between two flat plates and applying a shearing force. A similar method (called the double-shear sandwich) is to mount two samples vertically between three plates, as in Figure 5–23, and to apply a shear strain by pulling the central plate.

Shear stress can also be produced by torsional loading of cylindrical test pieces (Figure 5–24). The strain in a straight cylinder is not uniform, being maximum at the surface and zero at the central axis. This can be

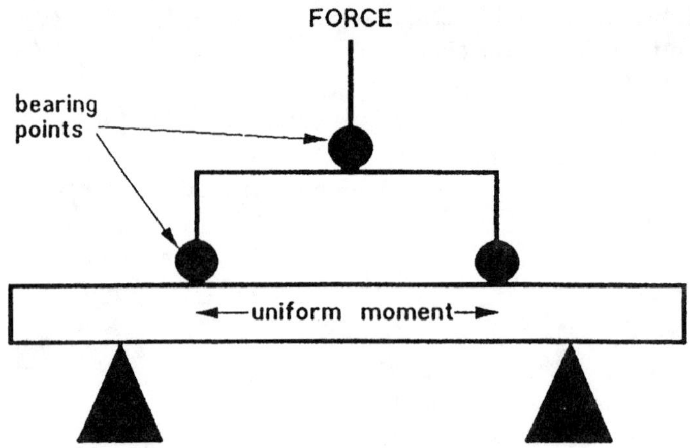

Figure 5–20 Typical Four-Point Bending Geometry. *Source:* J.F.V. Vincent, ed., *Biomechanics—Materials: A Practical Approach*, © 1992, by permission of Oxford University Press.

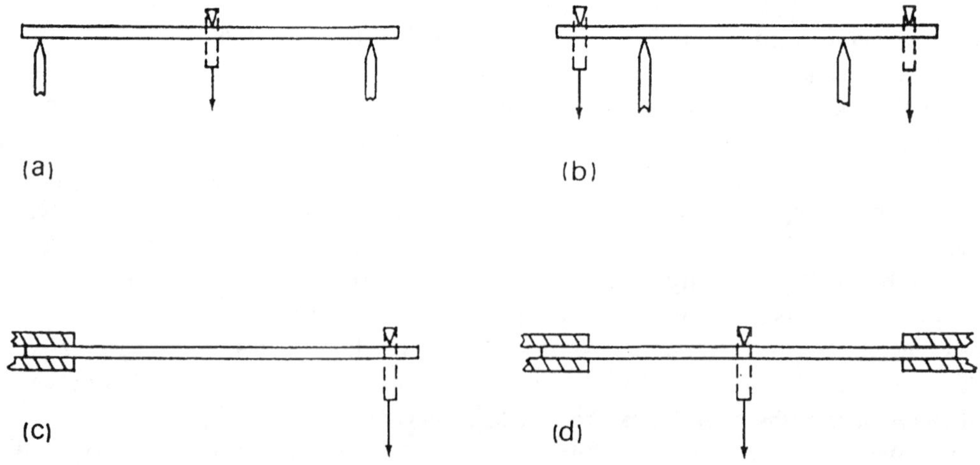

Figure 5–21 Methods Used in Bending. (a) Three-point bending, (b) four-point bending, (c) simple cantilever, (d) double cantilever. *Source:* Reprinted with permission from R.W. Whorlow, *Rheological Techniques*, 2nd ed., © 1992, Ellis Horwood Ltd.

	SECTION	SECOND MOMENT OF AREA (I)
	SOLID CYLINDER	$\pi r^4/4$
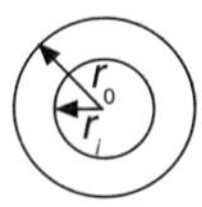	CYLINDRICAL TUBE	$\pi/4\ (r_0^4 - r_i^4)$
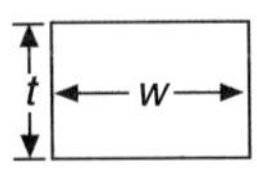	SOLID RECTANGLE SOLID SQUARE	$wt^3/12$ $t^4/12$
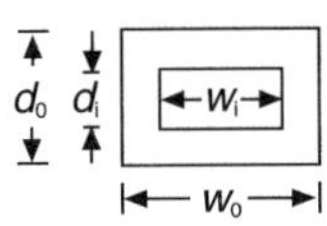	RECTANGULAR TUBE	$1/12\ (w_0 t_0^3 - w_i t_i^3)$
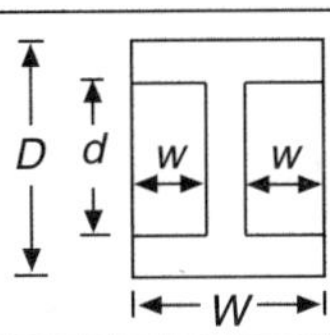	I-SECTION BEAM	$WD^3/12 - 2(wd^3)/12$

Figure 5–22 Cross-Section Area Values. *Source:* Reprinted with permission from *Structural Materials,* G. Weidmann et al., eds., pp. 17 and 22, © 1990, The Open University Rights Office/Butterworths Scientific Ltd.

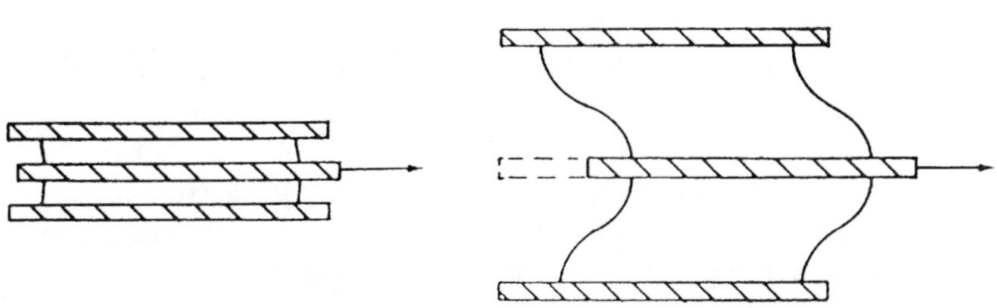

Figure 5–23 Double-Shear Sandwich. *Source:* Reprinted with permission from R.W. Whorlow, *Rheological Techniques*, 2nd ed., © 1992, Ellis Horwood Ltd.

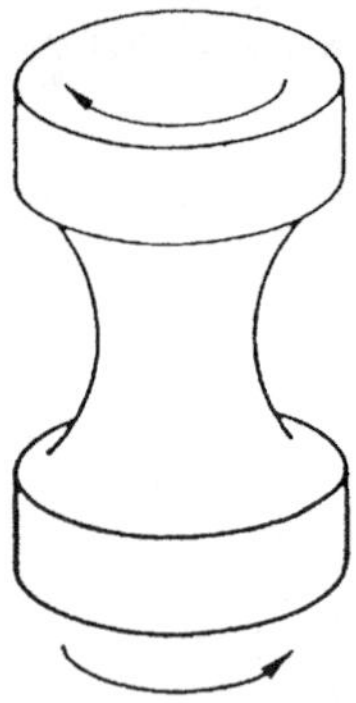

Figure 5–24 Dumbbell Specimen Geometry Used in Torsion Testing. *Source:* Reprinted with permission from R.W. Whorlow, *Rheological Techniques*, 2nd ed., © 1992, Ellis Horwood Ltd.

avoided by designing specially shaped test pieces such as a thin-walled hollow cylinder.

For materials that flow readily, shear stresses are commonly applied using standard viscometer geometries such as rotating parallel plates, cone and plate, and concentric cylinders as described in Figure 5–25. Shear between cone and plate geometry gives a uniform shear strain.

Extensional Flow

Most of the tests described previously have involved relatively small deformations. Many food materials undergo large deformations in practice during processing and eating, and many of these have a large extensional component (Figure 5–26). For example, extensional flows are important during swallowing and in sheeting of pastry and dough; converging and diverging flow, as in extrusion and pumping; spreading of soft solids such as butter, cheeses, and pastes; and expansion of bubbles in foams such as bread dough, cakes, and heat-extruded snacks. Unfortunately, most tests in flow are carried out in shear under small deformations, mainly because most conventional viscometers operate in shear, the equipment is readily available, and the technique is well established. However, measurements carried out in shear and using small deformations do not provide information about a material's behavior under large extension, mainly because in molecular terms extensional flows can be visualized as stretching, entanglement, and orientation of long-chain molecules during extensional flow, whereas in simple shear they remain coiled and can slip past each other. The molecular stretching and entanglement cause extensional viscosity to increase with strain and strain rate, a phenomenon known as *strain hardening*. From extensional studies on long-chain, high–molecular weight polymer melts, it is known that entirely different rheological properties are obtained in shear than in tension, especially if the polymer chains are branched (Cogswell, 1981; Padmanabhan, 1995). For example, the elongational viscosity of low-density (branched-chain) polyethylene melts increases with stress and strain rate (strain hardening), while the shear viscosity decreases with stress and strain rate (shear thinning), as shown in Figure 5–27. Therefore, there is an obvious need to perform measurements under conditions relevant to those experienced by the material in actual practice, and not simply because equipment is available.

Types of extensional flow measurements include simple uniaxial tension, fiber wind-up or spinning (Figure 5–28), converging flow, capillary extrusion, opposed jets, lubricated compression, and biaxial extension (Figure 5–29).

Biaxial Extension

In biaxial extension, a sample is stretched at equal rates in two perpendicular directions

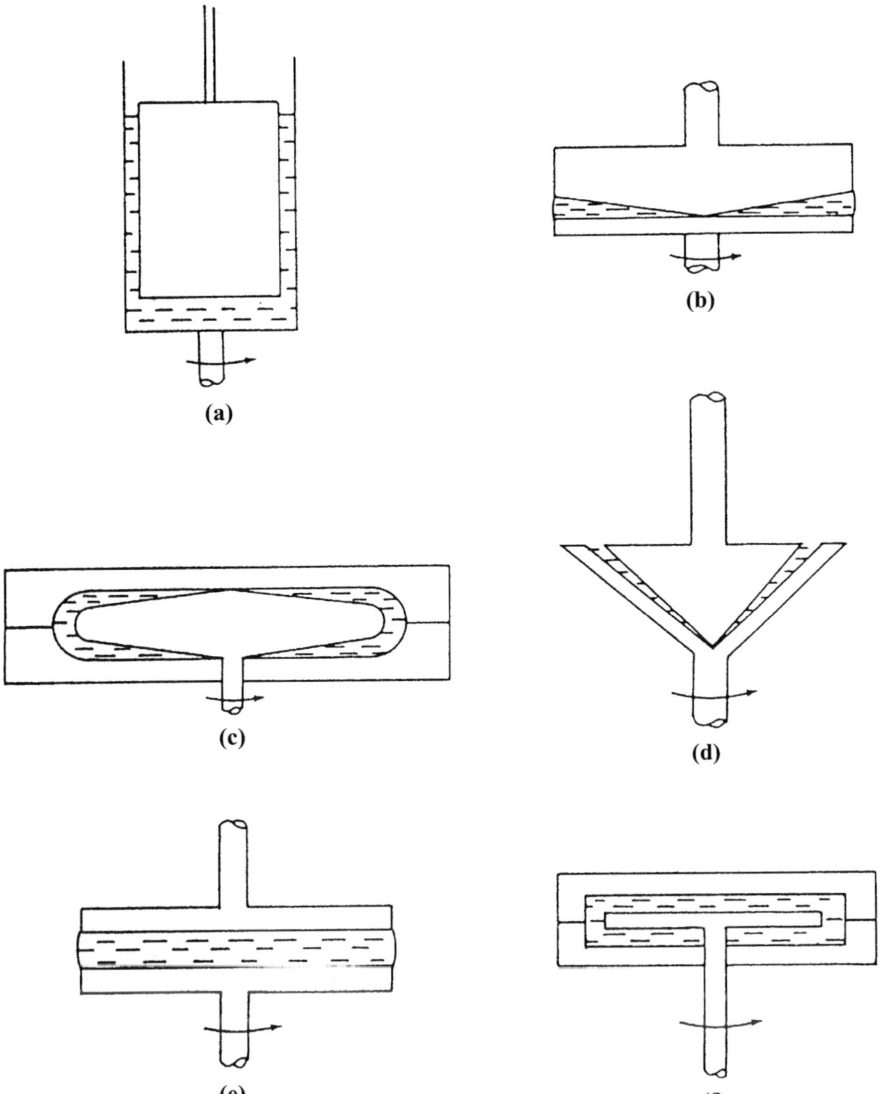

Figure 5–25 Common Geometries Used in Rotational Viscometers. (a) Concentric or coaxial cylinder, (b) cone and plate, (c) double cone and plate, (d) cone-cone, (e) parallel plate, (f) enclosed plate. *Source:* Reprinted with permission from R.W. Whorlow, *Rheological Techniques*, 2nd ed., © 1992, Ellis Horwood Ltd.

(x and y) in one plane, as in a spherical balloon (Figure 5–30). Assuming that volume remains constant, the strains in three perpendicular directions should all sum to zero (Dieter, 1986):

$$\varepsilon_x + \varepsilon_y + \varepsilon_z = 0$$
$$\varepsilon_z = -(\varepsilon_x + \varepsilon_y)$$

and assuming that $\varepsilon_x = \varepsilon_y$,

$$\varepsilon_z = -2\varepsilon_x = -2\varepsilon_y$$

Hence, the strain in the thickness (z) direction

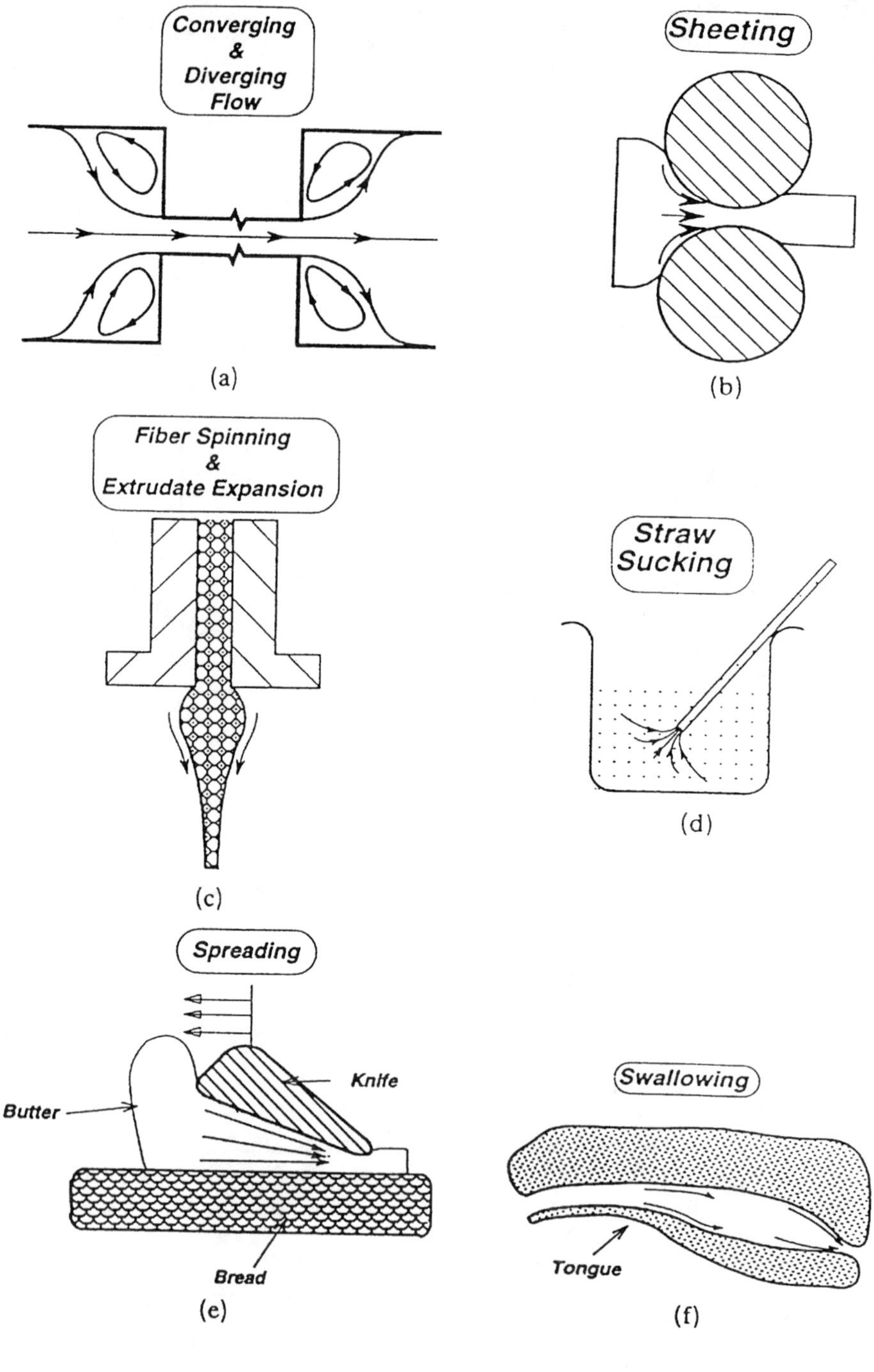

Figure 5–26 Typical Extensional Flows in Foods during Processing and Eating. *Source:* Reprinted from *Journal of Food Engineering,* Volume 25, M. Padmanabhan, Measurement of Extensional Viscosity of Visloelastic Foods, Page No's. 311–327, Copyright (1995), with permission from Elsevier Science.

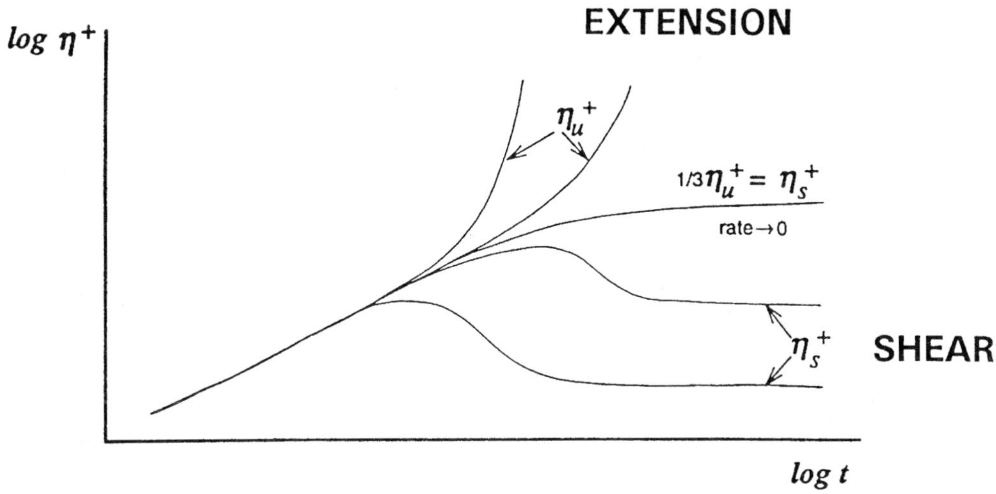

Figure 5–27 Schematic Comparison of the Differences between Transient Shear (η_s^+) and Transient Uniaxial Extensional (η_u^+) Viscosities Compared to the Theoretical Steady-State Relationship (η_u^+) = 3(η_s^+). *Source:* Reprinted from *Journal of Food Engineering,* Volume 25, M. Padmanabhan, Measurement of Extensional Viscosity of Visloelastic Foods, Page No's. 311–327, Copyright (1995), with permission from Elsevier Science.

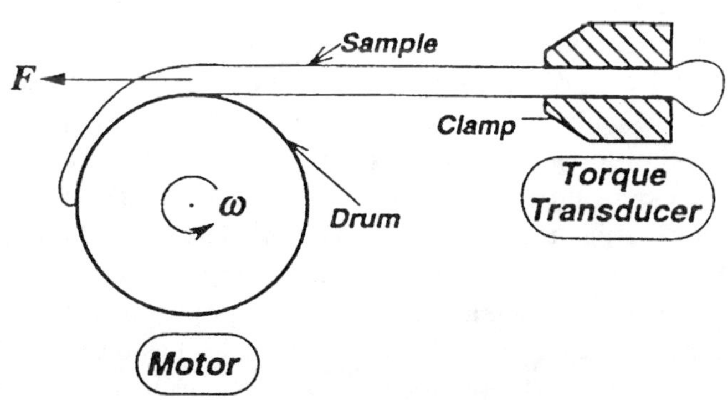

Figure 5–28 Extensional Flow Measurements: Fiber Wind-up Technique. *Source:* Reprinted from *Journal of Food Engineering,* Volume 25, M. Padmanabhan, Measurement of Extensional Viscosity of Visloelastic Foods, Page No's. 311–327, Copyright (1995), with permission from Elsevier Science.

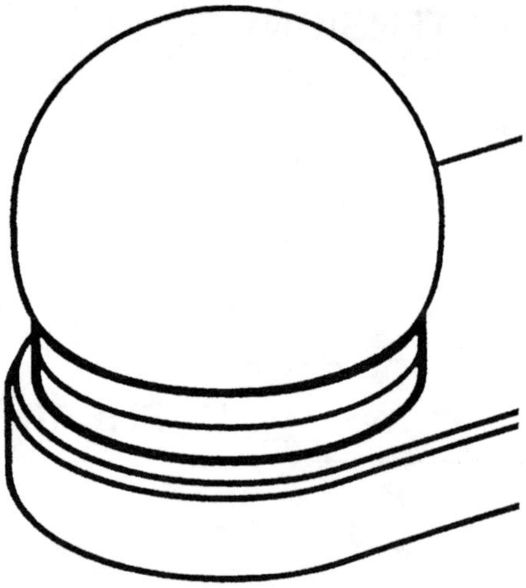

Figure 5–29 Extensional Flow Measurements: Inflation Technique. Courtesy of Stable Micro Systems Ltd., 1998, Godalming, Surrey, United Kingdom.

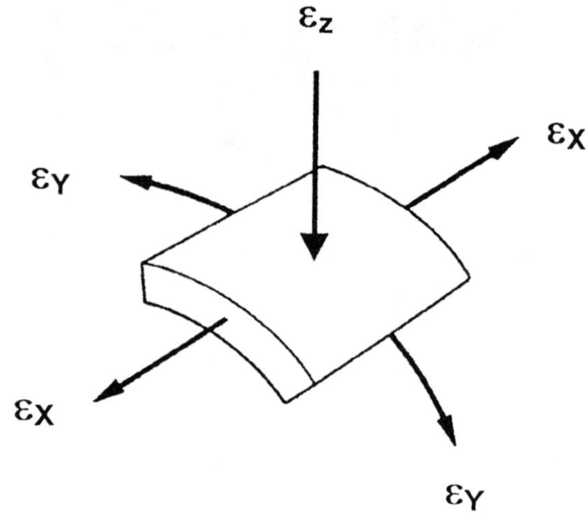

Figure 5–30 Biaxial Extension of a Sample Stretched at Equal Rates in Two Perpendicular Directions (ε_x and ε_y).

is twice the strain in any planar direction. We can then define an extensional planar strain (ε_x):

$$\varepsilon_x = \ln(l/l_o) = -\tfrac{1}{2}\varepsilon_z = -\tfrac{1}{2}\ln(t/t_o)$$
or
$$\varepsilon_z = 2\ln(l/l_o) = \ln(t/t_o)$$

One of the simplest and most widely used methods for measuring biaxial extension properties of biological materials, polymer melts, and rubbers has been the inflation technique. Treloar (1944) showed a good approximation to equal biaxial extension near the pole of an expanding bubble of rubber. In this technique, a sheet of the material is clamped between two flat plates, one of which has a circular aperture and the other a hole through which a controlled volume of air or liquid is pumped to inflate the sheet. Denson and Crady (1974), Denson and Gallo (1971), Joye, Poehlein, and Denson (1972), and Rhi-Sausi and Dealy (1981) used inflation to study the biaxial extensional properties of molten polymers. Shadwick (1992) reviewed the use of biaxial testing on arteries of invertebrates, worms, guts, and cow teats. Vincent (1992) described a method for measuring biaxial properties of fruit skins. Inflation was first used as an empirical technique to measure wheat gluten and bread dough extensibility in the 1920s by Hankoczy (1920) and Chopin (1921). This method was later developed for use as a rheological tool by Launay, Bure, and Praden (1977), based on equations derived by Bloksma (1957). Dobraszczyk and Roberts (1994) used an inflation method to measure the fracture and biaxial extensional rheological properties of wheat doughs, and this new rheometer has been developed by Stable Micro Systems (SMS) as a commercial test system for wheat flour doughs, called the D/R Dough Inflation System (Dobraszczyk, 1997a). (See Figure 5–31.)

Stress and strain can be derived directly from a knowledge of the pressure (P) and height of the bubble during inflation (Whorlow, 1992):

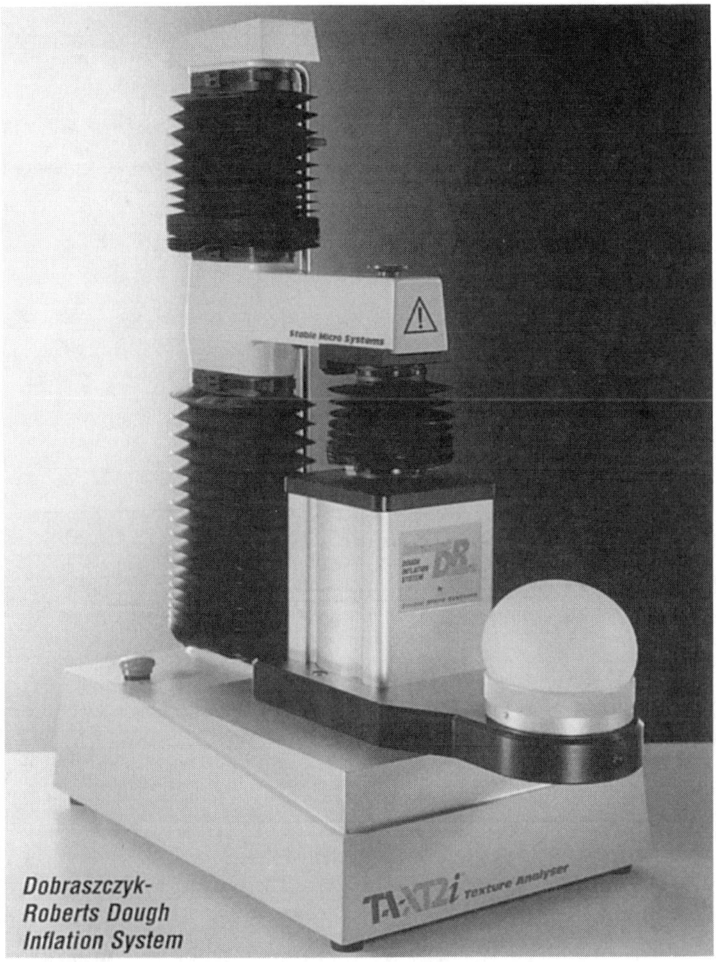

Figure 5–31 Biaxial Inflation of Wheat Flour Dough Using the Stable Micro Systems D/R Dough Inflation System. Courtesy of Stable Micro Systems Ltd., 1998, Godalming, Surrey, United Kingdom.

$$\sigma = PR/2h$$

$$\varepsilon = -\tfrac{1}{2}\ln(t/t_0) = \ln(\alpha R/w)$$

By measuring the height of the bubble (x), the parameters R and α and the bubble wall thickness (h) at any height can be calculated (Figure 5–32):

$$R = w^2 + (x^2/2x)$$

$$h = h_0 w/\alpha R$$

Bloksma (1957) independently calculated stress and strain in an expanding dough bubble using bubble height and pressure. Dobraszczyk and Roberts (1994) used a bubble inflation method to measure the biaxial extensional rheological properties of wheat doughs, using a laser ranger to measure the height of an expanding dough bubble and hence calculate stress and strain up to failure.

Other methods of biaxial extension measurement include compression between flat plates using lubricated surfaces, which produces purely extensional flow provided that

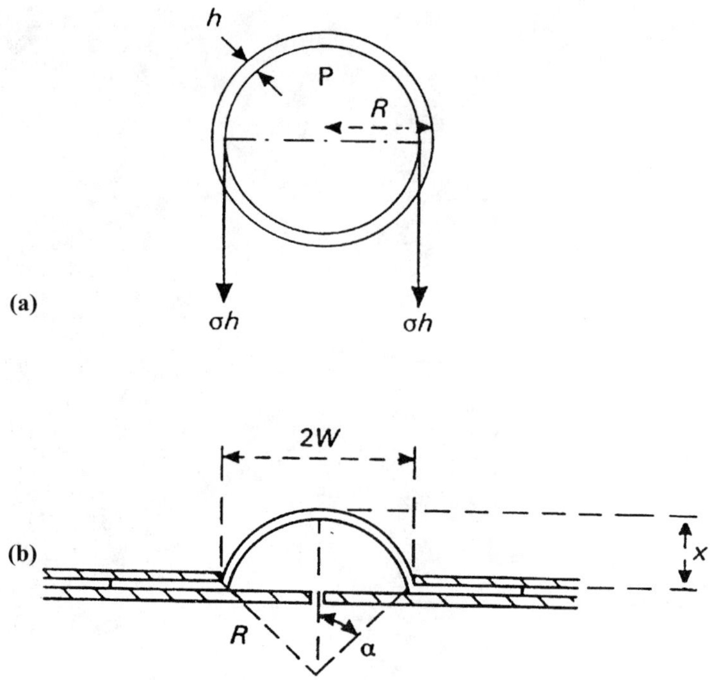

Figure 5–32 Biaxial Inflation of a Film. (a) Forces acting on the sphere; (b) geometry of the bubble during inflation. *Source:* Reprinted with permission from R.W. Whorlow, *Rheological Techniques*, 2nd ed., © 1992, Ellis Horwood, Ltd.

no friction occurs (Chatraei, Macosko, & Winter, 1981). Huang and Kokini (1993) and Kokelaar et al. (1996) measured the extensional rheological properties of wheat doughs using lubricated compression. Meissner (1987) developed a rotating clamp technique for measuring both uniaxial and biaxial elongational properties of molten polymers, and a more recent modification using conveyor belts has been used to measure the extensional properties of bread doughs (Conde-Petit, Escher, Halliger, & Schweizer, 1994; Meissner & Hostettler, 1994).

STRENGTH, TOUGHNESS, AND FRACTURE

Strength and toughness are often confused, and it is important to differentiate between them. Strength is defined as the maximum stress an object will withstand before it breaks, whereas toughness is the resistance to cracking. In terms of a force-extension relationship (Figure 5–33), strength is simply the maximum force divided by the cross-sectional area. Toughness is defined as the energy required to propagate a fracture by a given crack area, generally derived from the area under a force-extension curve. Hence, a strong material need not necessarily be tough. For example, ceramics and glass are strong—they break at a high force—but they are not as tough as material such as wood, which breaks at a lower force but at a much greater deformation (Figure 5–34). Tendon and muscle are even weaker, but they have a high toughness due to their large extensibility (hence the effect of bulls in china shops!).

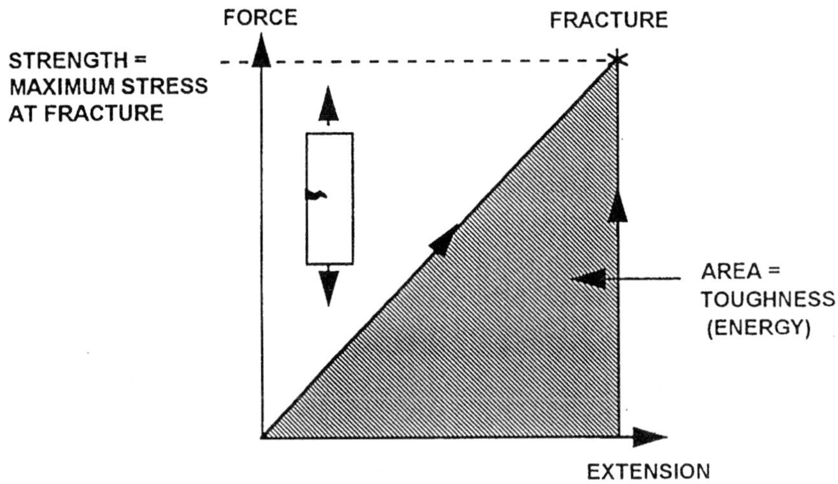

Figure 5–33 Nominal Force-Extension Plot Showing Definitions of Strength and Toughness

Stress Concentrations

An object will break much more readily if the available force is concentrated on relatively few bonds, such as the sharp tip of a crack. This produces a high local stress intensity or stress concentration. The stress intensity necessary to cause a material to break is more or less typical of the material and the processes of fracture: the stress intensity factor is often called the *fracture toughness* and has units of Pa • $m^{0.5}$. However, when a material fractures, the two new surfaces have an energy—surface energy—associated with them in the same way that a liquid has a surface energy. This introduces a second requirement for fracture to occur. Not only must there be sufficient force, concentrated in a small zone, but there must be a supply of strain energy sufficient to propagate the development of new surfaces. This energy requirement is also typical of the material, leading to a second fracture parameter—the work of fracture, measured in J • m^{-2}.

The stress concentration factor K around an elliptical flaw (Figure 5–35) is defined by

$$K = 1 + 2(a/b)$$

As b decreases, the flaw becomes a sharp crack, and the stress concentration increases. For sharp cracks, the stress concentration factor can be as high as 200. For $b \rightarrow 0$, $K \rightarrow \infty$, which suggests that a material with very sharp cracks would be unable to withstand any applied forces. To answer this problem, the science of fracture mechanics was born, which developed the ideas that a stable sharp subcritical crack can exist in a material and that such cracks become unstable only when it is energetically favorable for them to grow.

Fracture Mechanics

Fracture mechanics was originally developed in the 1920s by A.A. Griffith to describe and predict the occurrence of fracture in brittle materials such as glass. He was among the first to recognize that flaws act as stress concentrators and as the sites for fracture initiation in brittle solids. Many materials have a much lower ultimate strength than predicted by theory, based on the measured strength of

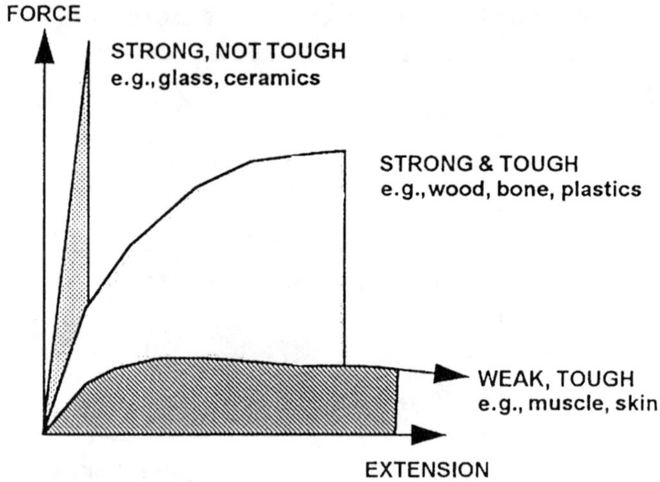

Figure 5–34 Nominal Force-Extension Plot Showing Several Materials of Contrasting Strength and Toughness

forces between atoms. Glass, for example, can be stronger than steel on a weight basis, yet in practice its brittleness inhibits its use in most structural applications. Griffith (1921) postulated that all materials contain minute cracks or flaws and that some materials are more sensitive to the presence of these cracks (notch sensitive) than others. Part of the reason has to do with amplification or concentration of stress around these cracks and how the material deals with that concentration. In a brittle material, stress is concentrated locally around the crack tip, building up very high stresses and an excess of energy that makes it easy for the crack to propagate. An example is cutting ceramic tiles or glass, where the trick is to start with scoring a sharp crack and then to propagate that crack by applying a load. In a ductile material, the stress around the crack is limited by plastic flow, giving a zone of intense deformation around the crack tip that effectively blunts the crack tip and prevents it from growing.

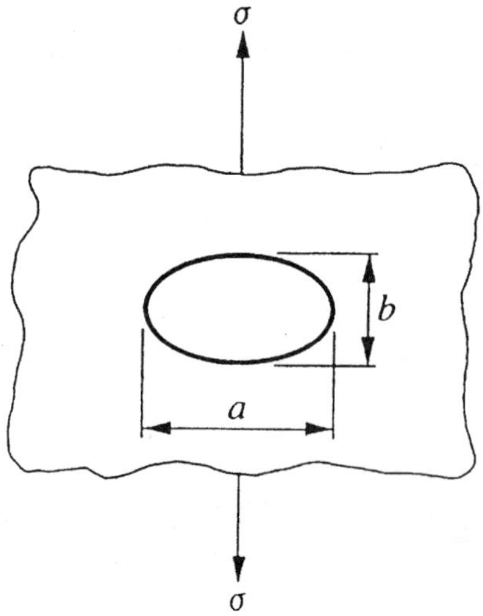

Figure 5–35 Stress Concentration around an Elliptical Flaw. *Source:* Benham and Crawford, *Mechanics of Engineering Materials*. Reprinted by permission of Addison Wesley Longman Ltd.

For a crack to grow, energy must be supplied to it from the surrounding stressed material, where energy is stored as elastic strain energy. Fracture occurs when the rate at which strain energy is released per unit crack area exceeds the rate at which surface energy is absorbed in creating new fracture surfaces. Resistance to crack growth is then defined as the work of fracture, which is the amount of energy required to propagate a crack by unit area. Figure 5–36 represents the two energy terms as a function of crack length. As the crack grows, surface energy (U_s) increases linearly with crack length (surface area being an area function), and strain energy (U_e) increases as the square of the crack length (strain energy being dependent on the volume of material being strained). The total energy is shown as the sum of the two energy contributions. A crack can grow only if the total energy of the system decreases (dU_{TOT} 0) and this point defines a critical crack length (a_c) at point A in Figure 5–36, where a crack changes from being stable to unstable when an increment in crack growth causes more strain energy to be released than is absorbed by the formation of new surfaces. The critical crack length defines the largest crack that can be present under a given stress.

Figure 5–37 shows a plate of unit thickness, under a uniformly applied tensile stress, containing a crack. Strain energy is stored elastically in the plate except for the shaded area extending above and below the crack in which the material is unstressed. Let U_e be the total elastic strain energy released due to the formation of the crack (represented by the shaded area in Figure 5–37), and U_s the surface energy of the newly formed fracture surfaces. The critical value for crack propagation occurs when the strain energy released per unit crack area (dU_e/da) is equal to or exceeds the energy required to create new fracture surfaces (dU_s/da), or

$$\frac{dU_e}{da} \geq \frac{dU_s}{da} = Gc$$

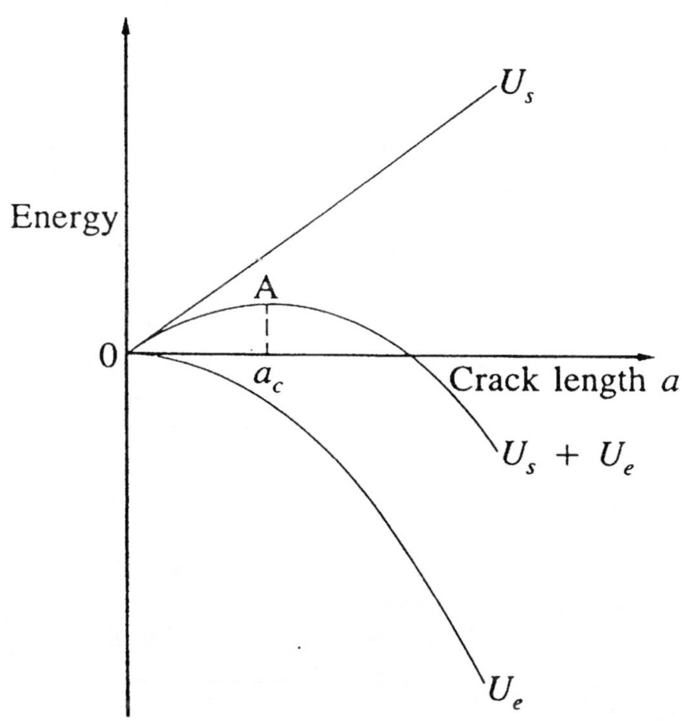

Figure 5–36 Variation of Surface (U_s) and Elastic (U_e) Strain Energy as a Function of Crack Length. *Source:* Benham and Crawford, *Mechanics of Engineering Materials*. Reprinted by permission of Addison Wesley Longman Ltd.

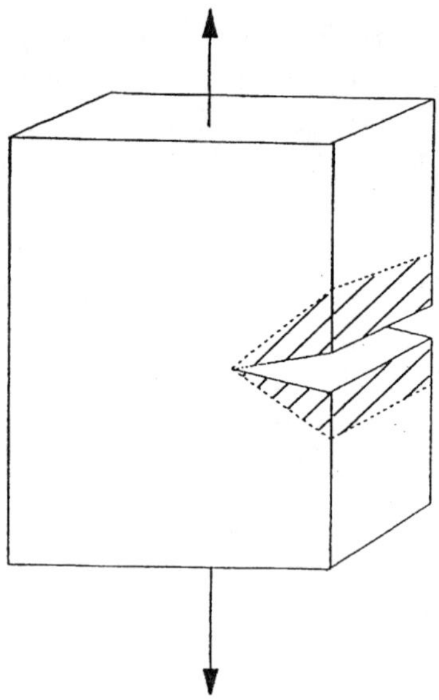

Figure 5–37 Plate of Unit Thickness, under a Uniform Tensile Stress, Containing a Crack. *Source:* Reprinted with permission from B.J. Dobraszczyk, Fracture Mechanics of Vitreous and Mealy Wheat Endosperm, *Journal of Cereal Science*, Vol. 19, p. 274, © 1994, Academic Press Ltd.

where Gc is the critical strain energy release rate, toughness or work of fracture, a fundamental material property, with units of J • m^{-2}. The strain energy released by crack propagation can be obtained by measuring the shaded area due to fracture under the force-extension curve in Figure 5–38.

Griffith showed that for unit thickness, the value of work of fracture (Gc) is given by

$$Gc = \frac{\pi\sigma^2 a_c}{E}$$

Thus, the critical energy for fracture (Gc) is dependent on the product of the applied stress (σ), the elastic modulus of the material (E), and a_c, a characteristic length that is dependent on the sample geometry. The traditional notion that fracture occurs at some critical stress can be seen to be seriously inadequate, since failure stress depends on the geometry and mode of cracking, fracture energy, elasticity, and size and geometry of the material. Work of fracture—that is, the strain energy released per unit area of fracture—and not the fracture stress is found to be the critical material property governing fracture.

The preceding energy analysis assumes linear elastic brittle behavior. Many food materials, however, do not conform to these conditions, but deform by plastic flow, called *ductile behavior*. This makes the analysis of fracture by the work area method described above much more difficult, because in this case not all of the strain energy dissipation is due to fracture (U_F); rather, energy can be channeled into other dissipative mechanisms such as yield and plastic flow (U_{PL}), buckling (U_B), and debonding (U_{DB}) and delamination (U_{DL}) around fibers and particles within the food. Clearly, then, the total area under the force-extension curve as described in Figure 5–38 will contain the combined contributions from other energy losses:

$$U_{TOT} = U_F + U_{PL} + U_B + U_{DB} +$$

Some workers take the total work area under a force displacement curve after cracking of a plastic material and divide by crack area to obtain an apparent work of fracture, but in reality this term is erroneous because it may contain energy-dissipating mechanisms other than fracture. The true fracture energy must be obtained by partitioning out the total energy release into flow and fracture components (Atkins & Mai, 1985; Dobraszczyk, Atkins, Jeronimidis, & Purslow, 1987).

Fracture and the Brittle-Ductile Transition

Food materials operate mostly in the ductile or rubbery region above the so-called

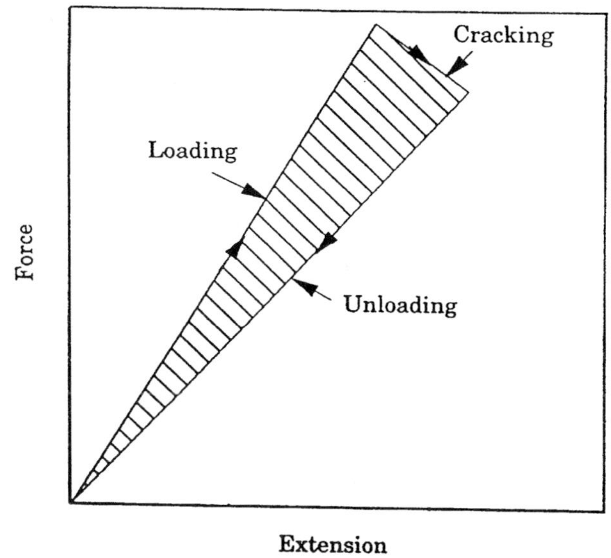

Figure 5–38 Force Extension Curve Showing Strain Energy Released (dU_e) Due to an Increment of Crack Extension (d*a*). *Source:* Reprinted with permission from B.J. Dobraszczyk, Fracture Mechanics of Vitreous and Mealy Wheat Endosperm, *Journal of Cereal Science*, Vol. 19, p. 274, © 1994, Academic Press Ltd.

---

glass transition or move in and out of the brittle-ductile region depending on the surrounding conditions. For example, many cereal-based products such as biscuits, wafers, snacks, pastry, and breakfast cereals are expected by consumers to be "crisp" or "crunchy" or to have a brittle fracture. When these products absorb moisture, either from the air or from being in close proximity with products with a higher moisture content (such as jam, fruit fillings, or raisins) they move into the ductile region, commonly referred to as "soft" or "soggy," where fracture can occur together with extensive plastic flow remote from the crack (Nicholls, Appelqvist, Davies, Ingman, & Lillford, 1995).

Because many foods operate in the plastic or ductile region, the large extent of the plastic energy dissipation mechanisms has important implications in terms of the speed or time over which deformation occurs. The fracture behavior of most foods depends on the rate of deformation, since plastic and viscous flow is highly rate dependent. Brittle fracture, typical of the so-called glassy state, is independent of rate. In terms of instrumental texture testing, the extent of brittle and ductile behavior is important because brittle fracture is highly dependent on the geometry of the sample, whereas ductile failure is more or less independent of shape but highly dependent on rate and temperature (Figure 5–39). Factors that control the brittle-ductile transition are rate and temperature (which principally affect the modulus and yield stress), moisture, additives such as sugar or fat, geometry, and the size of cracks in the material. Most materials will become less stiff as the temperature

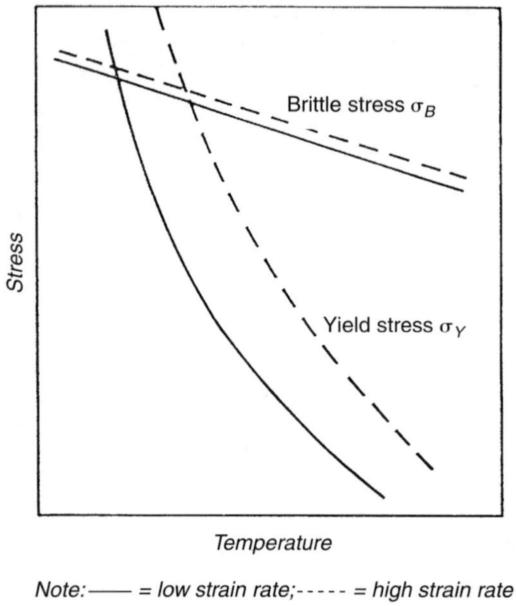

Figure 5–39 Effect of Temperature and Strain Rate on Brittle Fracture Stress and Yield Stress. *Source:* I.M. Ward, *Mechanical Properties of Solid Polymers*, p. 333, © 1971. Copyright John Wiley & Sons Limited. Reproduced with permission.

rises, and at some point they will go through the so-called glass transition, where they pass from the brittle elastic "glassy" state to the ductile flow region. Dobraszczyk et al. (1987) observed the effect of temperature on the brittle-ductile transition in frozen meat using dynamic rheological analysis and fracture testing and showed that the brittle-ductile transition temperature varied with rate of testing, geometry, and size. Cryo-scanning electron micrographs of fracture surfaces of frozen meat at various temperatures (Figure 5–40) and the associated stress-extension curves for fracture in tension (Figure 5–41) show that at –30°C the fracture surface is relatively smooth and glassy, indicative of a fast, brittle fracture. At –20°C there is some indication of energy dissipation in fiber debonding and pullout, and at higher temperatures (–8°C) the fracture surfaces are rough and distorted, showing individual fibers and bundles, indicating plastic flow and ductile deformation.

It is worth pointing out here that the glass transition should not be confused with a brittle-ductile transition. The so-called glass transition is defined by a temperature at which a measured specific heat shows a sudden change or transition, usually measured by differential scanning calorimetry (DSC). This has been one of the main techniques used in studying glass transitions in foods (Levine & Slade, 1993). However, when such transitions are measured with other techniques such as dynamic mechanical spectroscopy, X-ray diffraction, or nuclear magnetic resonance (NMR), the temperature at which a transition is seen to occur can often differ by as much as 25°C (Kalichevsky Jaroszkiewicz, Ablett, Blanshard, & Lillford, 1992; Nicholls et al., 1995). This has often been misinterpreted by researchers as indicating multiple transitions, whereas they are simply different phenomena. Mechanical methods measure changes in modulus, which are dependent on rate as well as temperature, and fracture tests can measure the extent of brittle or ductile behavior. DSC is not very sensitive to such transitions in foods, mainly because the glass transition is spread out due to the multicomponent nature of most foods and also is dependent on the heating rate (Blanshard, 1995). Mechanical tests are more sensitive because they measure changes in modulus, which has a major effect on brittle fracture. Fracture tests are probably the most sensitive to changes in brittle-ductile behavior and the most related to the changes that consumers perceive in practice when assessing textural properties of foods.

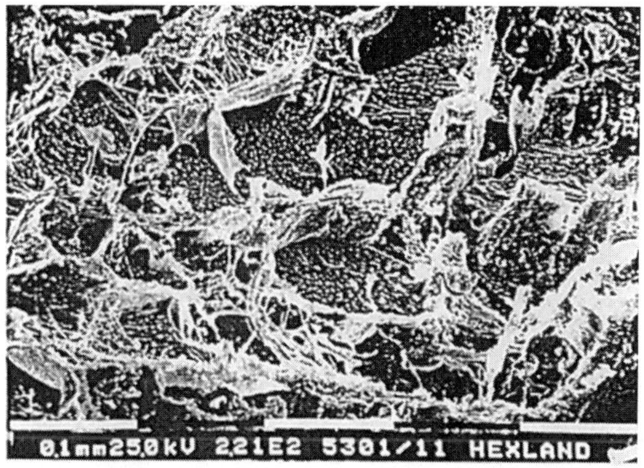

(a)

(b)

(c)

Figure 5–40 Cryo-Scanning Electron Microscope (SEM) Photographs of Fracture Surfaces of Frozen Meat at Various Temperatures: (a) –30°C, (b) –20°C, (c) –8°C. *Source:* Reprinted from *Meat Science*, Vol. 21, B.J. Dobraszczyk et al., Fracture Toughness of Frozen Meat, pp. 25–49, Copyright (1987), with permission from Elsevier Science.

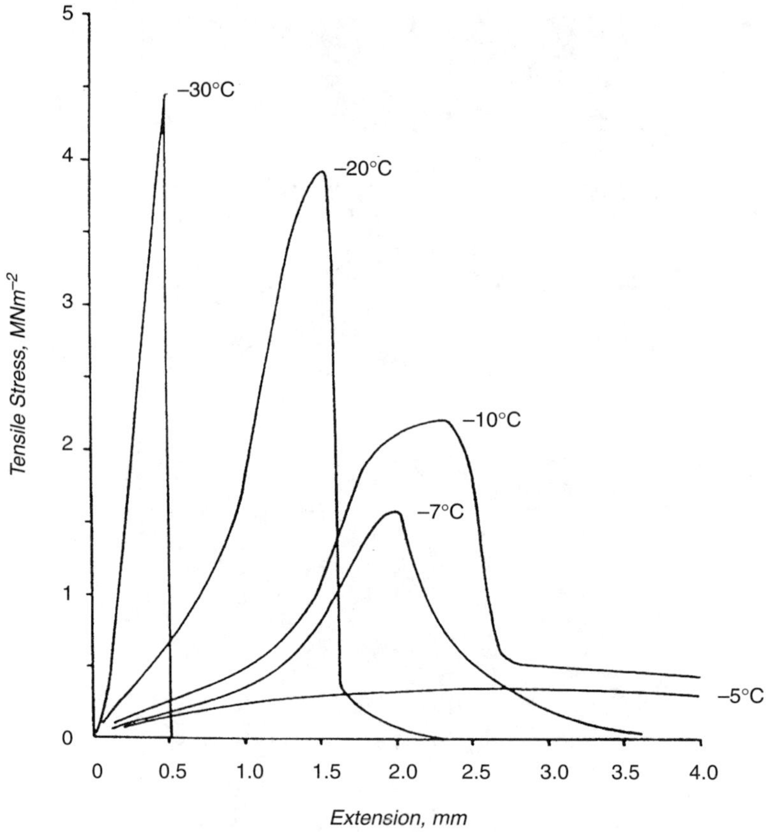

Figure 5–41 Tensile Stress-Extension Fracture Curves for Frozen Meat at Various Temperatures. *Source:* Reprinted from *Meat Science*, Vol. 21, B.J. Dobraszczyk et al., Fracture Toughness of Frozen Meat, pp. 25–49, Copyright (1987), with permission from Elsevier Science.

Size in particular has an important and often poorly understood effect on fracture. It is much easier to break large objects than small objects, not only because there is more probability of finding a large crack in a large object but also because there will be more volume relative to crack area in which to store excess strain energy. If the object is large, it is more likely that the sample dimensions will be large in comparison to any plastic zone at the crack tip so that elastic conditions will prevail in the bulk of the sample, which favors brittle fracture. In a small object, the size of the plastic zone will approach those of the sample dimensions, making plastic deformation more likely and inhibiting brittle fracture (Figure 5–42). This has important implications for size reduction processes such as grinding, milling, and comminution, which play a large part in food processing and texture. The critical size (a_c) at which a transition from elastic brittle behavior to ductile plastic flow occurs—the brittle-ductile transition—can be estimated by substituting the yield stress (σ_y) into the equation

$$a_c = \frac{\alpha E G c}{\sigma_y^2}$$

where α = geometric factor dependent on the size and shape of the specimen.

If $a_c < \alpha EGc/\sigma_y^2$ then ductile yielding will occur before fracture; if $a_c > \alpha EGc/\sigma_y^2$, then brittle fracture will be preferred (Atkins & Mai, 1986). It is therefore possible to predict the brittle-ductile transition from a knowledge of the material properties and geometry of a sample and also the effects of parameters such as rate, temperature, and geometry.

Methods of Measuring Fracture

There are three basic ways of breaking something: mode 1 (tension or crack opening), mode 2 (in-plane shear), and mode 3 (out-of-plane shear) (see Figure 5–43). But when an animal is eating something, it wants to use as little energy as possible preparing the food for digestion, so it will want its teeth to crack and chew. In general, the front teeth can cut into soft food materials like a knife, but they break brittle food materials (carrots, apples, etc.) with a wedging action that is therefore mode 1 fracture. Mode 2 fracture is uncommon (although the teeth of the aardvark, which eats ants, look as if they work in this mode). Much more commonly, the back teeth fracture food in mode 3, which is the way in which scissors cut into material. The ridges on cow teeth function this way, as do carnassial teeth in carnivores. In each instance, the adaptation is for cutting through a fibrous material that is insensitive to notches and therefore cannot readily have a crack propagating through. These different modes of fracture can be distinguished by the fracture surface that is produced. Unfortunately, when you examine the fracture surface of a chewed food item, you commonly find that it has been fractured in a mixture of all three modes. The only one you can be sure of is mode 1 produced by the incisors!

There are many tests designed to measure fracture properties, some of which are shown in Figure 5–44, but it is not possible to describe all the available testing geometries here. Atkins and Mai (1985) have a useful appendix covering a large number of these geometries. The ones we have chosen have been applied particularly to food materials. The main points to bear in mind when selecting suitable tests for food materials are:

1. Fracture mechanics was originally developed for brittle, elastic materials, and, as we have seen, foods rarely conform to these conditions. Foods tend to exhibit mixtures of elastic and plastic flow (i.e., viscoelastic behavior), so it should be remembered that any fracture test will be measuring not only fracture energy but energy dissipation due to plastic flow. Various ways to allow for this will be described later.
2. Most food materials are inhomogenous and frequently anisotropic, whereas fracture mechanics analyses assume that the material is homogenous and isotropic. The heterogeneity of food materials leads to complex fracture paths that are difficult to analyze.

However, even accounting for these problems, it is still possible to adapt many of the fracture tests designed for structural materials to foods.

Work Area Methods

This group of methods measures the energy required for fracture derived from the area enclosed by a force extension curve, as described in Figure 5–45. As mentioned previously, this is only strictly valid if most of the strain energy goes into propagating a single crack. If considerable plastic flow occurs remote from the crack tip, the area under the curve will also contain plastic work

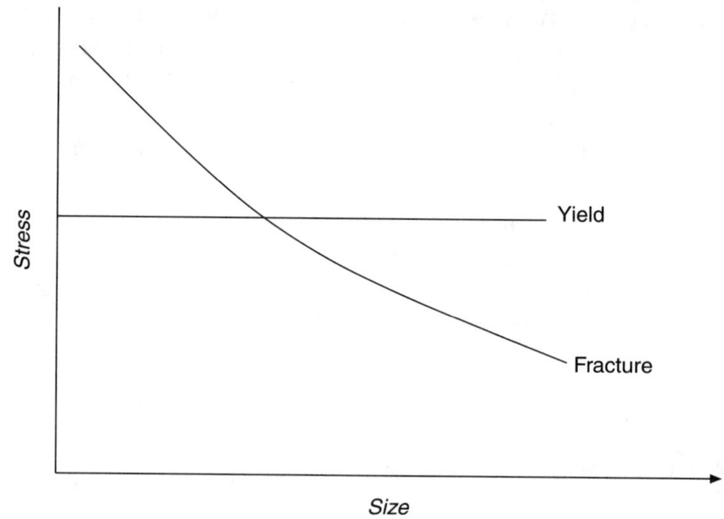

Figure 5–42 Effect of Size on Fracture. A large structure is more likely to fail by brittle fracture, while a small sample will yield before fracture. *Source:* Reprinted from *Meat Science*, Volume 21, B.J. Dobraszczyk et al., Fracture Toughness of Frozen Meat, Page No. 45, Copyright (1987), with permission from Elsevier Science.

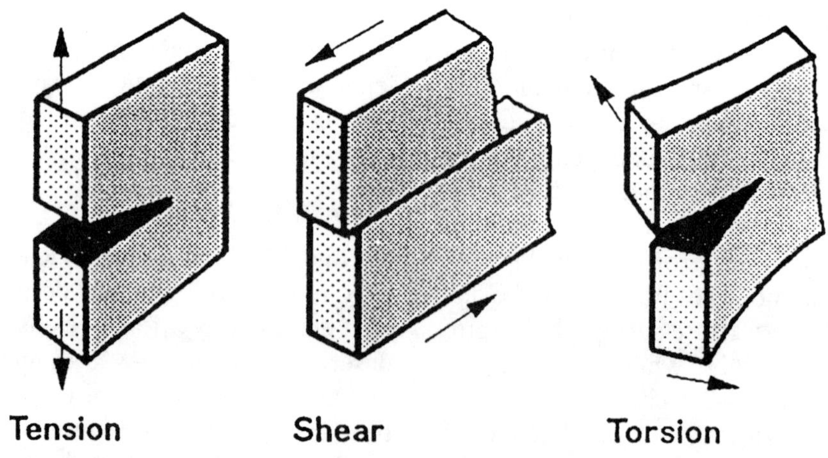

Figure 5–43 Three Modes of Fracture: Tension, Shear, and Torsion. *Source:* J.F.V. Vincent, ed., *Biomechanics—Materials: A Practical Approach*, © 1992, by permission of Oxford University Press.

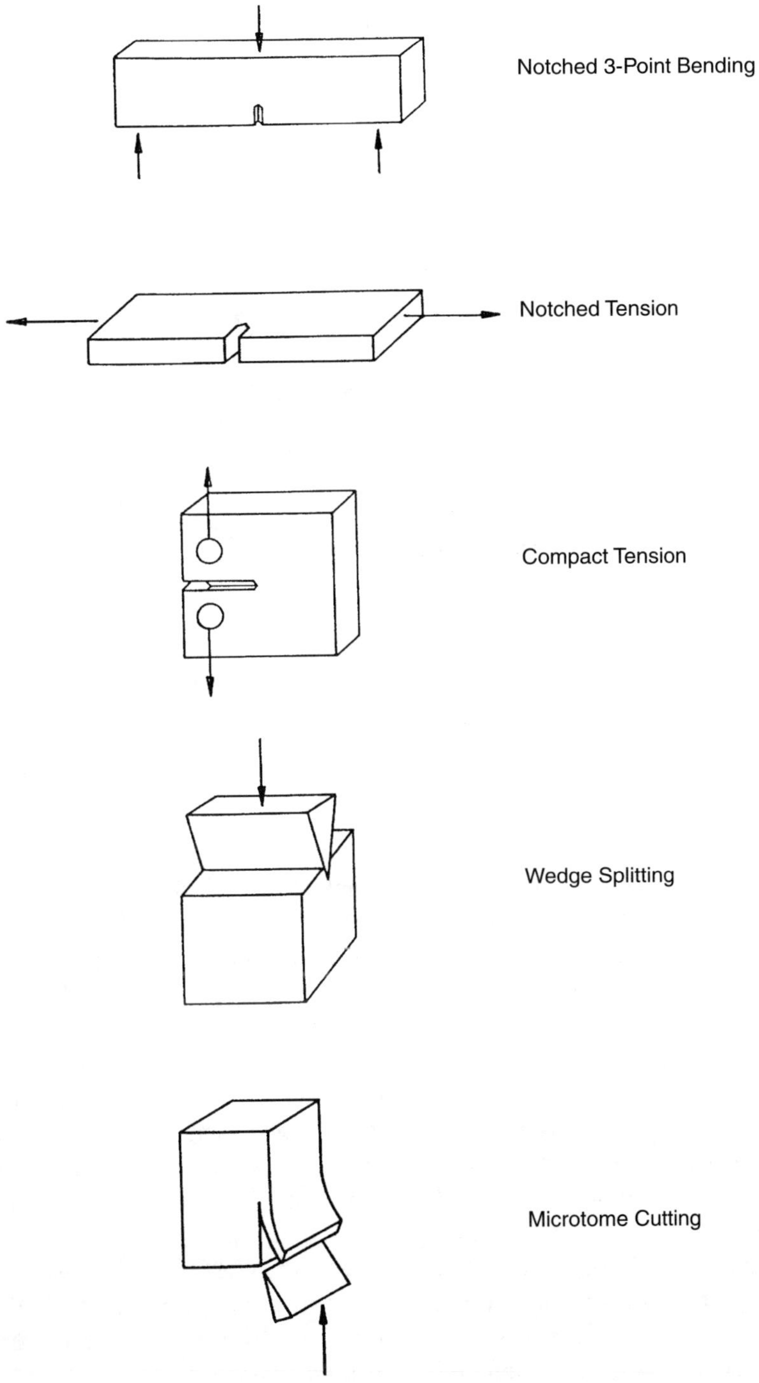

Figure 5–44 Test Geometries Commonly Used in Measuring Fracture Properties. *Source:* Reprinted from *Meat Science*, Vol. 21, B.J. Dobraszczyk et al., Fracture Toughness of Frozen Meat, pp. 25–49, Copyright (1987), with permission from Elsevier Science.

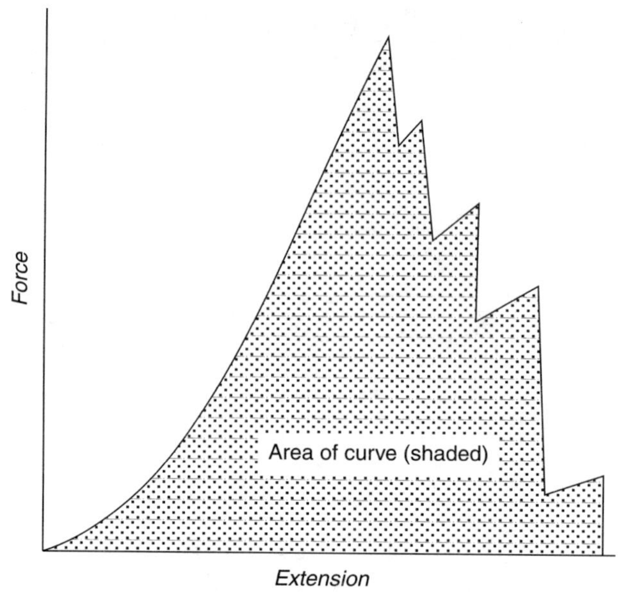

Figure 5–45 Work Area Method for Measuring Fracture Toughness (from the Area Enclosed under a Force-Extension Curve). *Source:* J.F.V. Vincent, ed., *Biomechanics—Materials: A Practical Approach*, © 1992, by permission of Oxford University Press.

of deformation. Alternatively, other energy-dissipating mechanisms may also be operating, such as debonding or delamination between fibers, particles, or surfaces and buckling.

Loading-Unloading Method

The loading-unloading method is particularly suitable for allowing for these problems. It has the advantage of being able to control and monitor crack growth and also to relate measured work areas to changes in crack growth and velocity or microstructure. A test piece, usually called a double-cantilever beam or compact tension sample, is notched at one side, and a groove is cut along both sides to guide the crack growth. The end of the notch is generally sharpened in some way (e.g., with a razor blade) to provide a starter crack (Figure 5–46).

The sample is slowly pulled apart in tension on a suitable testing machine until a crack starts to propagate by a small amount, and then is unloaded by reversing the direction of movement (Figure 5–47). The distance that the crack grows is measured using a traveling microscope. The sample is then loaded and unloaded several times until the crack has moved all the way across the sample. The area of each segment (A1–A4) in Figure 5–47 represents the work or energy used for each increment of crack growth, and if the energy (in Joules) is divided by the incremental crack area (in m^2) we can obtain the work of fracture or fracture toughness expressed in $J \cdot m^{-2}$.

If plastic deformation occurs, this will be apparent from the unloading lines not return-

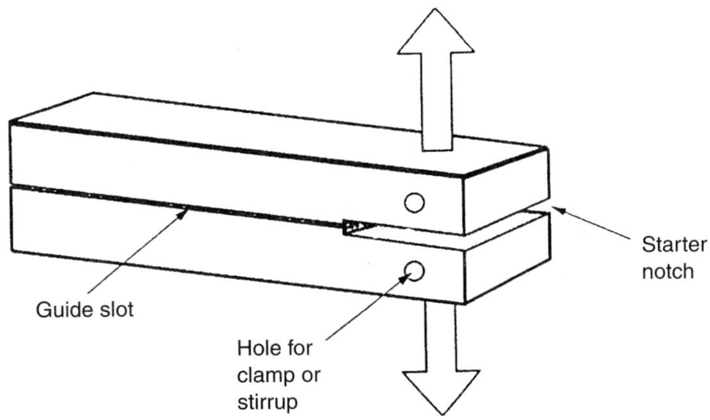

Figure 5–46 Double-Cantilever Beam (DCB) Test Geometry. *Source:* J.F.V. Vincent, ed., *Biomechanics—Materials: A Practical Approach*, © 1992, by permission of Oxford University Press.

ing to zero deformation but instead returning to a finite permanent deformation (Figure 5–48). The apparent fracture energy will also be much greater, indicating the toughening mechanism of plastic flow. Partition of the total work of deformation into plastic flow and fracture energy is possible by drawing a line parallel to the average original loading curve (Atkins & Mai, 1985). This method has been applied to the fracture of frozen meat (Dobraszczyk et al., 1987) and wheat endosperm (Dobraszczyk, 1994).

Trousers Tear Test

This painfully named test has been used for measuring the fracture properties of extensible materials such as rubber (Rivlin & Thomas, 1953) and soft animal tissues (e.g., rat skin, pig aorta, sea anenome mesoglea; Purslow 1983a, 1983b).

A strip of material has a starter crack cut longitudinally, and the "legs" of the trousers are pulled apart in a testing machine to extend the crack (Figure 5–49). The area under the force-deformation trace gives the work or energy to promote fracture (Figure 5–50).

Wedge Fracture

Strain energy can be concentrated by driving a wedge into a block of material so that a crack is propagated ahead of the wedge (Figure 5–51). This test has been adapted for use in measuring the fracture properties of foods such as cheese and raw and cooked fruits and vegetables by Vincent et al. (1991). A typical force-time trace during wedge penetration is shown in Figure 5–52.

This is how many food materials fracture in practice (in biting), and good correlations have been found between wedge and biting sensory tests (Vincent et al., 1991).

Cutting

A test related to wedge penetration is instrumented cutting. A sharp blade or wire forced through a material can propagate a crack ahead of the blade (Figure 5–53). An instrumented microtome developed by Atkins and Vincent (1984) has been used to measure the cutting force of fruit and vegetables and the fracture of wheat endosperm (Dobraszczyk, 1994). Similar experiments

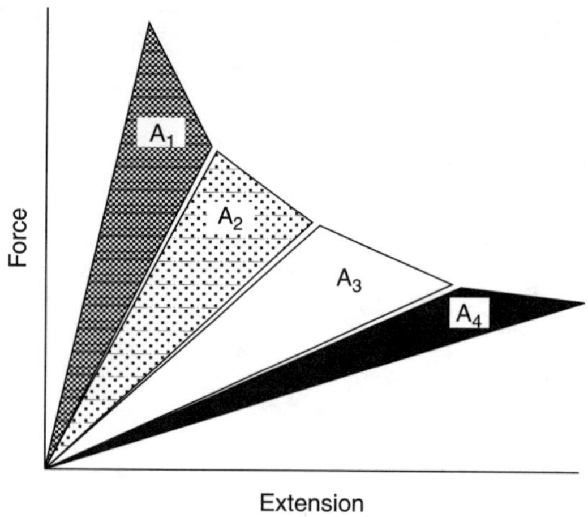

Figure 5–47 Associated Force-Extension Curve Showing Repeated Loading-Unloading Cycles. The shaded area of each segment (A1–A4) represents the energy for incremental crack growth. *Source:* J.F.V. Vincent, ed., *Biomechanics—Materials: A Practical Approach,* © 1992, by permission of Oxford University Press.

have used a wire instead of a blade to cut cheese (Luyten, van Vliet, & Walstra, 1991; Figure 5–54).

Although not absolutely necessary, a microtome with a load cell built into the knife mounting (Willis & Vincent, 1995) or the specimen holder (Ericson and Lindberg, 1996) is the simplest and best instrument for this test method. In the test, a block of the material is mounted in the chuck of the microtome, and slices are taken in the way one normally cuts histological sections. The measured force is made of three components: the force required to push the knife through the material, the work dissipated in the offcut as it is deformed over the knife (the offcut has to move out of the way to let the blade through), and the friction between the knife and the material. Friction can be accounted for either by lubricating the blade (which is done automatically if the material is a veg-

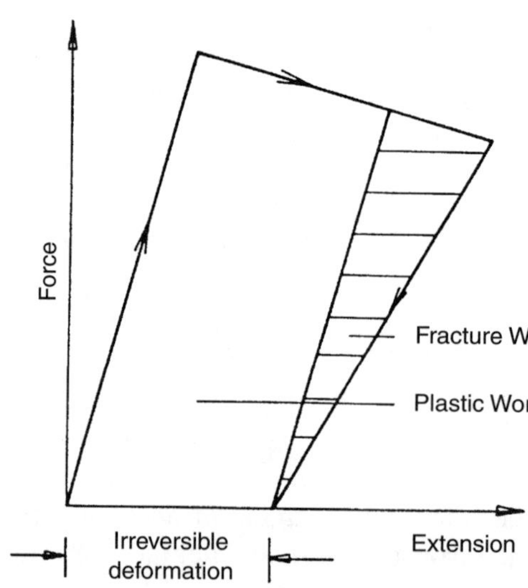

Figure 5–48 Loading-Unloading Trace for a Plastically Deforming Material, Showing Separation of Total Work Area into Plastic (ΔU_p) and Fracture Energy (ΔU_f) Components. *Source:* Reprinted from *Meat Science*, Volume 21, B.J. Dobraszczyk et al., Fracture Toughness of Frozen Meat, Page No's. 25–49, Copyright (1987), with permission from Elsevier Science.

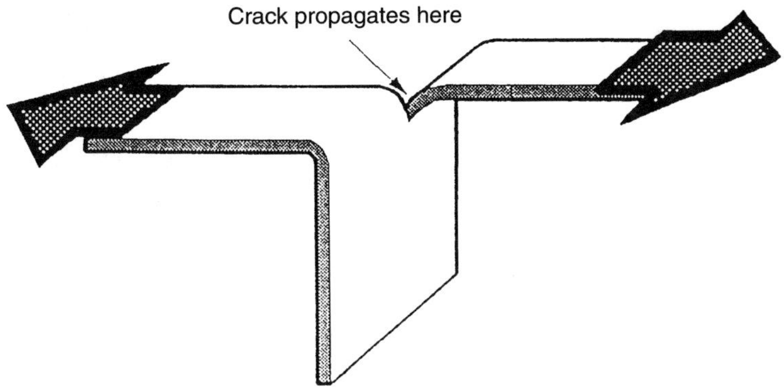

Figure 5–49 Trousers Tear Test. *Source:* J.F.V. Vincent, ed., *Biomechanics—Materials: A Practical Approach,* © 1992, by permission of Oxford University Press.

etable or fruit, since the cells release water as they are cut) or by allowing the specimen to come past the knife on the return stroke and monitoring the reverse force generated. The work dissipated in the offcut (the work of curling) is related to the shear strength of the material and can be eliminated by sectioning at different thicknesses of offcut and extrapolating to zero thickness in a graph of thickness versus force (Figure 5–55). The work of cutting (in J • m^{-2}) is then obtained by dividing the force (in N) by the width of the specimen (in m) (Ericson & Lindberg, 1996; Vincent, 1991). With some materials, such as cheese, the amount of friction is so great between the knife and the specimen that it is better to use a

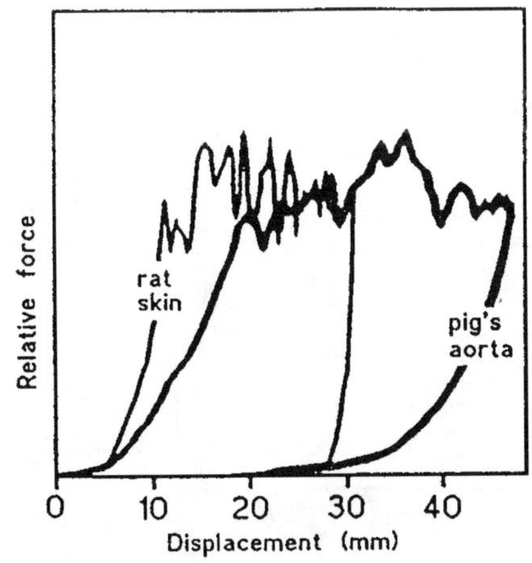

Figure 5–50 Force-Displacement Trace Associated with Trousers Tear Test. *Source:* J.F.V. Vincent, ed., *Biomechanics—Materials: A Practical Approach,* © 1992, by permission of Oxford University Press.

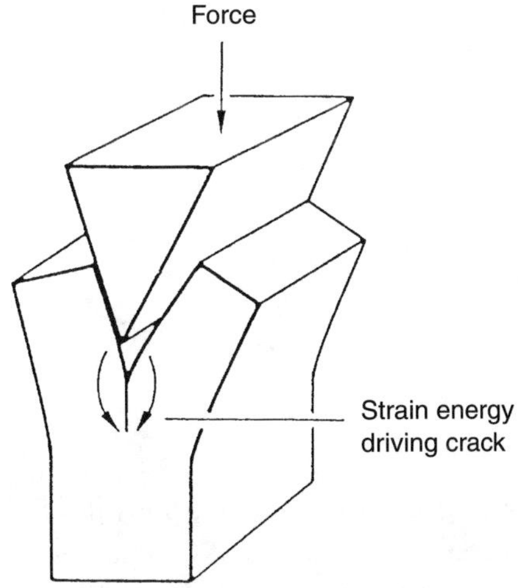

Figure 5–51 Wedge Fracture Test. *Source:* J.F.V. Vincent, ed., *Biomechanics—Materials: A Practical Approach,* © 1992, by permission of Oxford University Press.

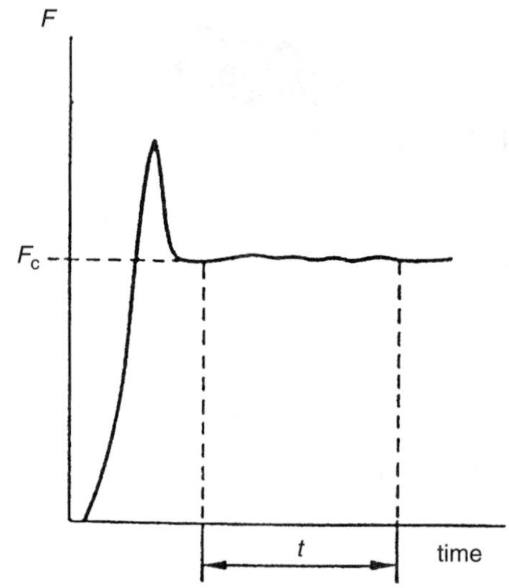

Figure 5–52 Force-Time Trace Associated with Wedge Fracture Test, Showing an Initial Peak in Force on Driving in the Wedge, followed by a Constant Force (F_c) on Crack Propagation ahead of the Wedge. *Source:* J.F.V. Vincent, ed., *Biomechanics—Materials: A Practical Approach*, © 1992, by permission of Oxford University Press.

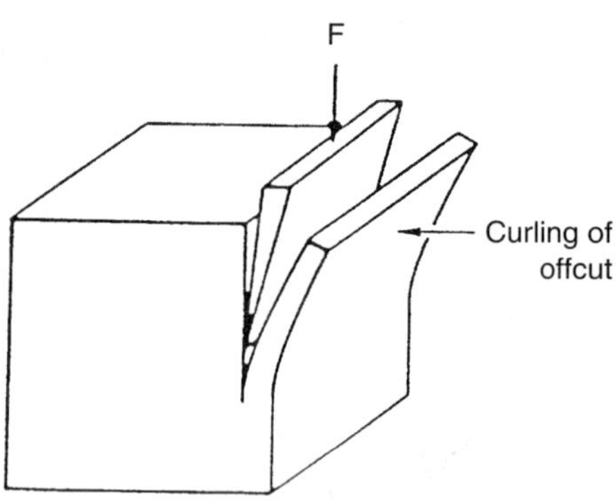

Figure 5–53 Instrumented Microtome Cutting Test Showing Blade Cutting into the Material with Force *F* and Curling of Offcut. *Source:* Reprinted from T. Van Vliet and H. Luyten, Fracture Mechanics of Solid Foods, in *New Physico-Chemical Techniques for the Characterization of Complex Food Systems*, E. Dickinson, ed., pp. 173–174, © 1995, Aspen Publishers, Inc.

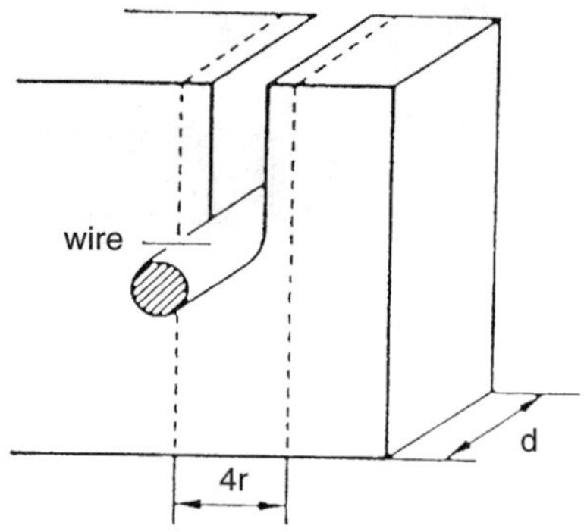

Figure 5–54 Instrumented Wire Cutting Test. *Source:* Reprinted from T. Van Vliet and H. Luyten, Fracture Mechanics of Solid Foods, in *New Physico-Chemical Techniques for the Characterization of Complex Food Systems*, E. Dickinson, ed., pp. 173–174, © 1995, Aspen Publishers, Inc.

wire. To obtain the cutting, one then uses wires of different diameters and extrapolates a graph of wire diameter versus force to zero diameter. The wire should be mounted tautly and not allowed to store strain energy in bending that cannot be released into the cut.

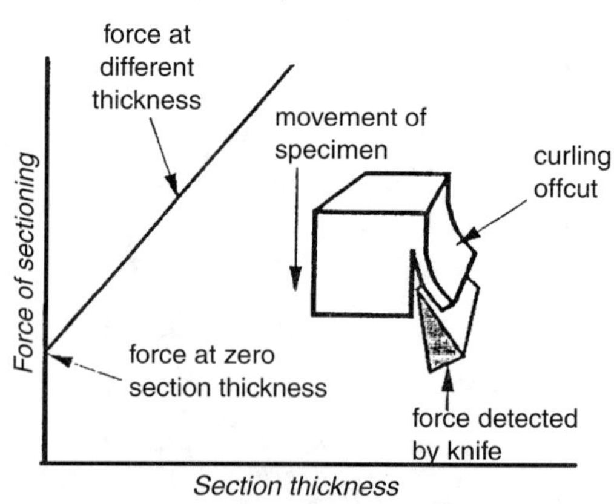

Figure 5–55 Factors Used in the Calculation of Toughness Using the Instrumented Microtome. *Source:* J.F.V. Vincent, ed., *Biomechanics—Materials: A Practical Approach*, © 1992, by permission of Oxford University Press.

Since the force recorded during cutting is representative of the area just ahead of the blade, information can be obtained about local variations in structure from variations in the force trace. This allows microdetermination of texture or inhomogeneities in structure of food materials that are often lost in tests that measure the mechanical properties of the bulk of the material. For example, in cutting food foams such as vegetables, it is possible to pick up variations in structures or size distributions in open-cell foams such as bread and cakes. The instrumented microtome is acting as a "mechanical microscope."

Instrumented Scissors

Lucas, Darvell, Lee, Yuen, and Choong (1995) measured the force of cutting using instrumented scissors. Cutting using scissors produces fracture in torsion (mode 3) and in many ways mimics the action of teeth in biting (Figure 5–56). This test has been used in measuring the fracture toughness of leaves (Lucas & Pereira, 1990) and the toughness of plant cell walls (Lucas et al., 1995) and grass (Vincent, 1992).

Single-Edge Notch: Bending and Tension

The single-edge notch bending (SENB) geometry shown in Figure 5–44 is widely used to determine the fracture of metals and plastics (American Society for Testing and Materials, 1984) and is one of the main test geometries to determine the stress intensity factor (Ewalds & Wanhill, 1986). Before testing, the sample is notched with a razor blade to give a sharp starter crack, with a dimension *a*. The sample can be loaded in three- or four-point bending. Assuming linear elastic behavior, the fracture toughness (G_c) can be derived from

$$G_c = \frac{U}{wt\Phi}$$

where U is the energy under the load-distance diagram between zero and crack initiation force, w is the width, t is the thickness, and Φ is a calibration factor, a function of (a/t), given in Williams (1984). Charalambides, Williams, and Chakrabarti (1995) obtained fracture toughness values for cheddar cheese in relation to maturing time using this test. Vincent (1982, 1983) used a single-edge test in tension to measure fracture properties of grasses.

CONCLUSION

With this armory of tests and theory, how should one approach a novel problem in food testing, whether it is a better quality control test or the necessity to find the cause of a particular problem? The hope, pious or not, has always been that materials science should be able to measure texture more discriminatively, more precisely, and more repeatably than people can. Recent results from our laboratory (obtained while we were preparing this chapter!) show that "hardness" of apples, measured by a sensory panel, can be tracked by the stress intensity factor with a correlation coefficient of about 0.95. This is good enough to say that perceived hardness *is* stress intensity—that is, the difficulty experienced in starting a crack in the material. As far as we know, this is the first time that a proper fracture mechanics parameter has been used to model "texture," although several people have shown that fracture mechanics is an important tool to describe the properties of food materials, mostly for cheese (Luyten et al., 1991; Vincent et al., 1991). From the sensory point of view, it means that the consumer is using a quite specific parameter on which to make the judgment about an aspect of texture. There is no reason (1) why other parameters

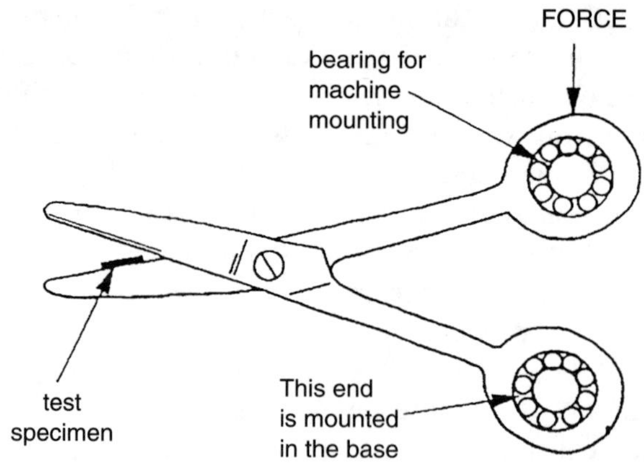

Figure 5–56 Instrumented Scissors Cutting Test. *Source:* J.F.V. Vincent, ed., *Biomechanics—Materials: A Practical Approach*, © 1992, by permission of Oxford University Press.

should not be found to behave in this way, and (2) why sensory scientists and panel leaders should not be told of this result (and related ideas in fracture mechanics) to characterize foods better. It seems to us that to deny them access to relevant information (so long as it is presented "neutrally")—that is, purely as information rather than interpretation—is not properly scientific. Sensory scientists or panel leaders can still choose to ignore the information. On the other hand, it may help them and their panels to arrive at a better assessment of the various factors contributing to texture or acceptability.

What of the "trade" (producers, wholesalers, retailers) for which the sensory panels are ostensibly working? Their requirement is for any type of test that leads to a better product. The true trial takes place in the marketplace, not the laboratory. The sensory panel, like the hangman's noose, concentrates the scientist's mind wonderfully. It is a tool, just like the tests of materials science, and serves its purpose only so long as it serves its masters. Let us hear no more, "Does the test replicate the conditions in the mouth?" The question *should* be, "Does the test enable us to produce a better and more reliable product that is texturally acceptable?"

REFERENCES

American Society for Testing and Materials. (1984) Standard E399-83: Standard test method for plane-strain fracture toughness of metallic materials. In *1984 annual book of ASTM standards* (Vol. 03.01, pp. 519–554). Philadelphia: Author.

Atkins, A.G., & Mai, Y.-W. (1985). *Elastic and plastic fracture: Metals, polymers, ceramics, composites and biological materials*. Chichester, UK: Ellis Horwood.

Atkins, A.G., & Mai, Y.-W. (1986). Review: Deformation transitions. *Journal of Materials Science, 21*, 1093–1110.

Atkins, A.G., & Vincent, J.F.V. (1984). An instrumented microtome for improved histological sections and the measurement of fracture toughness. *Journal of Materials Sciences Letters, 3*, 310–312.

Bagley, E.B., & Christianson, D.D. (1987). Stress relaxation of chemically leavened dough: Data reduction using the BKZ elastic fluid theory. *Journal of Rheology, 31*, 405–413.

Blanshard, J.M.V. (1995). The glass transition: Its nature and significance in food processing. In S.T. Beckett (Ed.), *Physicochemical aspects of food processing* (pp. 17–48). London: Blackie.

Bloksma, A.H. (1957). A calculation of the shape of the alveograms of some rheological model substances. *Cereal Chemistry, 34*, 126–136.

Brown, W.E., Eves, D., Ellison, M., & Braxton, D. (in press). Use of combined electromyography and kinesthesiology during mastication to chart the oral breakdown of foodstuffs: Relevance to the measurement of food texture. *Journal of Texture Studies*.

Casiraghi, E.M., Bagley, E.B., & Christianson, D.D. (1985). Behaviour of mozzarella, cheddar and processed cheese spread in lubricated and bonded uniaxial compression. *Journal of Texture Studies, 16*, 281–301.

Charalambides, M.N., Williams, J.G., & Chakrabarti, S. (1995). A study of the influence of ageing on the mechanical properties of cheddar cheeses. *Journal of Material Sciences, 30*, 3959–3967.

Chatraei, S.H., Macosko, C.W., & Winter, H.H. (1981). Lubricated squeezing flow: A new biaxial extensional rheometer. *Journal of Rheology, 25*, 433–443.

Chopin, M. (1921), Relations entre les proprietes mecaniques des pâtes de farines et la panification. *Bull. Soc. Encour. Ind. Natl,. 133*, 261.

Cogswell, F.N. (1981). *Polymer Melt Rheology*. London: George Godwin.

Conde-Petit, B., Escher, F., Halliger, A., & Schweizer, T. (1994). Absolutmessung des dehnverhaltens von brotteig mit kleinen probemengen. *Getreide, Mehl und Brot, 48*, 28–30.

Denson, C.D., & Crady, D.L. (1974). Measurements on the planar extensional viscosity of bulk polymers: The inflation of a thin rectangular polymer sheet. *Journal of Applied Polymer Science, 18*, 1611–1618.

Denson, C.D., & Gallo, R.J. (1971). Measurements on the biaxial extension viscosity of bulk polymers: The inflation of a thin polymer sheet. *Polymer Engineering Science, 11*, 174–176.

Dieter, G.E. (1986). *Mechanical metallurgy* (3rd ed.). New York: McGraw-Hill.

Dobraszczyk, B.J. (1994). Fracture mechanics of vitreous and mealy wheat endosperm. *Journal of Cereal Science, 19*, 273–282.

Dobraszczyk, B.J. (1997a). Development of a new dough inflation system to evaluate doughs. *Cereal Foods World, 42*, 516–519.

Dobraszczyk, B.J. (1997b). The rheological basis of dough stickiness. *Journal of Texture Studies, 28*, 139–162.

Dobraszczyk, B.J., Atkins, A.G., Jeronimidis, G., & Purslow, P.P. (1987). Fracture toughness of frozen meat. *Meat Science, 21*, 25–49.

Dobraszczyk, B.J., & Roberts, C.A. (1994). Strain hardening and dough gas cell-wall failure in biaxial extension. *Journal of Cereal Science, 20*, 265–274.

Ericson, M.L., & Lindberg, H. (1996). A method of measuring energy dissipation during crack propagation with an instrumented microtome. *Journal of Materials Science, 31*, 655–662.

Ewalds, H.L., & Wanhill, R.J.H. (1986). *Fracture mechanics*. London: Edward Arnold.

Gordon, J.E. (1978). *Structures—or why things don't fall down*. New York: Penguin.

Griffith, A.A. (1921). The phenomenon of rupture and flow in solids. *Philosophical Transactions of the Royal Society of London A, 221*, 163–198.

Hankoczy, J. (1920). Apparat fur kleberverwertung. *Z. Gesamte Getreidewiss, 12*, 57.

Heath, M.R. (1991). The basic mechanics of mastication: Man's adaptive success. In J.F.V. Vincent & P.J. Lillford (Eds.), *Feeding and the texture of food* (pp. 143–166). Cambridge, UK: Cambridge University Press.

Huang, H., & Kokini, J.L. (1993). Measurement of biaxial extensional viscosity of wheat flour doughs. *Journal of Rheology, 37*, 879–891.

Joye, D.D., Poehlein, G.W., & Denson, C.D. (1972). A bubble inflation technique for the measurement of viscoelastic properties in equal biaxial extensional flow. *Transactions of the Society of Rheology, 16*, 421–445.

Kalichevsky, K.T., Jaroszkiewicz, E.M., Ablett, S., Blanshard, J.M.V., & Lillford, P.J. (1992). The glass transition of amylopectin measured by DSC, DMTA and NMR. *Carbohydrate Polymers, 18*, 77–88.

Khan, A.A. (1989). *Mechanical properties of fruit and vegetables*. Unpublished Ph.D. dissertation, University of Reading, UK.

Kilcast, D., & Eves, A. (1991). Integrating texture and physiology: Techniques. In J.F.V. Vincent & P.J. Lillford (Eds.), *Feeding and the texture of food* (pp. 167–183). Cambridge, UK: Cambridge University Press.

Kokelaar, J.J., van Vliet, T., & Prins, A. (1996). Strain hardening and extensibility of flour and gluten doughs in relation to breadmaking performance. *Journal of Cereal Science, 24*, 199–214.

Launay, B., Bure, J., & Praden, J. (1977). Use of the Chopin alveograph as a rheological tool: I. Dough deformation measurements. *Cereal Chemistry, 54*, 1042–1048.

Levine, H, & Slade, H. (1993). The glassy state in applications for the food industry. In J.M.V. Blanshard and P.J. Lillford (Eds.), *The glassy state in foods* (pp. 333–373). Nottingham, UK: Nottingham University Press.

Lucas, P.W., Darvell, B.W., Lee, P.K.D., Yuen, T.D.B., & Choong, M.F. (1995). The toughness of plant cell walls. *Philosophical Transactions of the Royal Society of London B, 348*, 363–372.

Lucas, P.W., & Luke, D.A. (1983). Methods of analysing the breakdown of food in human mastication. *Archives of Oral Biology, 28*, 813–819.

Lucas, P.W., & Pereira, B. (1990). Estimation of the fracture toughness of leaves. *Functional Ecology, 4*, 819–822.

Luyten, H., van Vliet, T., & Walstra, P. (1991). Characterization of the consistency of Gouda cheese: Fracture properties. *Netherlands Milk Dairy Journal, 45*, 55–80.

Meissner, J. (1987). Polymer melt elongation: Methods, results, and recent developments. *Polymer Engineering and Science, 27*, 537–546.

Meissner, J., & Hostettler, J. (1994). A new elongational rheometer for polymer melts and other highly viscoelastic liquids. *Rheologica Acta, 33*, 1–21.

Nicholls, R.J., Appelqvist, I.A.M., Davies, A.P., Ingman, S.J., & Lillford, P.J. (1995). Glass transitions and the fracture behaviour of gluten and starches within the glassy state. *Journal of Cereal Science, 21*, 25–36.

Padmanabhan, M. (1995). Measurement of extensional viscosity of viscoelastic liquid foods. *Journal of Food Engineering, 25*, 311–327.

Purslow, P.P. (1983a). Measurement of the fracture toughness of extensible connective tissues. *Journal of Materials Science, 18*, 3591–3598.

Purslow, P.P. (1983b). Positional variations in fracture toughness, stiffness and strength of descending thoracic pig aorta. *Journal of Biomechanics, 16*, 947–953.

Rhi-Sausi, J., & Dealy, J.M. (1981). A biaxial extensiometer for molten plastics. *Polymer Engineering Science, 21*, 227–232.

Rivlin, R.S., & Thomas, A.G. (1953). Rupture of rubber I: Characteristic energy for tearing. *Journal of Polymer Science*, 10, 291–318.

Shadwick, R.E. (1992). Soft composites. In J.F.V. Vincent (Ed.), *Biomechanics—Materials: A practical approach* (pp. 133–163). Oxford, UK: Oxford University Press.

Shama, F., & Sherman, P. (1973). Identification of stimuli controlling the sensory evaluation of viscosity: II. Oral methods. *Journal of Texture Studies, 4*, 111–118.

Sherman, P. (1970). *Industrial rheology*. London Academic Press.

Shrive, N.G. (1997). Soft tissue strain measurement. In J.F. Orr & J.C. Shelton (Eds.), *Optical measurement methods in biomechanics* (pp. 154–172). London: Chapman & Hall.

Szczesniak, A.S. (1963). Classification of textural characteristics. *Journal of Food Science, 28*, 385–389.

Treloar, L.R.G. (1944). Strains in an inflated rubber sheet, and the mechanism of bursting. *Transactions of the Institute of Rubber Industry, 19*, 201–212.

Tschoegl, N.W., Rinde, J.A., & Smith, T.L. (1970). Rheological properties of wheat flour doughs II: Dependence of large deformation and rupture properties in simple tension on time, temperature and water absorption. *Rheologica Acta, 9*, 223–238.

Vincent, J.F.V. (1982). The mechanical design of grass. *Journal of Materials Science, 17*, 856–860.

Vincent, J.F.V. (1983). The influence of water content on the stiffness and fracture properties of grass leaves. *Grass Forage Science, 38*, 107–111.

Vincent, J.F.V. (1989). The relation between density and stiffness of apple flesh. *Journal of the Science of Food and Agriculture, 47*, 443–462.

Vincent, J.F.V. (1990a). Fracture in plants. *Advances in Botanical Research, 17*, 235–287.

Vincent, J.F.V. (1990b). *Structural biomaterials*. Princeton, NJ: Princeton University Press.

Vincent, J.F.V. (1991, May). Automating the microtome. *Microscopic Analysis*, pp. 19–21.

Vincent, J.F.V. (Ed.). (1992). *Biomechanics—Materials: A practical approach*. Oxford, UK: Oxford University Press.

Vincent, J.F.V. (in press). The quantification of crispness. *Journal of the Science of Food and Agriculture*.

Vincent, J.F.V., Jeronimidis, G., Khan, A.A., & Luyten, H. (1991). The wedge fracture test: A new method for measurement of food texture. *Journal of Texture Studies, 22*, 45–57.

Wainwright, S.A., Biggs, W.D., Currey, J.D., & Gosline, J.M. (1965). *Mechanical design in organisms*. London: Edward Arnold.

Waldron, K.W., Smith, A.C., Ng, A., Parker, M.L., & Parr, A.J. (1997). Thermal stability of mechanical properties of Chinese water chestnuts. In G. Jeronimidis & J.F.V. Vincent (Eds.), *Plant biomechanics 1997* (pp. 137–143). Reading, UK.

Whorlow, R.W. (1992). *Rheological techniques* (2nd ed,) Chichester. UK: Ellis Horwood.

Wikström, K. (1997). *Rheology of wheat flour dough at large deformations and the relation to baking quality and physical structure*. Unpublished Ph.D. dissertation, University of Lund, Sweden.

Williams, J.G. (1984). *Fracture mechanics of polymers*. Chichester, UK: Ellis Horwood.

Willis, A., & Vincent, J.F.V. (1995). Monitoring cutting forces with an instrumented histological microtome. *Journal of Microscopy*, *178*, 56–65.

CHAPTER 6

Starch-Based Foods

Andrew C. Smith

STARCH

Plants store energy by producing starch. The morphology of individual starch granules varies with the source, whether cereal, pulse, root, or tuber. Granule sizes vary from 2 to 100 μm (Greenwood, 1976). Starch granules have a well-known property of birefringence when viewed under polarized light. Granules vary in their composition of the polymers amylose and amylopectin and in protein, lipids, and other components. Amylose is a linear polymer, whereas amylopectin is highly branched (Figure 6–1). Crystallinity in starches is categorized as A type, found largely in cereals; B type, largely from tubers; and C type, which is from legume starches and is a mixture of A and B types. Amylose complexes with lipids to form a crystalline structure called the V type.

In excess water (typically >60% on a wet-weight basis), starch undergoes gelatinization at a temperature in the range 50 to 90°C, depending on botanical origin, marked by the loss of crystalline order (detected by loss of birefringence), swelling of the granule, and solubilization of amylose. A fluid is formed that comprises porous granules dominated by amylopectin in amylose, which forms a turbid viscoelastic paste or gel on cooling (Morris, 1990). There is a marked difference in rheology between a dispersion of native starch granules in water and the gelatinized paste or gel (Christianson & Bagley, 1983). This is often demonstrated in the Brabender Amylograph, which induces a temperature cycle on a dispersion of a starchy material (Anderson, Conway, Pfeifer, & Griffin, 1969; Gomez & Aguilera, 1984). Reduction of the water level below excess elevates the gelatinization temperature, which is clearly indicated by differential scanning calorimetry (Figure 6–2). Sugar addition to starch in place of water elevates the gelatinization temperature (Spies & Hoseney, 1982).

The situation during processing varies considerably. Starch is transformed at the molecular, crystalline, and granular levels (Blanshard, 1986). Heating in an excess of water may involve gelatinization of starch alone, and granule swelling and rupturing may occur. Other variables such as mechanical energy may cause disruption. In the extreme of high shear with high temperature and reduced water, which occurs in an extrusion cooker, the starch granule structure will be lost and some molecular degradation will occur. However, degradation to short oligosaccharides and glucose will not happen. The stages of granular disruption typical of various cereal products were highlighted by Greenwood (1976) (Figure 6–3).

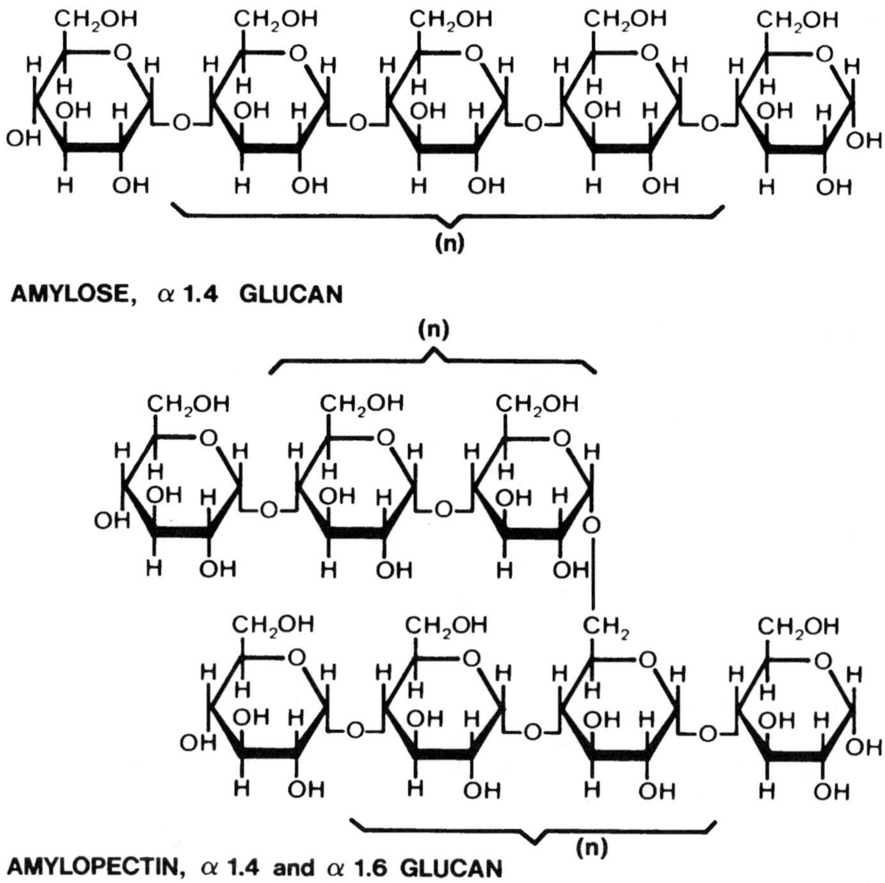

Figure 6–1 Amylopectin and Amylose Molecular Structures

In practice, starch-based foods are complex multicomponent systems containing starch, proteins, water, solutes, sugars, and lipids. Proteins are of fundamental importance in determining water absorption, reduction or oxidation, rheology, and gas retention (Blanshard, 1988). Gluten is the functionally important protein in wheat flour and comprises two largely insoluble protein groups, the glutenins and gliadins. Gluten has reactive side groups, and the effect of water and heat is to promote initially a time-dependent hydration and expansion leading to aggregation and gelation. As with starch, heat dramatically alters the rheology of proteins. Similarly, mechanical energy is an important factor in structure development in proteins as well as starch.

POLYMER SCIENCE APPROACH

Gelatinization

Gelatinization is a term peculiar to starch. It is a complex combination of the melting and glass transitions. Levine and Slade (1990) have championed the importance of a polymer science interpretation of starch and other food polymers and have maintained that the approach is fundamental in under-

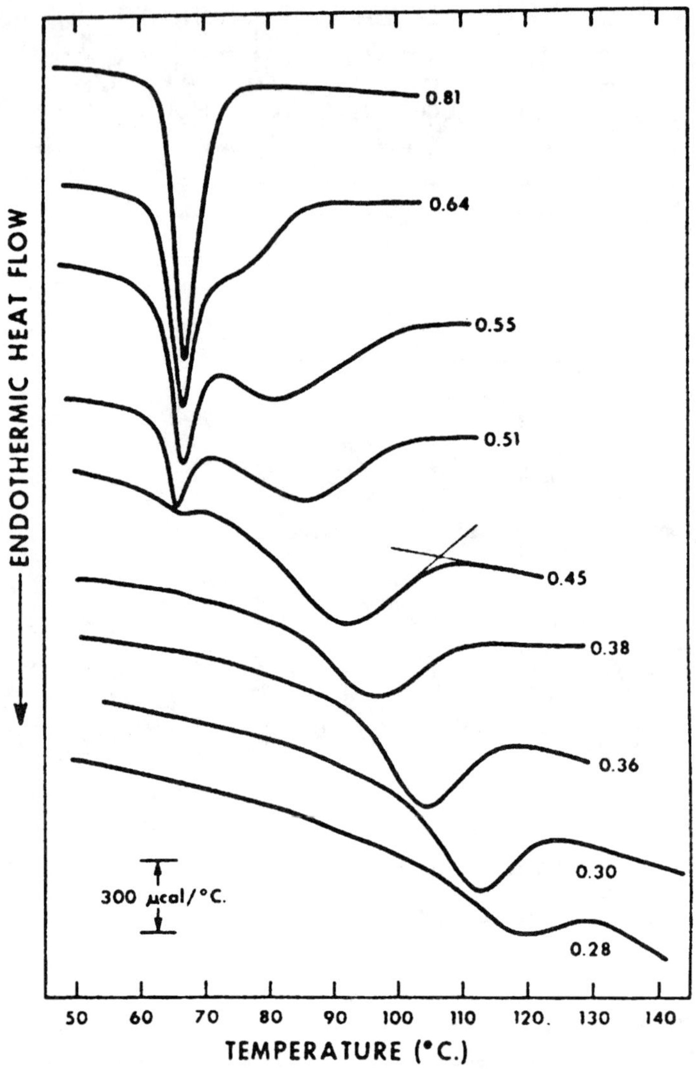

Figure 6–2 Differential Scanning Calorimetry of Starch at Different Water Contents (Expressed as Volume Fraction of Water). *Source: Biopolymers*, J.W. Donovan, Copyright © (1979, John Wiley & Sons, Inc.) Reprinted with permission of John Wiley & Sons, Inc.

standing structural, mechanical, and textural properties during processing and storage. Polymer science recognizes the melting and glass transitions, and the generality of the approach implies that they may also be used to describe food polymers. Early studies of starch gelatinization used polymer ideas of crystallinity and the prediction of changes in the melting point with amount of water (Lelievre, 1973).

Glass Transition

Melting points (Tm) of ice and the crystalline regions of polymers are important aspects of food polymer systems that affect rhe-

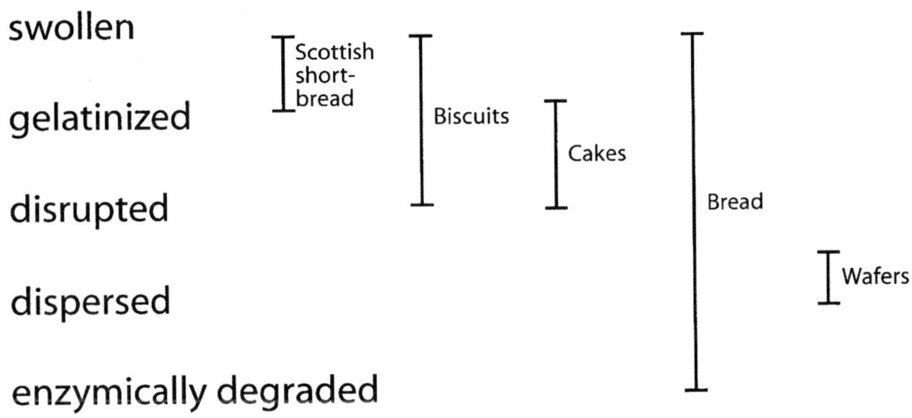

Figure 6–3 Starch Granule Disruption in Different Products. *Source:* Reprinted with permission from C.T. Greenwood, Starch, in *Advances in Cereal Science and Technology*, Y. Pomerantz, ed., pp. 119–157, © 1976, American Association of Cereal Chemists.

ology and particularly mechanical properties. Amorphous polymers have a glass transition temperature (Tg) that corresponds to an increase in volume and molecular mobility. This onset of molecular mobility with increasing temperature corresponds to a change from the glassy to the rubbery state (Slade & Levine, 1993) and has most often been studied using differential scanning calorimetry. This description immediately impinges on rheology and has implications for the stress-strain behavior of polymers. In addition to stress-strain measurements, thermomechanical analysis allows measurement of real (storage, E') and imaginary (loss, E'') moduli. The low-strain modulus undergoes a decrease at the glass transition, and the ratio of imaginary to real moduli (tan δ, defined as E''/E') shows a maximum. Mechanical spectroscopy can also show other lower temperature transitions with more minor changes in modulus and peaks in tan δ (Kalichevsky, Blanshard, & Marsh, 1993; Ward, 1983).

Another interesting element of these transitions is their change with water content. Water addition to a dry polymer reduces melting and glass transitions (Zeleznak & Hoseney, 1987). With reference to starch and cereals, Figure 6–4 shows the decrease in their glass transitions with increasing water content. In polymer science terms, this is referred to as *plasticization* of the polymer by a small molecule. Again, it is noteworthy that water is the most effective plasticizer of the starch polymers. Proteins have also been amenable to this type of description (Hoseney, Zeleznak, & Lai, 1986), as have the more complex mixtures of amylopectin, protein, and sugars (Kalichevsky & Blanshard, 1992, 1993; Kalichevsky, Blanshard, & Marsh, 1993; Kalichevsky, Jaroszkiewicz, Ablett, Blanshard, & Lillford, 1992; Kalichevsky, Jaroszkiewicz, & Blanshard, 1992a, 1992b, 1993). The pioneering work of Blanshard has established the glass transitions of these systems and revealed the complexity that can arise in mixtures such as phase separation, degree of miscibility, and competition for water. Even more complex systems are processed cereals and their mixtures with sugars and other commercial additives, which show that the same concepts can be applied (Georget & Smith, 1995, 1996a, 1996b) (Figure 6–5).

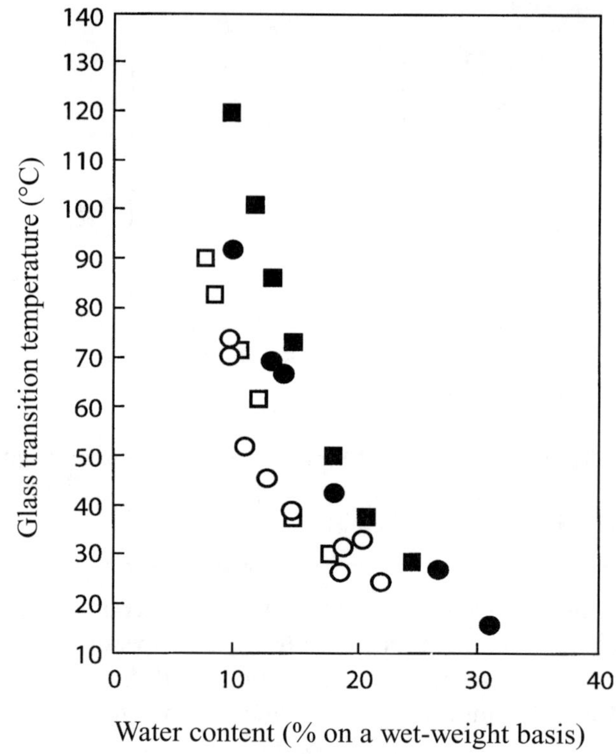

Figure 6–4 Glass Transition of Amylopectin (■), Gluten (□), Flaked Wheat (●), and Flaked Wheat Mixture (○) as a Function of Water Content. *Source:* Reprinted from *Carbohydrate Polymers,* Volume 28, D.M.R. Georget and A.C. Smith, Mechanical Properties of Wheat Flake Components, Page No's. 305–311, Copyright (1995), with kind permission from Elsevier Science Ltd., The Boulevard, Langford Lane, Kidlington 0X5 1GB, UK.

Sub-Tg Aging

When a polymer is in the glassy state, the initial conclusion might be that the material is "stable." While this is generally true as evidenced by various phenomena (Slade & Levine, 1991) such as caking of amorphous powders, collapse in freeze-dried products, loss of encapsulated volatiles, enzymic activity, and Maillard browning reactions, recent evidence suggests that changes do occur. We know from polymer science that a system must be taken into the glassy state infinitely slowly to achieve equilibrium. The fact that material can be cooled or dried either slowly or quickly leads to a wide range of nonequilibrium states. Theories of enthalpy relaxation that are universal for polymers (Hutchinson, 1995) have been applied to food materials such as starches (Shogren, 1992) and proteins (Lawton & Wu, 1993).

Polymers crystallize in the temperature (T) range Tg < T < Tm; that is, when they are in the rubbery state but below the melting point. The rate of crystal nucleation rises asymptotically toward Tg, and the rate of crystal growth rises similarly toward Tm. Consequently, the crystallization rate is a maximum midway between Tg and Tm (Figure 6–6). There is a temperature for maximum rate of crystallization that may be manipulated by composition—for example, water or sugar

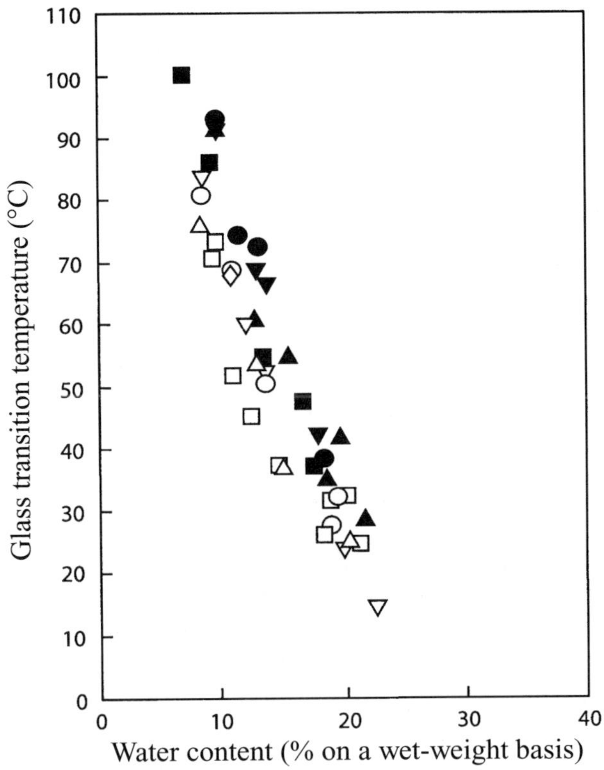

Figure 6–5 Glass Transition of Flaked-Wheat Mixtures as a Function of Water Content. i) No added Sucrose—wheat, NaCl, malt (■); wheat, malt (●); wheat, NaCl (▲); wheat (▼). ii) Added sucrose [ratio wheat : sucrose 6:1]—wheat, malt, NaCl, sucrose (□); wheat, malt, sucrose (○); wheat, NaCl, sucrose (△); wheat, sucrose (▽). *Source:* Reprinted from *Carbohydrate Polymers,* Volume 28, D.M.R. Georget and A.C. Smith, Mechanical Properties of Wheat Flake Components, Page No's. 305–311, Copyright (1995), with kind permission from Elsevier Science Ltd., The Boulevard, Langford Lane, Kidlington 0X5 1GB, UK.

addition. In the case of starch polymers, this process is usually termed *retrogradation* and it has important ramifications for product staling. (see below). Sugars are of particular interest in their effect on gelatinization temperatures and retrogradation rates (Cairns, Miles, & Morris, 1991; I'Anson et al., 1991).

PROCESSING

One of the reasons for processing is to add value by changing structure. Processing of starchy materials may be generalized into a number of stages (Smith, 1991). First, the ingredients are assembled, mixed, and prepared. Mixing may be carried out in a wide range of equipment, and the nature of the strain, temperature, and time history affects the process response and ultimately impinges on product attributes. For many processes, mixing results in a high-viscosity dough or dispersion and affects protein structure, particularly gluten development. The process variables of mixing operations are combinations of temperature, shear/stress, time, and ingredient composition. The raw ingredients

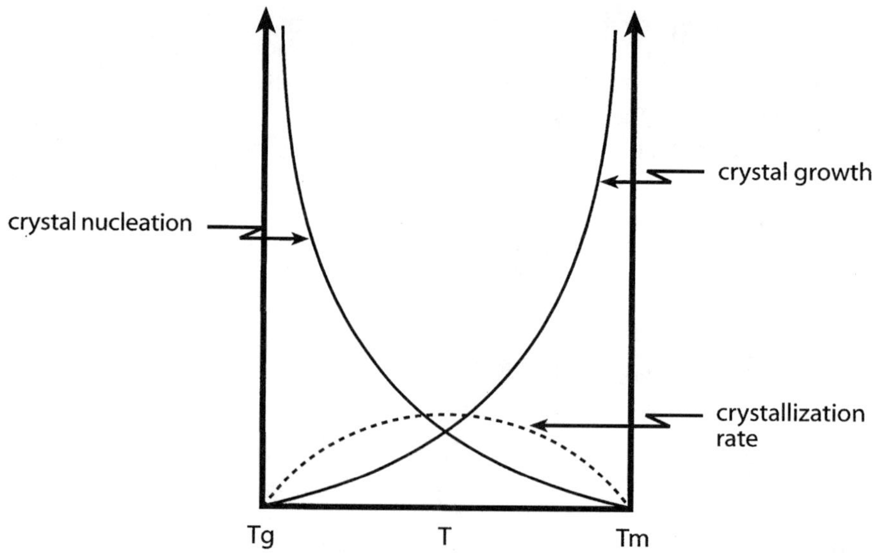

Figure 6–6 Schematic Crystallization Nucleation and Growth as a Function of Temperature. Tg is the glass transition temperature, and Tm is the melting temperature. *Source:* Reprinted from *Trends in Food Science and Technology,* Volume 1, V.J. Morris, Starch Gelatin Retroeradation, Page No. 6, Copyright (1990), with kind permission from Elsevier Science Ltd., The Boulevard, Langford Lane, Kidlington OX5 1GB, UK.

are hydrated and components incorporated as necessary for structure creation. "Structure" may be developed by a number of different approaches.

Structural change occurs at the microscopic level. It involves biopolymer melting or specifically starch gelatinization, protein denaturation, and complexing of these major functional components with each other, sugars, and fats. Marked rheological changes also accompany these changes, as described in a number of studies (Bloksma, 1980; Giesekus, 1983; Remsen & Clark, 1978). Gas is released from air inclusion, baking powder breakdown, or fermentation, as in the case of the enzymes in yeast breaking down starch into sugars and thence to carbon dioxide and ethanol. Additionally, vaporization of water and thermal expansion of trapped gas occur (Bloksma, 1986). The interaction of the matrix and the gas release process is complex but vital to the production of the product structure. Further manipulation of structure may be achieved by introduction of different amounts of various other materials such as bran.

The final stage is the stabilization of the structure or completion of the structuring process, principally through exhaustion of active ingredients, attainment of some pseudoequilibrium condition, and cooling and drying, associated with induction of the glassy state. Removal of water and temperature change affect the volume that may induce distortion and moisture gradients. Figure 6–7 summarizes the mobility changes associated with product formation. This is a schematic description, and examples are given in the following sections.

Baked Products

Bread baking is a traditional process in which heat, water, enzyme activity, starch,

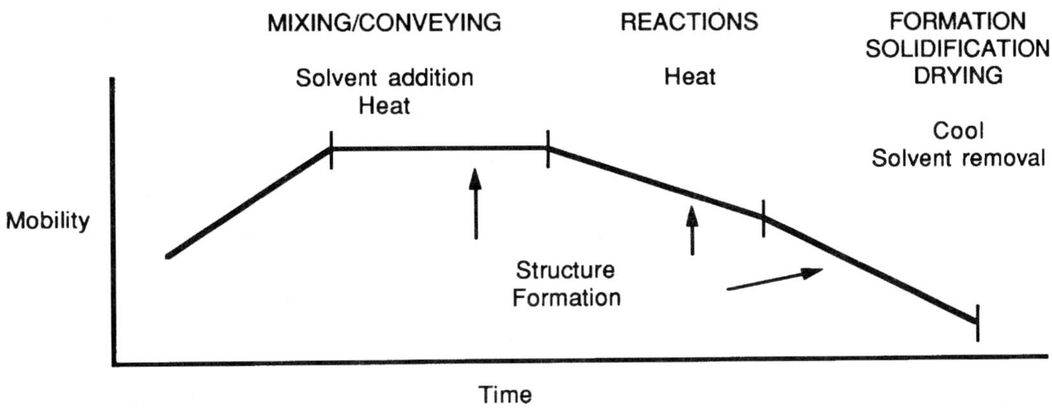

Figure 6–7 Schematic of Mobility Changes in Processing

and protein properties are the principal factors that ultimately determine bread quality (Yin & Walker, 1995). One of the major physical changes during baking is an increase in volume. This occurs through heat increasing the pressure of the gas within the cells, which include carbon dioxide, water vapor, and vaporized ethanol. Overall, baking processes make use of a number of ways of structuring, including air entrainment and gas release from enzyme action and chemical reaction (Blanshard, 1988).

In baking, starch competes for water with other components, particularly gluten. Amylolytic enzymes are active to produce maltose for yeast action, while proteolytic enzymes act on protein chains. Consequently, baked products differ in the degree of starch gelatinization depending on heat and associated water loss. In bread, the majority of starch granules have lost their birefringence, whereas low-moisture formulations such as biscuits may contain substantial amounts of ungelatinized starch (Blanshard, 1988). The starch granule may therefore contribute to the product architecture in different ways, as discussed above. The granule may remain substantially intact, it may swell and gain elastic properties, or it may gelatinize to form an amylopectin-filled matrix based on amylose with contributions from other components such as gluten, sugars, or fats. In baked products based on wheat, the gluten has a dominant role, and the effect of added sugars and fats is often to inhibit or disrupt the gluten network, leading to "softer" products (Lindley, 1988).

Sugars present in solid or liquid form affect structure of baked products as the matrix or in modifying the nucleation of gas bubbles. Sugars interact with proteins and starch and modify mechanical properties—for example, stiffening at low temperature and reducing the rubbery modulus (Kalichevsky, Blanshard, & Marsh, 1993). An increase in sugar disrupts the gluten network, which alters the structure and mechanical properties of biscuits. Sugars can solidify as crystals or glasses depending on local conditions and confer particular mechanical properties on the composite. The particle size of solid sugar as an ingredient is important for both processing and finished-product properties (Lindley, 1988). Finally, there have been studies of substituting sucrose with other sugars such as glucose and fructose.

Fats are present in baked products, where they are active in the dough preparation. Fats are hydrophobic and therefore interact with starch and proteins through limiting water or wetting and inhibit formation of a gluten network (Blanshard, 1986). Vegetable and animal fats behave differently in that the crystals of vegetable fat are more square and uniform and are better able to surround air cells to stabilize them (Jewell & Heathcock, 1988). In layered pastry, a lamination effect may be generated by construction of fat layers, which melt and form steam, lifting the rest of the dough (Bloksma, 1986). This confers anisotropic mechanical properties, whereby layers are readily separated.

To illustrate the influence of the processing route on the mechanical properties of the product as it is formed, cake batters may be heated in a rheometer. Initially, the storage modulus falls with increasing temperature from 30°C to between 70 and 80°C, corresponding to starch gelatinization and protein unfolding. The batter now has gel properties, and its storage modulus rises. Manipulation of the heating rate affects the amount of water evaporated and the ultimate storage modulus or, more generally, its stiffness (Dea, Richardson, & Ross-Murphy, 1984) (Figure 6–8). Ablett, Attenburrow, & Lillford (1986) have discussed the properties of wheat doughs in terms of the description of the gluten network as a rubber interacting with water as a swelling agent.

Breakfast Cereals

Cereal flakes such as those used in breakfast cereals may be produced from grains, grits, and flours of maize, wheat, oats, and rice. Addition of malt, salt, and sweeteners is usual. Pressure cooking is traditionally used to cook the milled or bumped grains, which raises their temperature and water content such that they may be plastically deformed by rolling. This is followed by drying and toasting to remove water, introduce some bubble structure, and change color and flavor. Extrusion cooking was introduced to replace the pressure cooker in a later version of this process. Degermed maize grits are transformed into a continuous rope of slightly expanded maize, which again is warm and moist enough to be plastically deformed by cutting into short pieces followed by rolling and toasting. The temperature trajectory associated with flake production is shown in Figure 6–9 (Parker & Smith, 1993). It shows that temperature is increased at stages in the process where the material needs to be rubbery enough to be deformed. Similar protocols are followed for wheat flakes.

Puffing of whole grains is the basis of another series of products (Fast, 1990) using gun-puffing technology, and again, extrusion technology has more recently been used to directly expand a cereal "melt." Shredded cereals are produced by rolling pressure-cooked grains or extruded dough between smooth and grooved rolls so as to produce shreds, which are then built up in layers and cut to produce a biscuit that is then baked.

Wafers

Wafers are seldom eaten alone and are often combined with foods having a contrasting texture, such as chocolate and ice cream (Beckett, Livings, & Schroeder, 1994). Wafers are formed from flour and water pastes or dispersions with small amounts of butter, sugar, salt, and sodium bicarbonate, which are mixed and then confined in preheated molds. Yeast may be used in place or in addition to bicarbonate. Pressure is generated by water evaporating to steam, and the wafer moisture content falls to a low level, typically 2%. Stevens (1976) reported the gelatiniza-

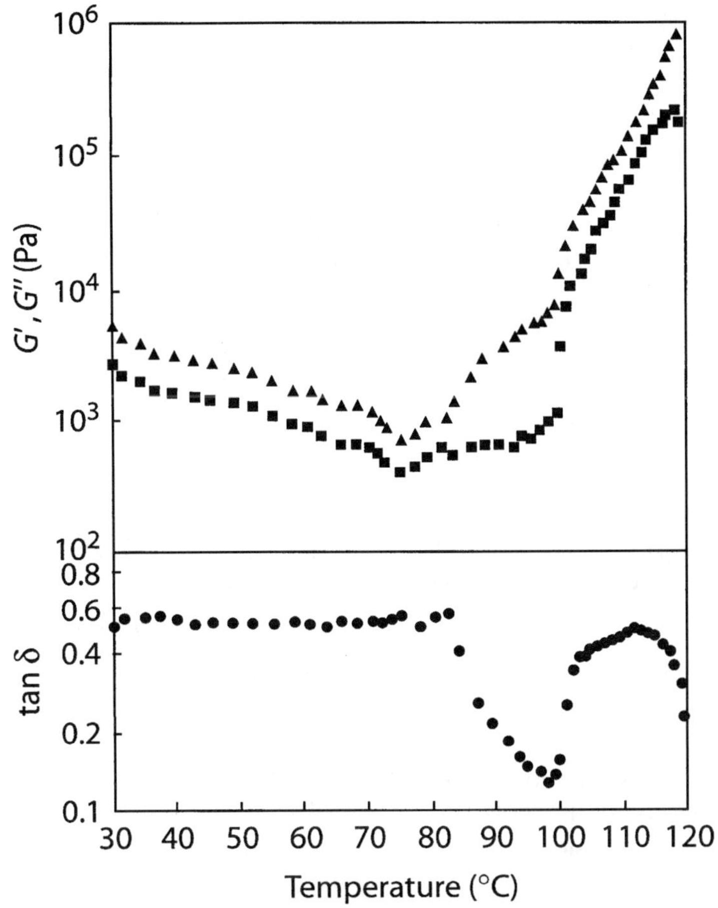

Figure 6–8 Mechanical Spectra of Batters as a Function of Temperature Relating to Cake Structure Formation. Heating rate = 3°C min^{-1}. G' (▲); G" (■); tan δ (●). *Source:* Reprinted with permission from G.D. Phillips et al., eds., *Gums and Stabilisers for the Food Industry, Vol. 11,* © 1984, S. Ross-Murphy.

tion of starch in wafers, while Huang, Luebbers, Lindamood, and Hansen (1989) found intact granules in ice cream wafer cones, which they described as starch granule–filled matrices that were porous.

Pasta

Pasta is produced from durum wheat semolina by preparation of a dough, extrusion, drying, and cooking (Cantarelli, 1986). Hydration of the wheat semolina is regulated by the grain size and uniformity of the feed, the dispersion of water, and the nature of the mixing process. High levels of moisture produce less damage to the gluten network and better starch retention. Lower extrusion temperatures produce less damage to the protein, yielding better hydration on cooking. The product leaving the die may be described as a gel containing dispersed starch granules in a gluten network. Other pasta and noodle products can also be fabricated from different raw materials, such as the flours of maize, barley, cassava, rice, and legumes (Pagani, 1986). Cooking quality is considered the most important organoleptic parameter. Cooked pasta

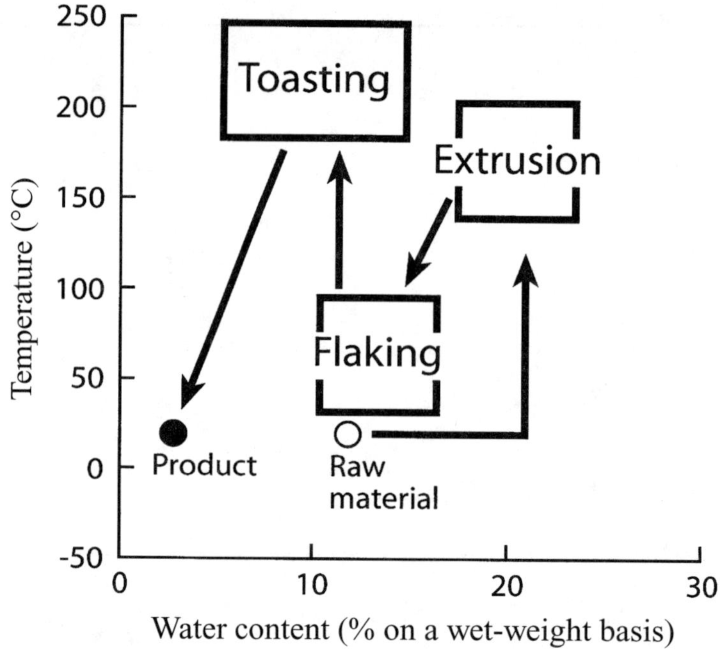

Figure 6–9 Temperature–Water Content Trajectory for the Extrusion Processing of Cereal Flakes

becomes sticky if the protein network is not strong enough to contain the gelatinized starch. The drying step is considered to be the most important unit operation in pasta manufacture, and a drying regime must be devised that avoids surface cracking (Andrieu & Stamatopoulos, 1986).

Snacks and Crispbread

Sheet and cut operations are one end use of the mixed doughs (Levine & Drew, 1990). A dough is rolled into a sheet and then cut out as desired shapes, which are then cooked by drying and baking or frying. Extrusion may provide additional mixing and pumps the dough through appropriately shaped dies. This is the basis of pasta and collet extrusion, which is the forerunner of many extrusion-cooked directly expanded products.

Expanded extrusion-cooked products are traditionally based on cereals and exploit the superheating of water under the elevated pressure conditions in the extruder (Smith, 1992). The superheated water turns to steam as the pressure drops to atmospheric when the material leaves the extruder barrel through the dies. The matrix, which is rubbery, due to the elevated temperature and moisture content, may expand. The expansion process depends on the driving force for expansion and how soon the material passes through its glass transition, thus "freezing" the structure. The extruder, and in particular the twin screw version, may achieve good mixing of the ingredients and the material microstructural degradation (see above). The expansion is enhanced through starch granule degradation and is reduced by additives that prematurely break up the expanding matrix film. Other variations are possible through additives that act as further nucleating agents, increasing the numbers of pores. Related means of causing expansion are frying and microwave heating, both of which can rapidly heat a water-containing matrix, causing a change from glassy to rubbery state, followed by expansion, cooling, and return to the glassy state, albeit with an induced lower density, porous structure.

In addition to aeration, structure creation may encompass the fabrication of composite structures by the combination of different feed streams—for example, in the production of filled extrudates using coextrusion. The extrusion foaming process may conveniently be divided into three contributions. The first is a nucleation stage during which vapor-filled bubbles form within the melt. The second is the growth of these bubbles as water vapor diffuses from the melt. Finally cooling and drying halt the growth, which may be followed by the collapse of some bubbles as the gas within them cools and contracts. It is the second bubble growth stage that has been studied most extensively. In considering the foaming of synthetic polymer melts, Han (1981) cited the growth rate in a gas-charged melt as depending on solubility and diffusivity of the gas in the melt, the interfacial tension between the gas and the melt, the melt viscosity and elasticity, the melt temperature, and the gas concentration. According to this model, water has a dual role, acting as a plasticizer to reduce the melt viscosity and as a blowing agent. The bulk density of foams increased with moisture content, whereas it might have been expected both to lower the melt viscosity and to provide a greater concentration of blowing agent. At a given moisture content, the microstructure of the maize grits is disrupted for the low-density, which are also the high-solubility extrudates (Kirby, Ollett, Parker, & Smith, 1988).

Foam formation may be achieved by means other than the phase change of water. Lai, Guetzlaff, and Hoseney (1989) have compared the extrudate structures using steam release and baking powder inclusion; Ferdinand, Clark, and Smith (1992) injected carbon dioxide directly into an extruder at low temperature conditions; and Rizvi, Mulvaney, and Sokhey (1995) combined supercritical carbon dioxide with extrusion.

STORAGE AND STABILITY

The removal of a significant amount of water such that the products are in the glassy state can lead to instabilities in the product. As mentioned in the schematic description above, removal of water may result in shrinkage and/or distortion of the product. Surface drying can result in acute moisture gradients. In pasta drying, this problem is called "case hardening," and trapped water when heated may vaporize and break through the surface. This problem can be ameliorated by use of a better controlled drying regime, possibly using radiofrequency technology (Maurer, Tremblay, & Chadwick, 1971). A similar problem that may occur with biscuits is called "checking"; it can cause stress development sufficient to cause spontaneous breakage (Wade, 1987). Baking techniques that give a more even moisture distribution permit faster cooling of biscuits. Molding stresses are a phenomenon in plastic products (Powell, 1983). They result from the rapid cooling of polymers and may be traced to the criterion that polymers should be taken through their glass transition infinitely slowly for equilibrium.

The effect of storage changes may limit shelf life. Here, they may conveniently be divided according to moisture content. For example, Stollman and Lundgren (1987) stored loaves wrapped in aluminium foil at 20°C and found changes in sensory attributes, such as a decrease in brittleness and softness and an increase in dryness. For high moisture content products such as baked products, one of the first mechanisms to be identified was starch recrystallization or retrogradation. The rate of retrogradation depends on the amount of water present (Longton & LeGrys, 1981) and is attributable to the amylopectin component, particularly the outermost short branches, which have a low melting point (Morris, 1990). A polymer science approach to crystallization was discussed above and revealed maximum crystallization midway between the Tg and Tm. The conclusion from this approach is that to slow down the rate at which bread goes stale, it should be stored below its Tg.

Other storage changes may be attributed to water content or water activity changes (e.g., Ablett et al., 1986). Low-moisture products whose Tg is below ambient would normally be described as stable. The stiffness, strength, and toughness of cereal products are highly dependent on water content, however, and any large ingress of water that depresses the glass transition temperature below the storage temperature will reduce stiffness and strength and increase toughness, as discussed below. Even if the product stays in its glassy state, there is a recent body of evidence that the properties are not absolutely time invariant. This may be understood in polymeric terms in that no product has entered the glassy state sufficiently slowly to be stable. Shogren (1992) and Thiewes and Steeneken (1997) described the changes in differential scanning calorimetry and associated these with enthalpy relaxation. Livings, Donald, and Smith (1993) also described the aging of confectionery wafers and changes in structure that occur.

STRUCTURE

Cell structure of bread has been studied in relation to texture by Day and Rogers (1996), who scanned slices of bread and processed the images by Fourier transform analysis. They calculated the "fineness" of the cells and found a good correlation with bakers' evaluations. Many cereal products are porous and may be described as solid foams. The size, shape, and distribution of the pores or bubbles is, however, different and will affect the mechanical properties and the texture. Structure is an important determinant of mechanical properties and texture, as shown in the image analysis studies of extrudates by Smolarz, van Hecke, and Bouvier (1989) and Barrett and Peleg (1992).

In the case of extruded foams, the foam structure is nonuniform. In many cases, the air cells are smaller near the surfaces, where surface tension and cooling produce edge effects, whereas centrally the pores are more uniform and bubblelike. The foams can vary in the thickness of the pore walls and in the dimensions of the pores. Manipulation of structure can be achieved by inclusion of inert particles such as bran or chalk, which affect nucleation or can break the structure prematurely. The shapes of pores can also vary, and this may be traced back to the time scale and mobility of the foaming process, which can be understood in terms of the glass transition. For example, if the foaming process has sufficient time, the pores deform and pack as polyhedral cells, whereas if the time scale is short or the structure becomes frozen while the foaming process continues, the foam will comprise spherical gas bubbles (Warburton, Donald, & Smith, 1990). In the context of extrusion cooking, a recent model of foaming by Fan, Mitchell, and Blanshard (1993) includes the contribution of matrix viscosity and glass transition and their dependence on process variables such as temperature and moisture content.

Although cereal foams are nonuniform, it is desirable to find a mathematical framework to describe foams. This would be useful if it could relate mechanical properties to structure and give a predictive capacity for at least one part of the texture "jigsaw." A framework for describing foam structure has been given by Gibson and Ashby (1988).

They modeled foams as three-dimensional arrays of identical cubic cells that were open or closed (Figure 6–10). The application of the Ashby model to extruded foam structures has been reviewed (Smith, 1989).

In addition to density and expansion, foamed solids may be characterized by cell shape and size. Some studies were directed to measuring the two-dimensional cell size distributions from low-power electron micrographs of multiple sections of extrudates (Hayter, Smith, & Richmond, 1986). With increasing density, the cells generally change from spherical to polyhedral (Warburton et al., 1990). Polyhedral cells appear to be a consequence of bubbles growing together and reaching an equilibrium structure, with some drawing of material from the cell faces to the cell edges before growth stops. Spherical cells tend to represent a rapidly set structure of growing bubbles. The volume fraction of material in the cell edges, ϕ, is given by

Equation 1

$$\phi = \frac{t_e^2}{t_e^2 + Z_f t_f l / \langle n \rangle}$$

where l is the cell size, t_e is the cell edge thickness, t_f is the cell face thickness, $\langle n \rangle$ is the mean number of edges per face, and Z_f is the face connectivity. ϕ is typically ~0.5 for polyhedral cells and ~0.02 for spherical cells (Warburton, Donald, & Smith, 1992).

Structural features have been inferred from mechanical properties studies on extrudates such as localized puncture tests (Hayter & Smith, 1988) and by Fourier transformation and calculation of the apparent fractal dimension from stress-strain curves (Barrett, Normand, Peleg, & Ross, 1992).

MECHANICAL PROPERTIES

In addition to structure, mechanical properties are seen as the principal building blocks of physical texture. The mechanical properties of starch-based foods are based on measures of stiffness, strength, and toughness. Studies of starch-based materials have been both on the product or research prototypes and on the properties of the matrix material after removal of structure. Although there is overlap with the literature on instrumental texture, some studies have focused on mechanical property measurement of actual or prototype products in engineering units. Some examples will be mentioned here. Andrieu and Stamatopoulos (1986) found a

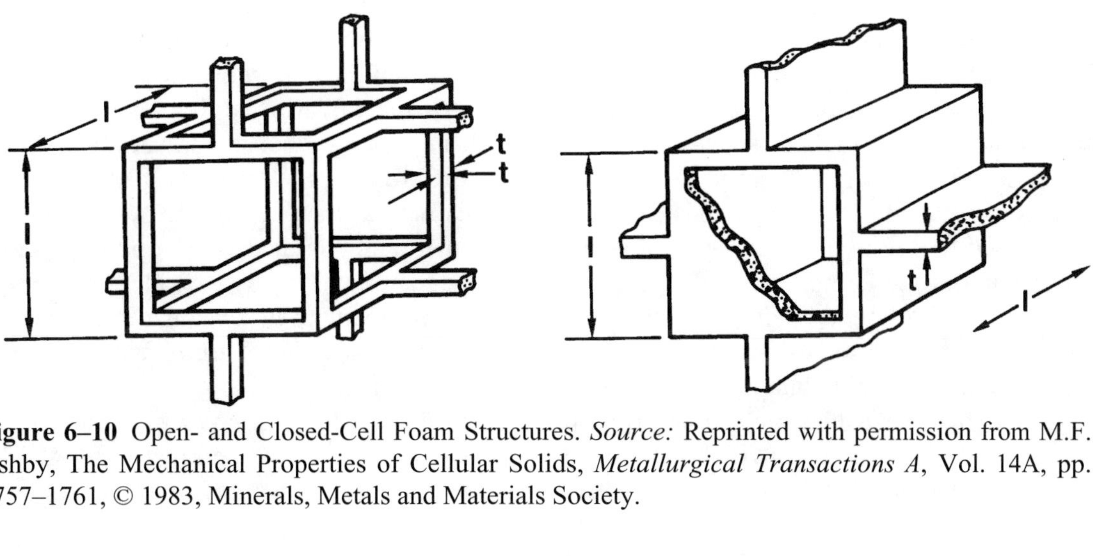

Figure 6–10 Open- and Closed-Cell Foam Structures. *Source:* Reprinted with permission from M.F. Ashby, The Mechanical Properties of Cellular Solids, *Metallurgical Transactions A*, Vol. 14A, pp. 1757–1761, © 1983, Minerals, Metals and Materials Society.

decrease in the Young's modulus of durum wheat pasta with increasing water activity, which was typical of later studies on pressed blocks of gluten or starch. Attenburrow and Davies (1993) reported a decrease in the Young's modulus of ice cream wafers with increasing moisture content. Attenburrow, Taylor, Goodband, and Lillford (1989) tested sponge cakes based on flour, sugar, egg, and emulsifier. They obtained sponges of different bulk densities by varying whisking time and adding baking powder. Cube samples were cut and equilibrated to different humidities and then tested in compression, using a universal testing machine to measure stiffness and strength. Cubic specimens of different breads were compressed between lubricated parallel plates, and the stress-strain relationship was described using three- or four-parameter models (Nussinovitch, Steffens, & Chinachoti, 1991; Swyngedau, Nussinovitch, Roy, Peleg, & Huang, 1991). Cylindrical extrudates have been tested in tension, compression, and flexure using a universal testing machine to measure stiffness and strength (Halek & Chang, 1992; Hutchinson, Siodlak, & Smith, 1987; van Hecke, Allaf, & Bouvier, 1995) and in impact using instrumented (Hayter et al., 1986) and Charpy (Kirby & Smith, 1988) pendula for strength and toughness, respectively. Georget, Parker, and Smith (1995) obtained stiffness and strength data for breakfast cereal flakes by central loading after hydration to flatten them in the rubbery state and subsequent drying as described below.

The other set of published studies has been carried out to define the base material properties without the complication of the gross structure, in particular their complex foam structure. In these studies, the specimens of simple geometry were obtained by hot pressing, or extrusion. Mechanical properties in the context of the glass transition have been studied particularly in wheat starch and amylopectin (Attenburrow, Davies, Goodband, & Ingman, 1992; Kalichevsky, Jaroszkiewicz, Ablett, et al., 1992; Nicholls, Appelqvist, Davies, Ingman, & Lillford, 1995; Ollett, Parker, & Smith, 1991) and gluten (Attenburrow, Barnes, Davies, & Ingman, 1990; Attenburrow et al., 1992; Hoseney et al., 1986; Kalichevsky, Jaroszkiewicz, & Blanshard, 1992a; Levine & Slade, 1993; Nicholls et al., 1995) as a function of water content (Figures 6–11 and 6–12). The comparison of the mixture of both components (Kalichevsky & Blanshard, 1992), as well as the effect of sugars, emulsifiers, and lipids addition to gluten or starch has been investigated (Kalichevsky & Blanshard, 1993; Kalichevsky, Jaroszkiewicz, & Blanshard, 1992a, 1992b, 1993; Ollett et al., 1991).

Ollett et al. (1991) studied the mechanical properties of unfoamed extrudates, using starch and glucose, which were then conditioned to different moisture contents. Glassy, bubble-free strips were produced from which three-point bend specimens were cut. The modulus and strength undergo an S-shaped variation with moisture content in the absence of glucose, which indicates plasticization of the starch by water. With the addition of glucose, the fall in modulus occurred at a lower moisture content, corresponding to a depression of the glass transition of starch due to monomer plasticization. The magnitude of the fall in modulus also decreased with increasing glucose content. The modulus difference between glassy and rubbery states is less than for a purely amorphous polymer and is indicative of a partially crystalline material. The increase in moisture was also accompanied by a change in the failure mode, with a general trend from fracture through tearing to yielding.

More complex mixtures have also been considered, such as breakfast cereals. The combination of wheat and other constituents in the process, principally water and sugar, alters the glass transition (Figure 6–5) and the mechanical properties of the matrix (Figure 6–11) (Georget & Smith, 1995, 1996a, 1996b). It is interesting that study of the brittle-ductile and glass transitions in food relates the macroscopic physical aspects of texture to mobility at the molecular level.

Breakfast cereal flake materials, comprising a control formulation and examples where one or more of the components had been subtracted, were milled and then molded as bars and conditioned to different water contents. A hot press technique was used to reconstitute the ground flakes as bar-shaped specimens to remove the geometry and structure effects and allow comparison of the matrix properties. Dynamic mechanical thermal analysis, together with some impact failure tests, was used to compare the response of multiple-component systems with that found for simpler one- and two-component materials that were also in the form of hot-pressed bars. In general, the decrease in

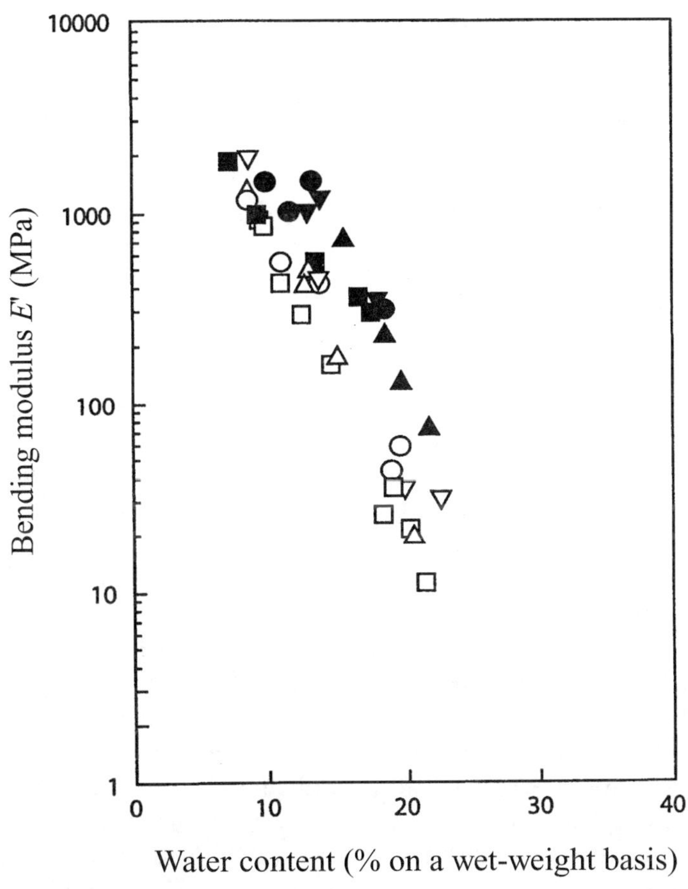

Note: Symbols defined as they are in Figure 6–5.

Figure 6–11 Bending Modulus E', as a Function of Water Content. *Source:* Reprinted from *Carbohydrate Polymers,* Vol. 28, D.M.R. Georget and A.C. Smith, Mechanical Properties of Wheat Flake Components, pp. 305–311, Copyright 1995, with kind permission from Elsevier Science Ltd., The Boulevard, Langford Lane, Kidlington OX5 1GB, UK.

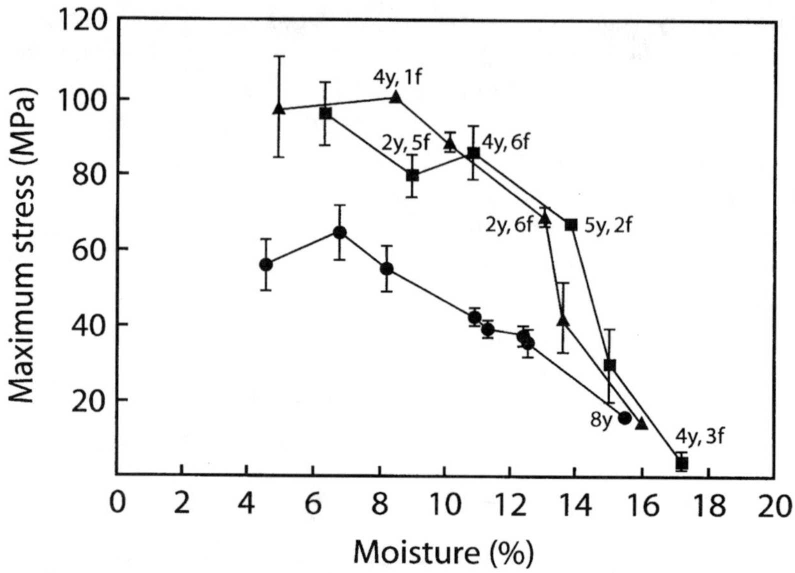

Figure 6–12 Fracture Strength of Samples in Three-Point Bend as a Function of Moisture Content (▲), Wheat Starch; (●), Gluten; and (■), Waxy Maize Starch. In some cases, not all samples failed by fracture, and these conditions are indicated adjacent to the data points by the numbers of samples failing by yielding (y) and fracture (f). *Source:* Reprinted with permission from R.J. Nichols et al., Glass Transitions and the Fracture Behaviour of Gluten and Starches Within the Glassy State, *Journal of Cereal Science*, Vol. 21, pp. 25–36, © 1995, Academic Press Ltd.

modulus is less well defined, and sugar peaks can be discerned. In related studies on flattened wheat flakes, a probe penetration technique was used (Georget et al., 1995), which showed a decrease of the Young's modulus with increasing water content and also enabled calculation of failure stress and strain. A summary of the stiffness and failure data on flakes, flattened flakes, and pressed blocks as a function of water content is given in Figure 6–13, which shows the toughness increasing and the modulus decreasing with increasing water content. Data for flakes that had been milled, sieved, and tested in compaction showed that the Heckel stress, which indicates the mechanical property of a powder, also falls with increasing water content.

These studies are useful in understanding models of foams. Gibson and Ashby used a two-dimensional analysis of honeycomb structures to relate the pore dimensions to mechanical properties (Figure 6–14). Their treatment originally considered the mechanical properties of idealized closed and open-celled foams that deform elastically, plastically, and by fracture. The original treatment considered three-dimensional cubic cells whose relative density was expressed in terms of the wall thickness and length. The deformation of the foams is understood from a two-dimensional treatment of a foam cell. Linear elastic deformation was derived by considering the cell walls to be bending beams, nonlinear behavior by treating the cell walls as Euler struts, and plastic yielding by the creation of plastic hinges at the cell wall intersections. The compressive strength of brittle foams is derived from the moment acting on the cell wall. In all cases, the mechanical property of the foam normalized by the

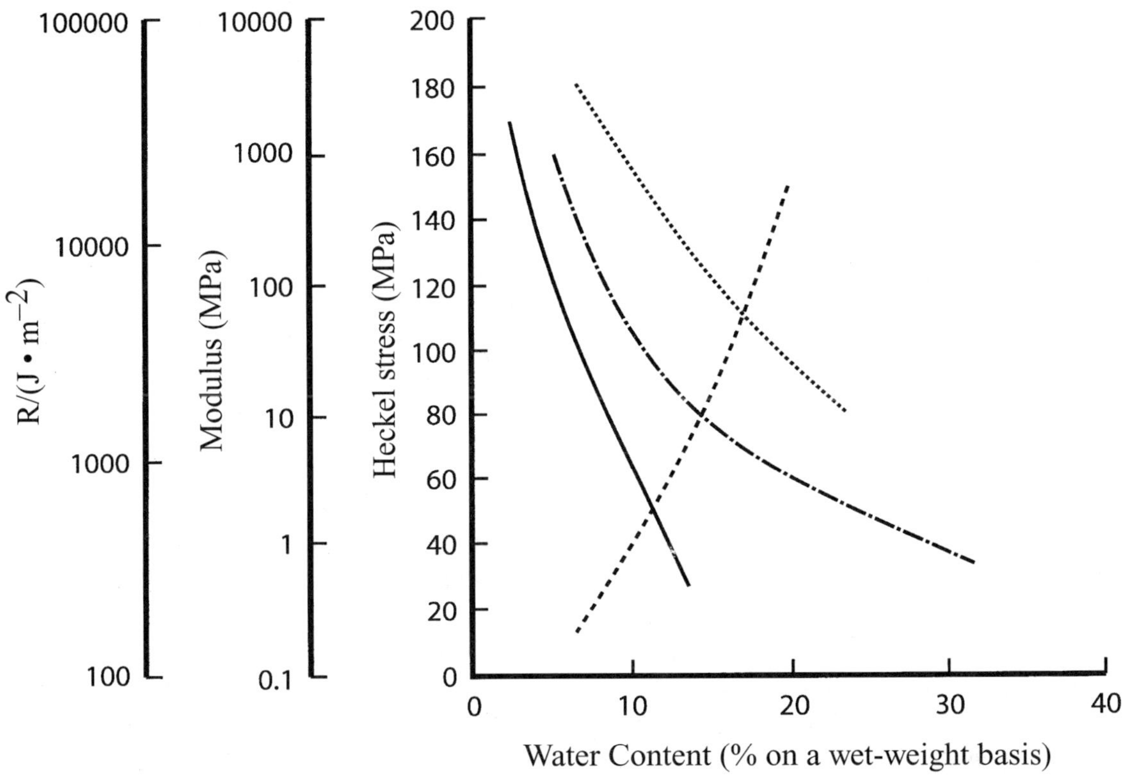

Figure 6–13 Mechanical Properties of Wheat Flakes. - - -, toughness, *R*; - · - · - · - ·, modulus of flattened flakes; ...., modulus of pressed milled flake material; ——, Heckel stress of milled flake powders tested in compaction.

wall value scales as a power of the normalized density (Figure 6–15) such as:

Equation 2

$$\sigma/\sigma_w = C\,(\rho/\rho_w)^n$$

Ashby provided evidence that the response of foamed polymers, ceramics, and metals obeyed these equations. Despite the simplifying assumptions of the treatment, it provides an elegant framework for the study of extruded foams. An immediate deviation from the model is apparent from the structure of real extrudates, which do not contain identical isotropic pores. In the case of foamed materials, σ_w in Equation 2 was identified with the bulk value. The analogous use of the mechanical property of an unfoamed food extrudate would be unrepresentative of a foam wall, not least because of the different processing conditions required in the two cases. A further, more general reservation about using the bulk value results from orientation of the material in the cell walls, which is known to affect the mechanical properties of plastics.

Initially, the basic relationship for modulus and strength of cereal extrudates with bulk density was tested on the assumption that the matrix properties were constant. Initial studies on extrudates used an approximation of

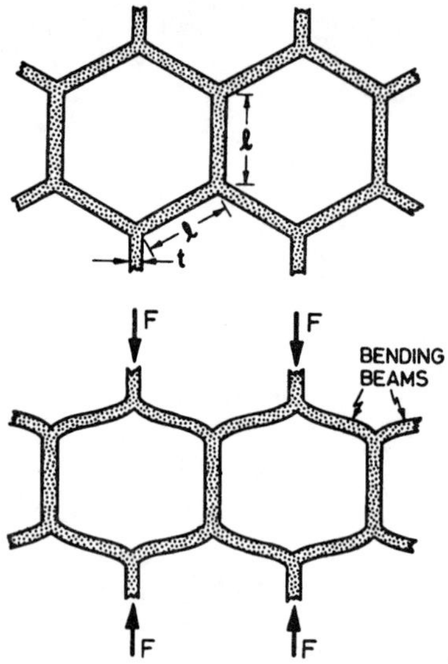

Figure 6–14 Two-Dimensional Honeycomb Model of Strength. *Source:* Reprinted with permission from M.F. Ashby, Mechanical Properties of Cellular Solids, *Metallurgical Transactions*, Vol. 14A, pp. 1757 and 1761, © 1983, Minerals, Metals and Materials Society.

Equation 2 for the case of constant wall properties, which reduces to

Equation 3

$$\sigma \propto \rho^n$$

This equation had been used earlier to describe the mechanical properties of foamed plastics. Rather than using texture-measuring techniques, studies have carried out compression, flexure, and tension tests according to British standards for plastics components. A typical relationship between strength and density is shown in Figure 6–16. The data are seen to fit Equation 3, and in general the modulus and strength of starchy foams obey this law in each of the testing modes. The power law exponent was higher for modulus than strength, in keeping with the model predictions, although the values were more typical of open-cell foams, whereas the foams

Deformation	**Open-cell Foams**	**Closed-Cell Foams**
Linear Elasticity	$\frac{E}{E_w} \propto \left[\frac{\rho}{\rho_w}\right]^2$	$\frac{E}{E_w} \propto \left[\frac{\rho}{\rho_w}\right]^3$
Elastic Collapse	$\frac{\sigma_e}{E_w} \propto \left[\frac{\rho}{\rho_w}\right]^2$	$\frac{\sigma_e}{E_w} \propto \left[\frac{\rho}{\rho_w}\right]^3$
Plastic Collapse	$\frac{\sigma_p}{\sigma_{pw}} \propto \left[\frac{\rho}{\rho_w}\right]^{3/2}$	$\frac{\sigma_p}{\sigma_{pw}} \propto \left[\frac{\rho}{\rho_w}\right]^2$
Brittle Crushing	$\frac{\sigma_f}{\sigma_{fw}} \propto \left[\frac{\rho}{\rho_w}\right]^{3/2}$	$\frac{\sigma_f}{\sigma_{fw}} \propto \left[\frac{\rho}{\rho_w}\right]^2$
Fracture Toughness	$\frac{K_{1c}}{\sigma_{fw}} \propto \left[\frac{\rho}{\rho_w}\right]^{3/2} \sqrt{1}$	$\frac{K_{1c}}{\sigma_{fw}} \propto \left[\frac{\rho}{\rho_w}\right]^2 \sqrt{1}$

Note: E = Young's modulus of foam; E_w = Young's modulus of foam wall material; σ_c = elastic collapse stress of foam; σ_p = plastic collapse stress of foam; σ_{pw} = yield strength of foam wall material; σ_f = crushing strength of foam; σ_{fw} = modulus of rupture of foam wall material; K_{lc} = fracture toughness of foam; l = foam cell size; ρ = density of foam; ρ_w = density of foam wall material.

Figure 6–15 Equations for Stiffness and Strength of Foams. *Source:* Reprinted with permission from M.F. Ashby, *Metallurgical Transactions A*, Vol. 14A, p. 1764, © 1983, ASM International.

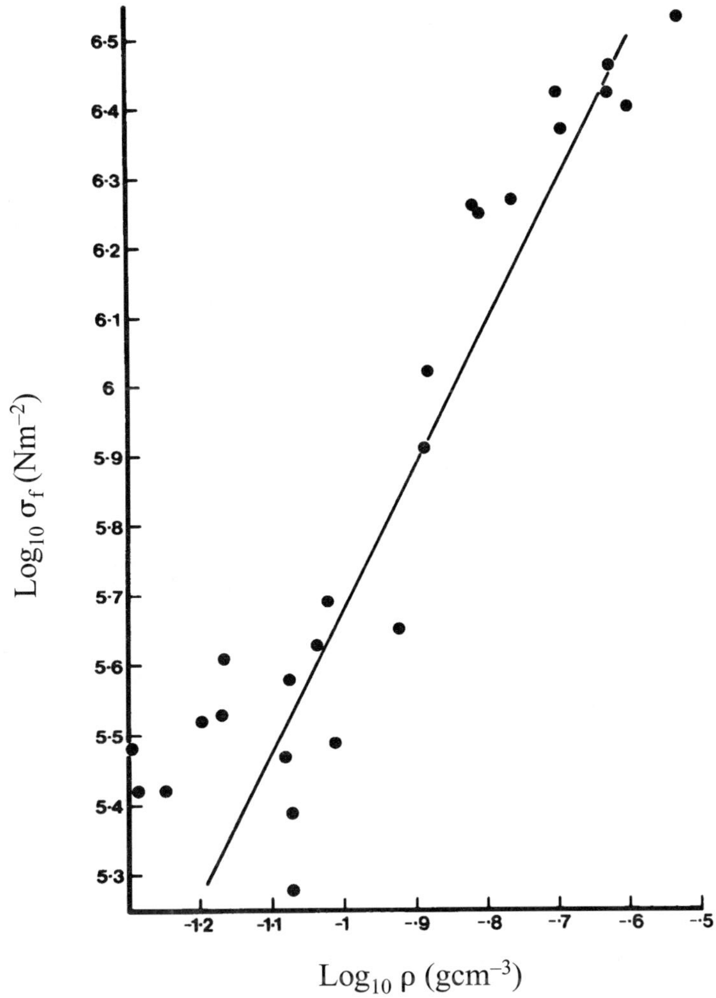

Figure 6–16 Flexural Strength, σ_f, of Extruded Maize as a Function of Bulk Density, ρ.

actually comprised closed cells. The strain rate for these measurements was typically $5 \times 10^{-3}\ s^{-1}$; hence, an instrumented pendulum technique was also used to enable high strain rate testing in compression. Although Ashby (1983) argued that often closed-cell foams behave like open-cell foams due to concentration of material in the cell edges, the experimental evidence points to extrudates being clearly of the closed cell type.

More recent treatments of Gibson and Ashby (1988) considered the different contributions to the deformation of closed-cell foams. The simple power law (Equation 2) applies for face bending only. In the more detailed treatment, it is recognized that the volume fraction of material in the cell edges, ϕ, contributes as for open-celled foams. The contribution to the modulus is therefore $\phi^2 (\rho/\rho_w)^2$ (as in the simple treatment in Figure 6–15). Bending of faces contributes $(1 - \phi)^3 (\rho/\rho_w)^3$ and stretching of faces contributes $(1 - \phi)(\rho/\rho_w)$.

Full details are given in Gibson and Ashby's excellent book (1988). This refinement of the model, which recognizes the open-cell contribution to closed-cell foams, actually fits the experimentally observed power laws better than the simple closed-cell equation (Figure 6–15). Recently, Gibson and Ashby (1988) have extended their work to include anisotropic pores and liquid-filled foams.

In experiments carried out using a Charpy pendulum, the energy was measured to break notched samples. This allowed calculation of the critical strain energy release rate G_c from K_{lc} and E in Figure 6–15:

Equation 4

$$G_c \propto (K_{lc})^2/E \propto l \times (\sigma_w^{\ 2}/E_w) \times (\rho/\rho_w)$$

For constant wall properties, Equation 4 reduces to

Equation 5

$$G_c \propto \rho l$$

This highlights the problem of defining a single pore size for an extrudate. Experimentally, G_c increases with density, as is also found for polyurethane (Kirby & Smith, 1988). In the case of toughness, the critical strain energy release rate, G_c, was related to the density and the pore size.

While the general application of the Gibson and Ashby models has been pointed out by Jeronimidis (1988), most applications have concerned extrudates. In addition to the above work in this laboratory, van Hecke et al. (1995) and Halek and Chang (1992) have considered the applicability of the Gibson and Ashby models. Halek and Chang (1992) used the hardness from Texture Profile Analysis of maize extrudates, which increased as relative density to a power of 1.4. van Hecke et al. (1995) found power laws of 1.6 and 3.1 for wheat flour mixtures extrusion cooked into flat breads, depending on composition.

Blanshard (1988) cited the work of Wheeler and Gedney on expanded gluten-starch systems as obeying these equations. Another early test of the models by Attenburrow et al. (1989) employed compressive tests on well-defined specimens of sponge cakes. Interestingly, this work comprised open-cell foams rather than the closed-cells of the extrudates, and the authors similarly tested out the scaling laws (Equation 3), assuming that the matrix properties were constant. They found general agreement with theory, although in some cases the strength was more sensitive to density than theory predicted. Amemiya and Menjivar (1992) applied these models to cookies and crackers. However, Nussinovitch et al. (1991) found no evidence for a relationship between tensile force and density for different commercial breads. It should be pointed out that a number of authors have related mechanical properties to porosity. Mizukoshi (1985) used a relationship for shear modulus for a solid filled with spherical holes defined by MacKenzic (1950), which reduced to a quadratic dependence on porosity. Good agreement with the model was found for cakes of different density produced by altering the mixer vacuum.

It is clear from Equation 2 that the foam wall mechanical properties are important in determining the foam properties. One way of causing a variation in wall properties is to vary their water content, as would occur in postextrusion conditioning. Equation 2 may be rewritten in terms of volumes:

Equation 6

$$\sigma/\sigma_w \propto (V_w/V)^n$$

where V is the volume of the foam and V_w is the volume of the walls. A variation in moisture content will change the volume ratio depending on swelling of the foam structure. It will also profoundly affect σ_w (as shown in Figures 6–11 and 6–12). Similar behavior in the foam wall will cause the foam's mechanical properties to fall dramatically at a certain moisture content. This is what is observed in practice. Hutchinson, Mantle, and Smith (1989) equilibrated wheat flour and starch foams to different water contents and found that the effect of water content could be superimposed on the power law relationship between foam strength or stiffness and foam density (Figure 6–17). The density will increase in proportion to water content but will also vary with volume changes due to swelling as a result of imbibed water. The authors found that the strength or stiffness decreased with increasing density if this was due to water uptake, indicating dominance of the matrix properties with a response typical of starchy materials (Figures 6–11 and 6–12). Interestingly, Figure 6–17 emphasizes that the response is opposite to a density increase caused by different extrusion conditions. The implications for real processing, which is a combination of unit operations, is that one process may produce a particular density and implicitly defined values of stiffness and strength but that a downstream operation, such as drying, will perturb the density and mechanical properties. Figure 6–17 is a type of physical texture "map" that shows how particular combinations of structure and mechanical properties could be obtained from multiple processes. Generalizations of this approach could be used to achieve desired structures and mechanical properties of products that could ultimately underpin sensory texture. Attenburrow et al. (1989) found a similar decrease in strength and stiffness of sponge cakes with increasing water activity.

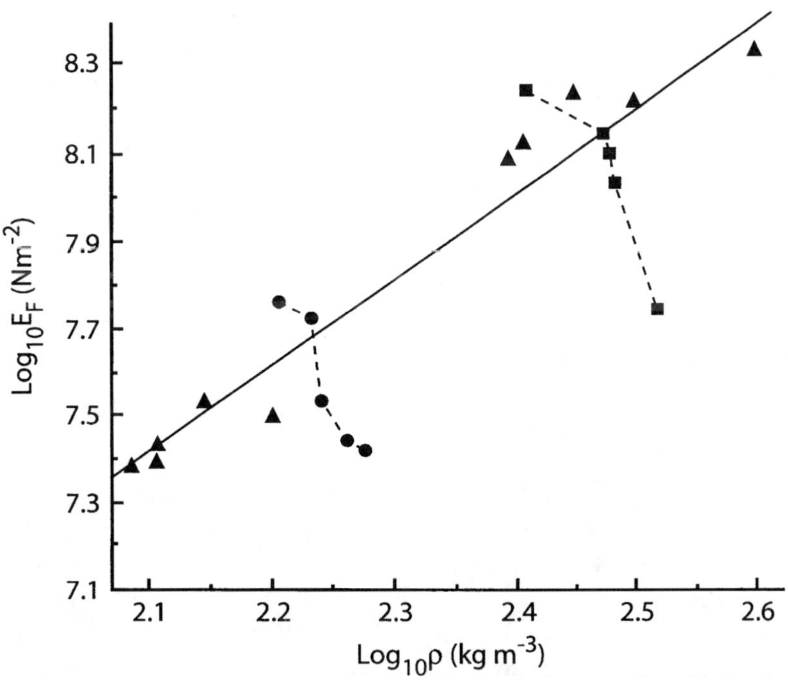

Figure 6–17 Flexural Modulus, E_F, as a Function of Density, ρ, for Extruded Wheat Flour. (▲) samples obtained by varying extrusion conditions; (●, ■) samples obtained by subsequent drying or hydration steps. *Source:* Reprinted with permission from R.J. Hutchinson, S.A. Mantle, and A.C. Smith, The Effect of Moisture Content on the Mechanical Properties of Extruded Food Foams, *Journal of Materials Science*, Vol. 24, pp. 3249–3253, © 1989, Chapman & Hall.

A related phenomenon is structural collapse (To & Flink, 1978) in which porous materials flow when the matrix is above its collapse temperature, which is related to the glass transition temperature and the alteration of mechanical properties. A low-density structure may then adopt a lower energy, higher density configuration with associated greater stiffness and strength, as predicted by Equation 2 or 6.

The role of structure in sensory texture has been discussed by Lillford (1989). In addition to the relationship between mechanical properties and porosity size described above, a greater reduction in particle size occurs with brittle foams; they then soften by hydration and to a lesser extent by the temperature rise in the mouth, and the particles are reassembled into a bolus prior to swallowing. Plastically deforming foams require larger strains before failure, and hence comminution is less likely to occur. While water plasticizes and reduces the stiffness and strength of food structures, the enzymic content of saliva will degrade starch molecules.

INSTRUMENTAL TEXTURE

Instrumental methods of texture measurement conform to three types: empirical, imitative, and fundamental (see Chapter 1 of this volume). Imitative instruments, such as the Volodkevich tenderometer, simulate the action of the jaw and teeth (Kilcast & Eves, 1991). The Brabender Amylograph for measuring the viscosity of pastes during a temperature locus also falls into the imitative category (Szczesniak, 1963).

Bread texture has been described using various instrumental tests ranging from those based on purpose-built instruments such as the Baker Compressimeter, which uses a plunger to depress a slice of bread through to Instron-based tests (Bourne, 1990). Texture Profile Analysis (see Chapter 1 of this volume) based on a "two-bite" compression test has been used to derive instrumental hardness, cohesiveness, elasticity, and chewiness, the latter being the product of the first three. For example, Brady and Mayer (1985) measured these parameters for breads and compared them with sensory parameters of the same name.

Guinot and Mathlouthi (1991) used a universal testing machine for measurement of sponge cake firmness as a function of additives and storage conditions. They tested in compression rectangular-section specimens cut from the center of the cakes. Huang et al. (1989) reported the testing of ice cream cone instrumental strength by placing a golf ball in the cone and using a compression tester to measure the force to crack the cone.

The Davis Shortometer was described some 75 years ago for brittle foods and is basically a three-point bend test suitable for biscuits and crackers (Bourne, 1990). Biscuit texture has been described by Brennan and Jowitt (1977). They used a number of bending, compression, and tensile tests on biscuits equilibrated in different humidity environments, as well as Charpy impact and an instrumental sawing test (the time taken for a circular saw to cut into a stack of biscuits to a constant depth), which is described in detail by Wade (1968). Interestingly, Wade (1968) considered the effect of biscuit aeration by comparing stacks of 20 biscuits of constant weight and found that stack height correlated with the time to cut through the biscuits. This technique was sensitive to flour protein content and the fat content of biscuit recipes.

For breakfast cereals, Szczesniak, Einstein, and Pabst (1974) described a shear cell for small samples and illustrated its use with specimens, which were tested as supplied and after a hydration step. Loh and Mannell (1990) provided a focused review of the determination of structure and mechanical properties of breakfast cereals. Sauvageot and Blond

(1991) tested single samples conditioned to different water contents and compressed by a moving blade. This yielded force-deformation data from which the initial slope, displacement to break, area under part of the curve, and number of peaks to break were obtained. A pin deformation test has been applied to testing breakfast cereal flakes. While the operation of this test is typical of many instrumental texture methods, the principal distinction here is that it is used to deflect a specimen of finite thickness rather than as an indentation test. There are circumstances when the test may be analyzed using classical mechanics to provide stresses, strains, and elastic moduli (Georget et al., 1995).

The effect of water content or water activity on the instrumental texture of cereal products has been the subject of a number of studies, ranging from studies of snacks (Katz & Labuza, 1981; Seymour & Hamann, 1988) to studies of biscuits (Zabik, Fierke, & Bristol, 1979) and breakfast cereals (Sauvageot & Blond, 1991; Szczesniak et al., 1974). The data in these publications are primarily expressed in terms of quantities derived from the force-displacement response and complement the results of rheological properties as a function of water content (see above).

ACOUSTIC EMISSION

Vickers and colleagues reviewed the evaluation of crispness and the hypothesis that auditory sensations are involved in the perception of crispness (Vickers, 1988; Vickers & Bourne, 1976). They concluded that the number of emitted sounds per unit biting distance and the loudness of the sounds changed with perceived crispness. One means of assessing sensory attributes such as crispness of baked foods is to measure the sounds produced during compression of foods. Studies have indicated that vibratory stimuli can lead to the distinguishing of crisp and crunchy foods (Vickers, 1985). Crisper sounds were higher in pitch and louder than the crunchier sounds. A good correlation between sensory crispness and acoustic emission has been reported (Attenburrow et al., 1990) (Figure 6–18). Figure 6–19 shows the acoustic emission from breaking a starch extrudate (Hayter, Kirby, & Smith, 1989).

SENSORY ATTRIBUTES

Some examples will be given below of studies that relate sensory attributes to structure, mechanical properties, and instrumental texture of starch-based foods. The practicalities of sensory textural evaluation in relation to baked products are given in Hansen and Setser (1990). Bramesco and Setser (1990) reviewed the application of sensory profiling to baked products. Attenburrow et al. (1989) found a significant correlation between sensory hardness assessed by a trained panel and either stiffness or strength of cakes (Figure 6–20). Similarly, Brennan and Jowitt (1977) found that their instrumental results above correlated well with sensory crispness and hardness. The Charpy impact test and the biscuit saw test were claimed to be the most reproducible. Sensory tests revealed that textural crispness was more sensitive but less reproducible than hardness to changes in moisture content.

Bread texture, as might be expected, has been extensively studied. Brady and Mayer (1985) examined rye and French breads and gave sensory and instrumental (using Texture Profile Analysis) definitions of hardness, cohesiveness, elasticity, and chewiness. Cohesiveness and chewiness were significantly correlated for rye bread but only chewiness for French bread. Stollman and Lundgren (1987) studied the sensory and instrumental texture of crumb, crust, and whole slices of white bread. They found a strong, negative, nonlinear relationship between softness and

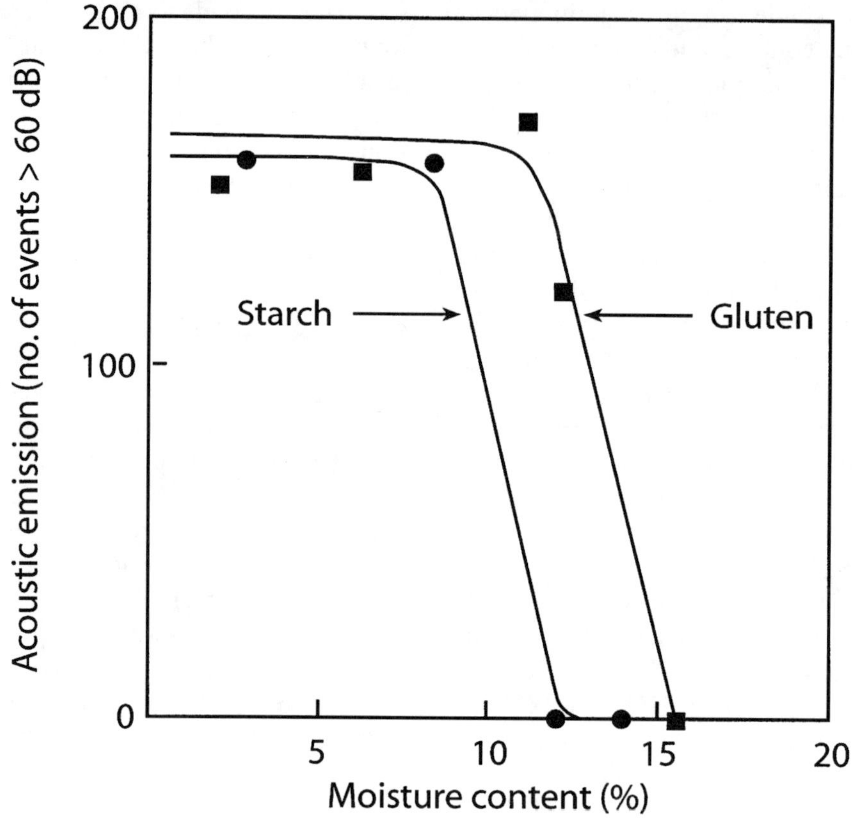

Figure 6–18 Acoustic Emission as a Function of Moisture Content from Starch and Gluten. *Source:* Reprinted with permission from G.E. Attenburrow, The Fracture Behaviour of Starch and Gluten in the Glassy State, *Journal of Cereal Science*, Vol. 16, pp. 1–12, © 1992, Academic Press Ltd.

compressive force for the bread crumb (Figure 6–21). The instrumental test was able to show significant differences between samples where none was found in the sensory tests, as also reported by Brady and Mayer (1985).

Sensory studies on extruded prototypes are represented by the studies of Barrett, Cardello, Lesher, and Taub (1994) and Hu, Huff, Heymann, and Hsieh (1996) on corn extrudates and Faller and Heymann (1996) on dehydrated potato. Barrett et al. considered sensory attributes of crispness, hardness, and crunchiness, which correlated with bulk density, although "perceived density" did not correlate with bulk density. The latter two studies considered the effects of additives. Hu et al. showed that soy fiber (Figure 6–22) and emulsifier significantly affected most of the sensory attributes studied. Faller and Heymann used generalized Procrustes analysis to show that harder samples were chewier but less easy to fracture. In comparing sensory and physical properties, they found that instrumental breaking strength was highly positively correlated with sensory chewiness and crispness and negatively correlated with sensory fracturability.

Bramesco and Setser (1990) considered moisture sorption in relation to mouth dryness, cohesiveness of cereal product types, and adhesion to teeth. They emphasized the

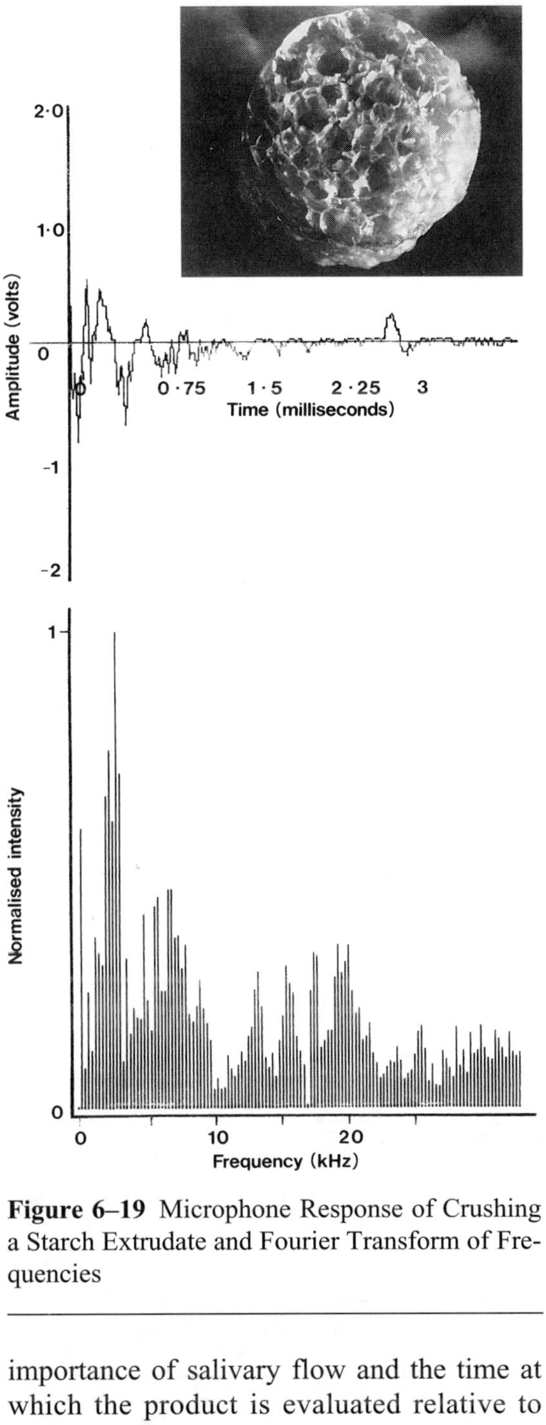

Figure 6–19 Microphone Response of Crushing a Starch Extrudate and Fourier Transform of Frequencies

importance of salivary flow and the time at which the product is evaluated relative to time in the mouth.

Finally, a comparison of the response of a starchy product with other foods is made in the context of a generalized three-dimensional model for in-mouth perceived texture involving degree of structure, degree of lubrication, and time (Hutchings & Lillford, 1988). As a food is changed in the mouth, it describes its own breakdown path. A dry sponge cake is easily broken down but cannot be swallowed immediately because of insufficient lubrication. A biscuit absorbs released saliva as well as moisture in the mouth; and consequently, the degree of lubrication decreases, then increases. In contrast, an orange loses liquid when deformed but after expression of the juice, the matrix cell wall material is perceived to be drier (Figure 6–23).

CONCLUSION

In the context of starch-based products, the starch structure may exist in a number of forms, ranging from native granules to degraded oligomers. For low-moisture starch products, the biopolymer approach to thermal events such as melting and glass-rubber transitions has reconciled much apocryphal data and given new insights into the prediction of phenomena such as staling and storage stability. It also indicates the propensity to brittle or ductile behavior of the matrix of a processed starch product.

The matrix is, however, only part of the story, since structure is the other key factor in understanding overall mechanical properties. At present, there are few predictive approaches applicable to the mechanical properties of a structured food product. Again the principles of materials science of composites indicate a way forward. Models of solid foams have been cited as a means of predicting mechanical properties of aerated products based on starch. Similarly, much may be learned from considering composites, which comprise a matrix that is stiffened with or toughened by inclusion of a second component. A consolidation of models at molecular and macroscopic scales offers a

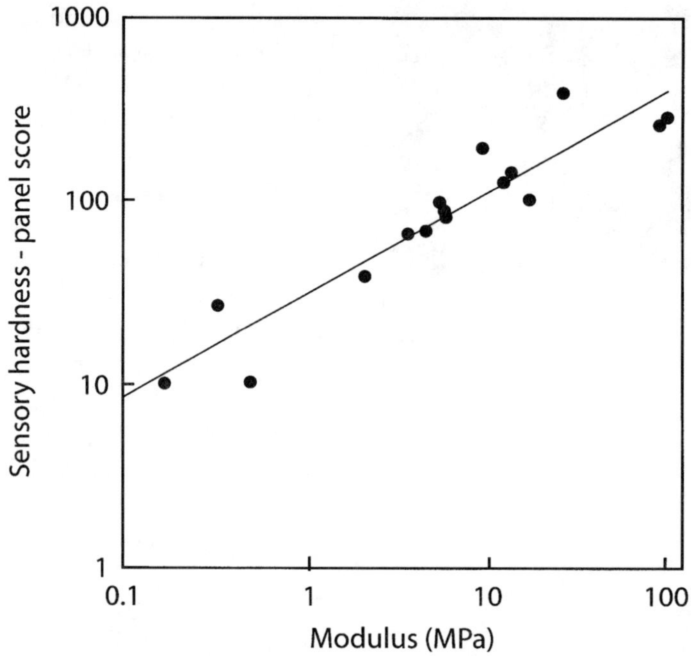

Figure 6–20 Correlation between Sensory Hardness and Mechanical Stiffness for Sponge Cakes. *Source:* Reprinted with permission from G.E. Attenburrow et al., Structure Mechanics and Texture of a Food Sponge, *Journal of Cereal Science*, Vol. 9, pp. 61–70, © 1989, Academic Press Ltd.

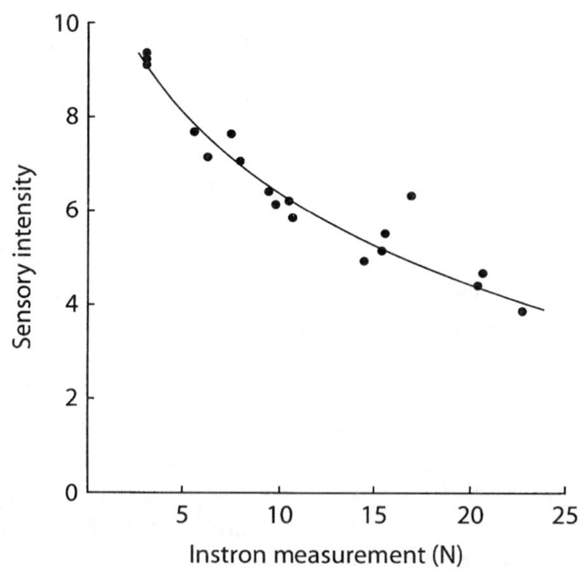

Figure 6–21 Sensory-Evaluated Softness and Compressive Force for Bread Crumb Using an Instron Universal Machine. *Source:* Reprinted with permission from V. Stollman and B. Lundgren, Texture Changes in White Bread: Effects of Processing and Storage, *Cereal Chemistry*, Vol. 64, No. 4, pp. 230–236, © 1987, American Association of Cereal Chemists.

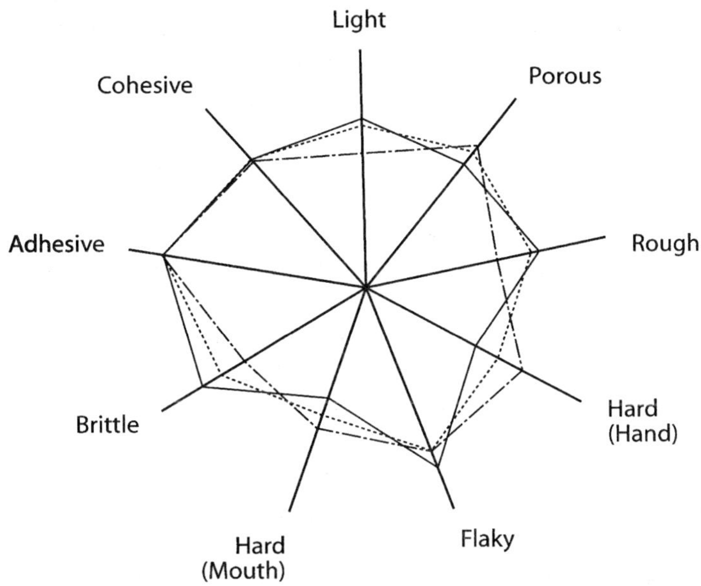

Figure 6–22 Polar Coordinate Graph of the Adjusted Means of the Panel Scores for the Nine Sensory Attributes for Corn-Based Extrudates. *Source:* Reprinted with permission from L. Hu, H.E. Huff, H. Heymann, and F. Hsieh, Effects of Emulsifier and Soy Fiber Addition on Sensory Properties of Corn Meal Extrudate, *Journal of Food Quality*, Vol. 19, pp. 57–77., © 1996, Food & Nutrition Press, Inc.

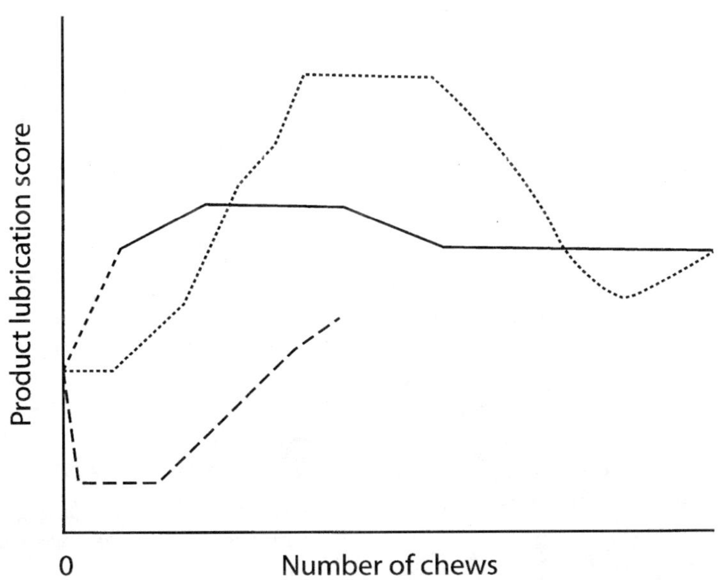

Figure 6–23 Changes in Perceived Product Lubrication during Progressive Chewing. Solid line (—), apple; dotted line (..........), orange; dashed line (- - - -), dry biscuit. *Source:* Reprinted with permission from J.B. Hutchings and P.J. Lillford, The Perception of Food Texture—The Philosophy of the Breakdown Path, *Journal of Texture Studies*, Vol. 19, pp. 103–115, © 1988, Food & Nutrition Press, Inc.

long-term approach to understanding and predicting physical and ultimately sensory texture.

The link between physical and sensory texture is, however, much more difficult to make, and the required mechanical properties in terms of toughness, stiffness, or strength under particular stress states and strain rates are not known or specified. It has also been pointed out that single-point measurements are rarely adequate. There is an argument in favor of instrumental tests that mimic sensory properties well, and their subsequent analysis in terms of basic mechanical properties is another approach (Bourne, 1994). It is perhaps apposite to cite Peleg (1983), who commented on the misleading similarity in language used by instrumentalists and panelists such that the same words are used but measure different properties.

Compared to proteins, starch is often seen as a simpler system, and much research effort has been directed to pure starch. However, the interactions of starch with other natural constituents or added components in processed cereals or tubers will affect their textural properties. The science to deal with complex mixtures and the heterogeneity in the structure of processed starchy materials remains a challenge to existing approaches.

REFERENCES

Ablett, S., Attenburrow, G.E., & Lillford, P.J. (1986). The significance of water in the baking process. In J.M.V. Blanshard, P.J. Frazier, & T. Galliard (Eds.), *Chemistry and physics of baking* (pp. 30–41). London: Royal Society of Chemistry.

Amemiya, J., & Menjivar, J.A. (1992, September). *Mechanical properties of cereal based food cellular systems*. Paper presented at the 77th annual meeting of the American Association of Cereal Chemists, Minneapolis, MN.

Anderson, R.A., Conway, H.F., Pfeifer, V.F., & Griffin, L.E.J. (1969). Gelatinisation of corn grits by roll- and extrusion-cooking. *Cereal Science Today, 14*, 4–7, 11–12.

Andrieu, J., & Stamatopoulos, A. (1986). Durum wheat pasta drying kinetics. *Food Science and Technology, 19*, 211–214.

Ashby, M.F. (1983). The mechanical properties of cellular solids. *Metallurgical Transactions A, 14*, 1755–1769.

Attenburrow, G., Barnes, D.J., Davies, A.P., & Ingman, S.J. (1990). Rheological properties of wheat gluten. *Journal of Cereal Science, 12*, 1–14.

Attenburrow, G., & Davies, A.P. (1993). The mechanical properties of cereal based foods. In J.M.V. Blanshard & P.J. Lillford (Eds.), *The glassy state in foods* (pp. 317–331). Nottingham, UK: Nottingham University Press.

Attenburrow, G., Davies, A.P., Goodband, R.M., & Ingman, S.J. (1992). The fracture behaviour of starch and gluten in the glassy state. *Journal of Cereal Science, 16*, 1–12.

Attenburrow, G.E., Taylor, L.J., Goodband, R.M., & Lillford, P.J. (1989). Structure, mechanics and texture of a food sponge. *Journal of Cereal Science, 9*, 61–70.

Barrett, A., Cardello, A.V., Lesher, L.L., & Taub, I.A. (1994). Cellularity, mechanical failure, and textural perception of corn meal extrudates. *Journal of Texture Studies, 25*, 77–95.

Barrett, A.M., Normand, M.D., Peleg, M., & Ross, E. (1992). Characterization of the jagged stress-strain relationships of puffed extrudates using the fast Fourier transform and fractal analysis. *Journal of Food Science, 57*, 227–232, 235.

Barrett, A.M., & Peleg, M. (1992). Extrudate cell structure-texture relationships. *Journal of Food Science, 57*, 1253–1257.

Beckett, S., Livings, S., & Schroeder, P. (1994). Edible wafers need physics. *Physics World, 7*(9), 49–53.

Blanshard, J.M.V. (1986). The significance of the structure and function of the starch granule in baked products. In J.M.V. Blanshard, P.J. Frazier, & T. Galliard (Eds.), *Chemistry and physics of baking* (pp. 1–13). London: Royal Society of Chemistry.

Blanshard, J.M.V. (1988). Elements of cereal product structure. In J.M.V. Blanshard & J.R. Mitchell (Eds.), *Food structure: Its creation and evaluation* (pp. 313–330). London: Butterworths.

Bloksma, A.H. (1980). Effect of heating rate on viscosity of wheat flour doughs. *Journal of Texture Studies, 10*, 261–269.

Bloksma, A.H. (1986). Rheological aspects of structural changes during baking. In J.M.V. Blanshard, P.J. Frazier, & T. Galliard (Eds.), *Chemistry and physics of baking* (pp. 170–178). London: Royal Society of Chemistry.

Bourne, M.C. (1990). Practical texture measurements of cereal foods. In H. Faridi. & J.M. Faubion (Eds.), *Dough rheology and baked product texture* (pp. 557–571). New York: Van Nostrand Reinhold.

Bourne, M.C. (1994). Converting from empirical to rheological tests on foods: It's a matter of time. *Cereal Foods World, 39*(1), 37–39.

Brady, P.L., & Mayer, S.M. (1985). Correlation of sensory and instrumental measures of bread texture. *Cereal Chemistry, 62,* 70–72.

Bramesco, N.P. & Setser, C.S. (1990). Application of sensory texture profiling to baked products: Some considerations for evaluation, definition of parameters and reference products. *Journal of Texture Studies, 21*, 235–251.

Brennan, J.G., & Jowitt, R. (1977). Some factors affecting the objective study of food texture. In G.G. Birch, J.G. Brennan, & K.J. Parker (Eds.), *Sensory properties of foods* (pp. 227–245). London: Applied Science.

Cairns, P., Miles, M.J., & Morris, V.J. (1991). Studies of the effect of sugars ribose, xylose and fructose on the retrogradation of wheat starch gels by X ray diffraction. *Carbohydrate Polymers, 16*, 355–365.

Cantarelli, C. (1986). Technological development in the pasta industry. In C. Mercier & C. Cantarelli (Eds.), *Pasta and extrusion cooked foods* (pp. 33–51). London: Elsevier.

Christianson, D.D., & Bagley, E.B. (1983). Apparent viscosities of dispersions of swollen cornstarch granules. *Cereal Chemistry, 60*, 116–121.

Day, D.D., & Rogers, D. (1996). Fourier-based texture measures with application to the analysis of the cell structure of baked products. *Digital Signal Processing, 6*, 138–144.

Dea, I.C.M., Richardson, R.K., & Ross-Murphy, S.B. (1984). Characterisation of rheological changes during the processing of food materials. In G.O. Phillips, D.J. Wedlock, & P.A. Williams (Eds.), *Gums and stabilisers for the food industry* Vol. 2, (pp. 357–366). Oxford, UK: Pergamon.

Donovan, J.W. (1979). Phase transitions of the starch-water system. *Biopolymers, 18*, 263–275.

Faller, J.Y., & Heymann, H. (1996). Sensory and physical properties of extruded potato puffs. *Journal of Sensory Studies, 11*, 227–245.

Fan, J., Mitchell, J.R., & Blanshard, J.M.V. (1993). A computer simulation of the dynamics of bubble growth and shrinkage during extrudate expansion. *Journal of Food Engineering, 24*, 337–356.

Fast, R.B. (1990). Manufacturing technology of ready-to-eat cereals. In R.B. Fast & E.F. Caldwell (Eds.), *Breakfast cereals and how they are made* (pp. 15–42). St. Paul, MN: American Association of Cereal Chemists.

Ferdinand, J.M., Clark, S.A., & Smith, A.C. (1992). Structure formation in extrusion cooked starch-sucrose mixtures by carbon dioxide injection. *Journal of Food Engineering, 16*, 283–291.

Georget, D.M.R., Parker, R., & Smith, A.C. (1995). An assessment of the pin deformation test for measurement of mechanical properties of breakfast cereal flakes. *Journal of Texture Studies, 26*, 161–174.

Georget, D.M.R., & Smith, A.C. (1995). The mechanical properties of wheatflake components. *Carbohydrate Polymers, 24*, 305–311.

Georget, D.M.R., & Smith, A.C. (1996a). The mechanical properties of processed cereals. *Journal of Thermal Analysis, 47*, 1377–1389.

Georget, D.M.R., & Smith, A.C. (1996b). A study of the effect of water content on the mechanical texture of breakfast wheat flakes. In G.R. Fenwick, C. Hedley, R.L. Richards, & S. Khokhar (Eds.), *Agri-food quality* (pp. 196–199). London: Royal Society of Chemistry.

Gibson, L.J., & Ashby, M.F. (1988). *Cellular solids: Structure and properties*. Oxford, UK: Pergamon.

Giesekus, H. (1983). Disperse systems: Dependence of rheological properties on the type of flow with implications for food rheology. In R. Jowitt, F. Escher, B. Hallström, H.F.T. Meffert, W.E.L. Spiess, & G. Vos (Eds.), *Physical properties of foods* (pp. 205–220). London: Applied Science.

Gomez, M.H., & Aguilera, J.M. (1984). A physico-chemical model for extrusion of corn starch. *Journal of Food Science, 49*, 40–43, 63.

Greenwood, C.T. (1976). Starch. In Y. Pomerantz (Ed.), *Advances in cereal science and technology* (pp. 119–157). St. Paul, MN: American Association of Cereal Chemists.

Guinot, P., & Mathlouthi, M. (1991). Instron measurement of sponge cake firmness: Effect of additives and storage conditions. *Journal of Science of Food and Agriculture, 54*, 413–420.

Halek, G.W., & Chang, K.L.B. (1992). Effect of extrusion operation variables on functionality of extrudates. In J.L. Kokini, C.-T. Ho., & M.V. Karwe (Eds.), *Food extrusion science and technology* (pp. 677–691). New York: Marcel Dekker.

Han, C.D. (1981). *Multiphase flow in polymer processing.* New York: Academic Press.

Hansen, L.M., & Setser, C.S. (1990). Texture evaluation of baked products using descriptive sensory analysis. In H. Faridi & J.M. Faubion (Eds.), *Dough rheology and baked product texture* (pp. 573–596). New York: Van Nostrand Reinhold.

Hayter, A.-L., Kirby, A.R., & Smith, A.C. (1989). The physical properties of extruded food foams. In H.G. Ang (Ed.), *Trends in food processing: 1* (pp. 265–274). Singapore: Singapore Institute of Food Science and Technology.

Hayter, A.-L., & Smith, A.C. (1988). The mechanical properties of extruded food foams. *Journal of Materials Science, 23*, 736–743.

Hayter, A.-L., Smith, A.C., & Richmond, P. (1986). The physical properties of extruded food foams. *Journal of Materials Science, 21*, 3729–3736.

Hoseney, R.C., Zeleznak, K.J., & Lai, C.S. (1986). Wheat gluten: A glassy polymer? *Cereal Chemistry, 63*, 285–286.

Hu, L., Huff, H.E., Heymann, H., & Hsieh, F. (1996). Effects of emulsifier and soy fiber addition on sensory properties of corn meal extrudate. *Journal of Food Quality, 19*, 57–77.

Huang, V.T., Luebbers, S.T., Lindamood, J.B., & Hansen, P.M.T. (1989). Ice cream cone baking: 2. Textural characteristics of rolled sugar cones. *Food Hydrocolloids, 3*, 41–55.

Hutchings, J.B., & Lillford, P.J. (1988). The perception of food texture: The philosophy of the breakdown path. *Journal of Texture Studies, 19*, 103–115.

Hutchinson, J.M. (1995). Physical aging of polymers. *Progress in Polymer Science, 20*, 703–760.

Hutchinson, R.J., Mantle, S.A., & Smith, A.C. (1989). The effect of moisture content on the mechanical properties of extruded food foams. *Journal of Materials Science, 24,* 3249–3253.

Hutchinson, R.J., Siodlak, G.D.E., & Smith, A.C. (1987). Influence of processing variables on the mechanical properties of extruded maize. *Journal of Materials Science, 22*, 3956–3962.

I'Anson, K.J., Miles, M.J., Morris, V.J., Bedford, L.S., Jarvis, D.A., & Marsh, R.A. (1991). The effects of added sugars on the retrogradation of wheat starch gels. *Journal of Cereal Science, 11*, 243–248.

Jeronimidis, G. (1988). Structure and properties of liquid and solid foams. In J.M.V. Blanshard & J.R. Mitchell (Eds.), *Food structure: Its creation and evaluation* (pp. 59–74). London: Butterworths.

Jewell, G.G., & Heathcock, J.F. (1988). Structured fat systems. In J.M.V. Blanshard & J.R. Mitchell (Eds.), *Food structure: Its creation and evaluation* (pp. 279–295). London: Butterworths.

Kalichevsky, K.T., & Blanshard, J.M.V. (1992). A study of the effect of water on the glass transition of 1:1 mixtures of amylopectin, casein and gluten using DSC and DMTA. *Carbohydrate Polymers, 19*, 271–278.

Kalichevsky, K.T, & Blanshard, J.M.V. (1993). The effect of fructose and water on the glass transition of amylopectin. *Carbohydrate Polymers, 20*, 107–113.

Kalichevsky, K.T., Blanshard, J.M.V., & Marsh, R.D.L. (1993). In J.M.V. Blanshard & P.J. Lillford (Eds.), *The glassy state in foods* (pp. 133–156). Nottingham, UK: Nottingham University Press.

Kalichevsky, K.T., Jaroszkiewicz, E.M., Ablett, S., Blanshard, J.M.V., & Lillford, P.J. (1992). The glass transition of amylopectin measured by DSC, DMTA and NMR. *Carbohydrate Polymers, 18*, 77–88.

Kalichevsky, K.T., Jaroszkiewicz, E.M., & Blanshard, J.M.V. (1992a). Glass transition of gluten. 1: Gluten and gluten-sugar mixtures. *International Journal of Biological Macromolecules, 14*, 257–266.

Kalichevsky, K.T., Jaroszkiewicz, E.M., & Blanshard, J.M.V. (1992b). Glass transition of gluten. 2: The effect of lipids and emulsifiers. *International Journal of Biological Macromolecules, 14*, 267–273.

Kalichevsky, K.T., Jaroszkiewicz, E.M., & Blanshard, J.M.V. (1993). A study of the glass transition of amylopectin-sugar mixtures. *Polymer, 34*, 346–358.

Katz, E., & Labuza, T.P. (1981). Effect of water activity on the sensory crispness and mechanical deformation of snack food products. *Journal of Food Science, 46*, 403–409.

Kilcast, D., & Eves, A. (1991). Integrating texture and physiology: Techniques. In J.F.V. Vincent & P.J. Lillford (Eds.), *Feeding and the texture of food* (pp. 167–183). Cambridge, UK: Cambridge University Press.

Kirby, A.R., Ollett, A.L., Parker, R., & Smith, A.C. (1988). An experimental study of screw configuration effects in the twin-screw extrusion-cooking of maize grits. *Journal of Food Engineering, 8*, 247–272.

Kirby, A.R., & Smith, A.C. (1988). Impact studies on extruded food foams. *Journal of Materials Science, 23*, 2251–2254.

Lai, C.S., Guetzlaff, J., & Hoseney, R.C. (1989). Role of sodium bicarbonate and trapped air in extrusion. *Cereal Chemistry, 66*, 69–73.

Lawton, J.W., & Wu, Y.V. (1993). Thermal behavior of annealed acetic acid-soluble wheat gluten. *Cereal Chemistry, 70*, 471–475.

Lelievre, J. (1973). Starch gelatinization. *Journal of Applied Polymer Science, 18*, 293–296.

Levine, H., & Slade, L. (1990). Influences of the glassy and rubbery states on the thermal, mechanical and structural properties of doughs and baked products. In H. Faridi & J.M. Faubion (Eds.), *Dough rheology and baked product texture* (pp. 157–330). New York: Van Nostrand Reinhold.

Levine, H., & Slade, L. (1993). The glassy state in applications for the food industry, with an emphasis on cookie and cracker production. In J.M.V. Blanshard & P.J. Lillford (Eds.), *The glassy state in foods* (pp. 333–373). Nottingham, UK: Nottingham University Press.

Levine, L., & Drew, B.A. (1990). Rheological and engineering aspects of the sheeting and laminating of doughs. In H. Faridi. & J.M. Faubion (Eds.) *Dough rheology and baked product texture* (pp. 513–555). New York: Van Nostrand Reinhold.

Lillford, P.J. (1989). Structure and properties of solid food foams. In A.J. Wilson (Ed.), *Foams: Physics, chemistry and structure* (pp. 149–166). Berlin: Springer Verlag.

Lindley, M.G. (1988). Structured sugar systems. In J.M.V. Blanshard and J.R. Mitchell (Eds.), *Food structure. Its creation and evaluation* (pp. 297 311). London: Butterworths.

Livings, S.J., Donald, A.M., & Smith, A.C. (1993). Ageing in confectionery wafers. In J.M.V. Blanshard & P.J. Lillford (Eds.), *The glassy state in foods* (pp. 507–511). Nottingham, UK: Nottingham University Press.

Loh, J., & Mannell, W. (1990). Application of rheology in the breakfast cereal industry. In H. Faridi & J.M. Faubion (Eds.), *Dough rheology and baked product texture* (pp. 405–420). New York: Van Nostrand Reinhold.

Longton, J., & LeGrys, G.A. (1981). Differential scanning calorimetry studies on the crystallinity of ageing wheat starch gels. *Die Staerke, 33*, 410–414.

MacKenzie, J.K. (1950). The elastic constants of a solid containing spherical holes. *Proceedings of the Physical Society, B63*, 2–11.

Maurer, R.L., Tremblay, M.R., & Chadwick, E.A. (1971). Microwave processing of pasta. *Food Technology, 25*, 1244–1249.

Mizukoshi, M. (1985). Model studies of cake baking. V: Cake shrinkage and shear modulus of cake batter during baking. *Cereal Chemistry, 62*, 242–246.

Morris, V.J. (1990). Starch gelation and retrogradation. *Trends in Food Science and Technology, 1*, 2–6.

Nicholls, R.J., Appelqvist, I.A.M., Davies, A.P., Ingman, S.J., & Lillford, P.J. (1995). Glass transitions and the fracture behaviour of gluten and starches within the glassy state. *Journal of Cereal Science, 21*, 25–36.

Nussinovitch, A., Steffens, M.S., & Chinachoti, P. (1991). Exponential model of the compressive stress-strain relationships of bread. *Food Science and Technology, 24*, 266–269.

Ollett, A.-L., Parker, R., & Smith, A.C. (1991). Deformation and fracture behaviour of wheat starch plasticised with water and glucose. *Journal of Materials Science, 26*, 1351–1356.

Pagani, M.A. (1986). Pasta products from non conventional raw materials. In C. Mercier & C. Cantarelli (Eds.), *Pasta and extrusion cooked foods* (pp. 52–68). London: Elsevier.

Parker, R., & Smith, A.C. (1993). The mechanical properties of starchy food materials at large strains and their ductile-brittle transitions. In J.M.V. Blanshard & P.J. Lillford (Eds.), *The glassy state in foods* (pp. 519–522). Nottingham, UK: Nottingham University Press.

Peleg, M. (1983). The semantics of rheology and texture. *Food Technology, 37*(11), 54–61.

Powell, P.C. (1983). *Engineering with polymers*. London: Chapman & Hall.

Remsen, C.H., & Clark, J.P. (1978). A viscosity model for a cooking dough. *Journal of Food Process Engineering, 2*, 39–64.

Rizvi, S.S.H., Mulvaney, S.J., & Sokhey, A.S. (1995). The combined application of supercritical fluid and extrusion technology. *Trends in Food Science and Technology, 6*, 232–240.

Sauvageot, F., & Blond, G. (1991). Effect of water activity on crispness of breakfast cereals. *Journal of Texture Studies, 22*, 423–442.

Seymour, S., & Hamann, D. (1988). Crispness and crunchiness of selected low moisture foods. *Journal of Texture Studies, 19*, 79–95.

Shogren, R.L. (1992). Effect of moisture content on the melting and subsequent physical aging of cornstarch. *Carbohydrate Polymers, 19*, 83–90.

Slade, L., & Levine, H. (1991). Beyond water activity: Recent advances based on an alternative approach to the assessment of food quality and safety *Critical Reviews in Food Science and Nutrition, 30*, 115–360.

Slade, L., & Levine, H. (1993). The glassy state phenomenon in food molecules. In J.M.V. Blanshard and P.J. Lillford (Eds.), *The glassy state in foods* (pp. 35–101). Nottingham, UK: Nottingham University Press.

Smith, A.C. (1989). Solid foams. In R.D. Bee, P. Richmond, & J. Mingins (Eds.), *Food colloids* (Special pub. No. 75, pp. 56–73). London: Royal Society of Chemistry.

Smith, A.C. (1991). Brittle textures in processed foods. In J.F.V. Vincent & P.J. Lillford (Eds.), *Feeding and the texture of food* (pp. 185–210). Cambridge, UK: Cambridge University Press.

Smith, A.C. (1992). Studies on the physical structure of starch-based materials in the extrusion cooking process. In J.L. Kokini, C.-T. Ho, & M.V. Karwe (Eds.), *Food extrusion science and technology* (pp. 573–618). New York: Marcel Dekker.

Smolarz, A., van Hecke, E., & Bouvier, J.M. (1989). Computerized image analysis and texture of extruded biscuits. *Journal of Texture Studies, 20*, 223–234.

Spies, R.D., & Hoseney, R.C. (1982). Effect of sugars on starch gelatinisation. *Cereal Chemistry, 59*, 128–131.

Stevens, D.J. (1976). The structure of wafer biscuit sheets and its relation to composition. *Die Staerke, 28*, 25–29.

Stollman, U., & Lundgren, B. (1987). Texture changes in white bread: Effects of processing and storage. *Cereal Chemistry, 64*, 230–236.

Swyngedau, S., Nussinovitch, A., Roy, I., Peleg, M., & Huang, V. (1991). Comparison of four models for the compressibility of breads and plastic foams. *Journal of Food Science, 56*, 756–759.

Szczesniak, A.S. (1963). Objective measurements of food texture. *Journal of Food Science, 28*, 410–420.

Szczesniak, A.S., Einstein, M., & Pabst, R.E. (1974). The texture tester, principles and selected applications. *Journal of Texture Studies, 5*, 299–316.

Thiewes, H.J., & Steeneken, P.A.M. (1997). The glass transition and sub-T_g endotherm of amorphous and native potato starch at low moisture content. *Carbohydrate Polymers, 32*, 123–130.

To, E.C., & Flink, J.M. (1978). Collapse, a structural transition in freeze-dried carbohydrates. I–III. *Journal of Food Technology, 13*, 551–594.

van Hecke, E., Allaf, K., & Bouvier, J.M. (1995). Texture and structure of crispy-puffed food products I: Mechanical properties in bending. *Journal of Texture Studies, 26*, 11–25.

Vickers, Z.M. (1985). The relationships of pitch, loudness and eating technique to judgements of the crispness and crunchiness of food sounds. *Journal of Texture Studies, 16*, 85–95.

Vickers, Z.M. (1988). Evaluation of crispness. In J.M.V. Blanshard & J.R. Mitchell (Eds.), *Food structure: Its creation and evaluation* (pp. 433–448). London: Butterworths,.

Vickers, Z.M., & Bourne, M.C. (1976). A psychoacoustical theory of crispness. *Journal of Food Science, 41*, 1158–1164.

Wade, P. (1968). A texture meter for the measurement of biscuit hardness. In *Rheology and the texture of foodstuffs* (No. 27, pp. 225–235). London: Society of Chemical Industry.

Wade, P. (1987). Biscuit baking by near-infrared radiation. *Journal of Food Engineering, 6*, 165–175.

Warburton, S.C., Donald, A.M., & Smith, A.C. (1990). The deformation of brittle starch foams. *Journal of Materials Science, 25*, 4001–4007.

Warburton, S.C., Donald, A.M., & Smith, A.C. (1992). Structure and mechanical properties of brittle starch foams. *Journal of Materials Science, 27*, 1469–1474.

Ward, I.M. (1983). *Mechanical properties of solid polymers.* New York: Wiley Interscience.

Yin, Y., & Walker, C.E. (1995). A quality comparison of breads baked by conventional versus nonconventional ovens: A review. *Journal of Science of Food and Agriculture, 67*, 283–291.

Zabik, M.E., Fierke, S.G., & Bristol, D.K. (1979). Humidity effects on textural characteristics of sugar-snap cookies. *Cereal Chemistry, 56*, 29–33.

Zeleznak, K.J., & Hoseney, R.C. (1987). The glass transition in starch. *Cereal Chemistry, 64*, 211–214.

CHAPTER 7

Fabricated Fat-Based Foods

Dick W. de Bruijne and Arjen Bot

Fat is an important ingredient in the human diet for both its nutritional and its savory contributions. Fat adds to the pleasure of eating of other food components: it generally improves mouth-feel, facilitates mastication, and adds flavor to food. The effect of fat on these processes is difficult to quantify because the experience is to a large extent subjective. An adequate understanding of the relations between the sensory effects of fat and the structure and rheology of fat-containing products is therefore of great importance. The need for understanding is even increasing nowadays because in the Western world the nutritional value of fat is considered more and more to be a burden rather than a benefit.

The caloric value of fat is roughly twice as high at that of the other major food components, proteins and carbohydrates (typically 38 kJ/g vs. 17 kJ/g). The intake of fat occurs through two categories of food products. The first group comprises products that are perceived by consumers as fat containing (e.g., margarines, dressing, salad oils, cooking oils, frying fats), often referred to as "visible" fat. The second category consists of products that are not thought to contain fat (e.g., meat, poultry, fish), sources of "invisible" fat. Visible and invisible fat consumption is about equal. U.S. consumption estimates for 1993 show a total annual fat consumption of 57.2 kg per capita (Chrysam, 1996).

The worldwide trend toward fat reduction in foods has meant that considerable effort is spent nowadays on trying to produce zero- or low-fat versions of the fabricated fatty foods, especially for products containing visible fat. This requires either mimicking the unique properties of fat by using nondigestible low-calorie analogues of fat or reducing the amount of fat needed for the product. It is clear that this can be done successfully only if detailed knowledge is available on the structure and the sensory properties of fat-containing products.

The aim of this chapter is to provide an overview of the various types of products in which fats act as a main structurant. Examples are margarines, halvarines, mayonnaises, and dressings. These products have in common that they usually contain both water and fat (or oil). Since fat and water are mutually insoluble, the products contain two distinct phases. The phases are inmiscible on a molecular scale but can be mixed on a longer length scale—say, above 10 µm. That mixture is called an *emulsion*. One uses the terms *o/w emulsion* if the oil phase components are dispersed in a continuous aqueous phase and *w/o emulsion* if the aqueous components are dispersed in the fat phase. Mayonnaise and dressings are examples of o/w emulsions, and margarine and chocolate are examples of w/o emulsions.

The word *emulsion* in the context of foods often refers to more complex systems than the prototype system consisting of liquid oil and liquid water, as is illustrated in Figure 7–1. Oil phases may be completely solid, as in the case of chocolate, or partially solid, as in the case of margarine at room temperature. In some cases, the aqueous phase of dressings is a dispersion of biopolymeric precipitates rather than a liquidlike solution of biopolymers. Such multiple dispersed structures are quite common in foods. Even in the case of pure oil and water, one has sometimes duplex emulsions: emulsified droplets turn out to be emulsions themselves on a smaller length scale.

The large differences in physical properties between o/w emulsions and w/o emulsions make it convenient to split the present chapter into two parts. The first part will deal with w/o emulsions. The text will refer predominantly to the structure and sensory properties of margarine, which is chosen as the typical example for such a system. A number of subsections will address the properties of other w/o emulsions, such as halvarines, chocolate, and low-fat and zero-fat spreads (although certain types of zero-fat spreads are actually w/w emulsions). The second part will discuss the properties of o/w emulsions. Here, mayonnaise is chosen as the typical example.

W/O EMULSIONS: MARGARINE, HALVARINES, CHOCOLATE, AND LOW-FAT AND ZERO-FAT SPREADS

Margarine was developed as a butter substitute by a French chemist, Hippolyte Mège-Mouriés, in 1869 (McGee, 1984). It consists of partially liquid fat phase in which water droplets are dispersed. The product has been very successful: the annual U.S. consumption was estimated to be 4 kg per capita in 1993 (Chrysam, 1996), making margarine responsible for approximately 7% of the total fat intake. This figure varies from country to country and depends strongly on the culinary

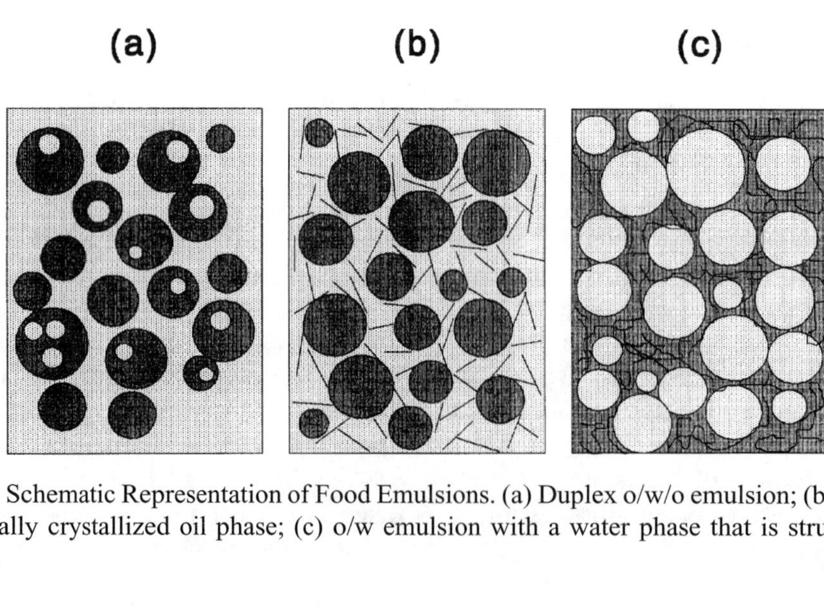

Figure 7–1 Schematic Representation of Food Emulsions. (a) Duplex o/w/o emulsion; (b) w/o emulsion with a partially crystallized oil phase; (c) o/w emulsion with a water phase that is structured using a biopolymer.

history and culture. For example, northern Europeans use more margarine per capita than Mediterranean people because the latter prefer the use of salad and cooking oils in many applications (Chrysam, 1996).

Margarine and butter both contain approximately 20% water. This water content is too low to affect either the product properties or the requirements for the production process. Therefore, margarine, butter, and shortenings (100% fat) and spreads with slightly reduced fat levels (about 60%) are considered here as one class of products.

Reduction of the fat level below typically 60% requires modification of the production process and product formulation to accommodate those larger amounts of water. An example of such a reduced-fat spread is halvarine. Somewhat surprisingly, chocolate also qualifies as a reduced-fat product (35–40%) in this classification scheme, but it should be realized that the production process is completely different from that of margarine. At still lower fat levels, the products are no longer fat continuous, but water continuous. The properties of the aqueous phase therefore dominate the product. The characteristics of the fat phase are of only minor importance then.

Structure

Margarine owes its consistency to agglomerates of crystalline fat dispersed in liquid oil. The chemical details of the fat triglyceride composition determine the amount of solid fat in the oil phase at a given temperature. Natural sources provide a wide range of triglycerides, containing different combinations of approximately 10 fatty acids. Chemical modification, such as interesterification and hydrogenation, increases the availability of certain triglycerides not occurring abundantly in nature. The chemistry of the fat crystals is relatively unimportant given a certain solid-fat content because the consistency of margarine is determined predominately by the percentage of crystalline fat in the oil.

Margarines show their characteristic melting behavior because the solid-fat content varies with temperature. The differences in melting behavior between triglycerides, both natural and modified, can be exploited to formulate blends with a desired melting behavior. In fact, the ability to blend available fats to products for which the consistency varies with temperature in a predefined way can be considered the essence of making margarines.

Water does not affect the product consistency and the manufacturing process very much. The role of water is mainly to accommodate water-soluble taste and flavor components, such as salt and acid, and to enable calorie reduction without detriment to other functional properties of the margarine.

Triglyceride fats, as well as monoglycerides and diglycerides, occur in three major classes of crystal modifications: α, β', and β. In all modifications, the triglyceride molecules take a chairlike conformation (with the glycerol group forming the seat) and are arranged in layers. A schematic representation is given in Figure 7–2. The fatty acid chains are oriented more or less perpendicularly to the layers. The tilt angle—that is, the angle between the direction of the fatty acid chains and the normal to the plane of the layers—varies for the different modifications. The tilt angle is zero for the α modification and increases for the β' and β modifications. Melting point and melting heat are lowest for the α modification and highest for the β modification; the three modifications differ about 10 to 15°C in melting point and about 20% in melting heat.

The packing of the molecules is rather loose in the α form. The chains are free to ro-

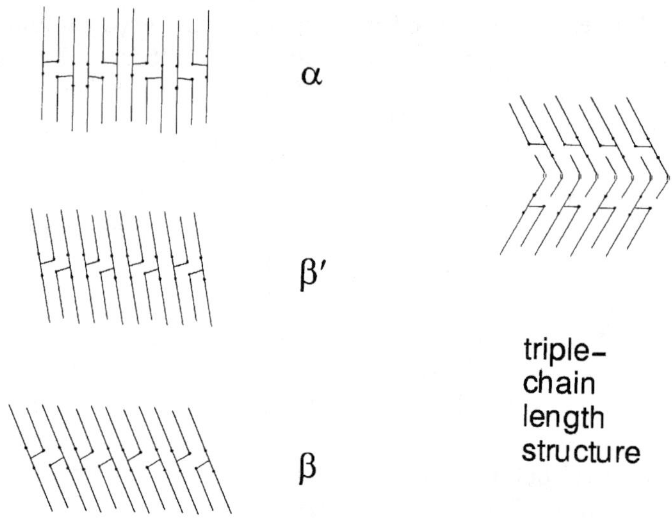

Figure 7–2 Schematic Representation of Crystal Modifications of Triglycerides. Left: α, β', and β triple–chain length modification. Right: example of triple–chain length modification.

tate, and as a result there is abundant opportunity to fit molecules of different shape and different chain length into the lattice. Each α crystal, therefore, usually contains a mixture of all the triglycerides present in the blend. To a lesser extent, this is true for β' crystals, in which far fewer foreign triglycerides are tolerated in the lattice. The β modification is the most perfect modification, in which usually only one type of triglyceride is allowed in the lattice. Thus, segregation often occurs if a crystal transforms from one modification to another. For example, mixtures of tristearin (trioctadecanoate) and tripalmitin (trihexadecanoate) segregate when β' crystals recrystallize into the β modification (Kellens & Reynaers, 1992).

The crystal structure of the stable β phase was determined in detail for a number of triglycerides, such as tricaprin (tridecanoate; Jensen & Mabis, 1963), trilaurin (tridodecanoate; Larsson, 1964), and tristearin (Skoda, Hoekstra, van Soest, Bennema, & van den Tempel, 1967). Two adjacent triglyceride molecules are stacked in such a way that both "seats" are positioned on top of each other. Crystallographic details on a length scale smaller than the fatty acid chains are still under discussion for the α and β' phases (Larsson, 1994). If β crystals are formed in margarines, this occurs by recrystallization at storage. The crystals may grow to such a size that they become perceivable at tasting and then lead to a defect known as "sandiness."

Mixed oleic-saturated glycerides such as POP, POS, and SOS (O = oleic acid = 9-*cis*-octadecenoic acid; P = palmitic acid = hexadecanoic acid; S = stearic acid = octadecanoic acid) tend to form triple-chain structures—that is, structures in which the triglyceride "chairs" are stacked with the chair backs against each other (Larsson, 1994). An example is sketched in Figure 7–2. These β' crystals are different from the regular β' crystals, and recrystallization may occur between these two β' modifications (Sato et al., 1989). A defect known as "graininess"—marginally harder spots in an otherwise smooth product—is a consequence of such recrystallization processes.

To obtain a margarinelike product structure, it appears to be essential to crystallize via the α modification. This requires rapid cooling. Slow cooling to moderate temperatures would lead to β'-crystal formation and inadequate soft and weak products. The production process for margarine is optimized with respect to cooling and crystallization rates.

Margarine is produced in a continuous process that comprises a sequence of separate cooling and crystallization steps. Cooling is most often achieved by means of scraped surface heat exchangers (A units), and crystallization occurs in vessels equipped with pin stirrers (C units). In the case of wrapper margarine, one needs an adequate (minimum) product consistency in the packing machine. The sequence then ends with an unstirred crystallizer (B unit). Although a number of variations in the process exist (Hoffmann, 1989), a typical production process is as follows. The solid fat and the temperature of the emulsion passing through the process line is shown in Figure 7–3.

A dosing pump feeds the mutually adjusted oil and water flows into the first unit, which usually is a heat exchanger. The phase ratio of water and oil is such that a w/o emulsion is formed. The drop size of that emulsion and of the final product is determined primarily by the intensity of the flow around the scraper blades (Batchelor, 1983).

The crystallization rate for each modification increases as the degree of supercooling increases. Since the degree of supercooling is higher with respect to the crystallization temperature of the β' modification than to the α modification crystallization temperature, one might expect that crystallization would occur preferentially in the β' modification. However, the crystallization rate for the α modification is so much higher than the crystallization of the other modifications that this outweighs the effects of supercooling. When the supercooling is great enough, crystallization will inevitably occur in the α phase, notwithstanding the much greater supercooling of the β' and β phase. The residence time in the scraped heat exchanger is typically 5 s. This is too short for the formation of β' crystals but long enough to obtain the equilibrium amount of α crystals (the state of crystallization at that instant is represented by the symbol A in Figure 7–3).

Subsequently, the blend passes through a crystallizer. The residence time in this unit is typically 0.5 to 1 min, which is long enough to obtain an adequate degree of β' (re)crystallization (C in Figure 7–3). The product temperature rises a few degrees in the crystallizer, as a consequence of stirring and of recrystallization from the α to the more stable β' modification. The blend is therefore cooled in a subsequent scraped surface heat exchanger (A2 in Figure 7–3).

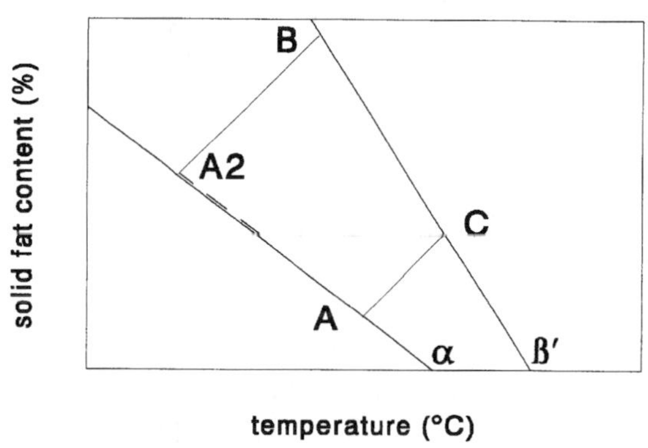

Figure 7–3 Schematic Representation of the Path of a Crystallizing Fat during Margarine Processing in a "Solids versus Temperature" Diagram. The fat melt is cooled to below the crystallization temperature of the modification. Subsequently, adiabatic crystallization to the β' modification leads to an increase in temperature, but new crystallites can be formed as long as the temperature remains below the β' crystallization temperature. Further cooling leads to the formation of new α and β' crystallites. Eventually, all α crystallites transform into β' crystallites.

The formation of β' crystals in the crystallizer and the scraped surface heat exchanger does not lead to any consistency of the emulsion. The emulsion remains fully fluid irrespective of the percentage of fat crystallized. This is due to the intense flow in those units, which immediately breaks down the structures formed. This fluid emulsion can be packed in tubs. The product consistency develops as a consequence of recrystallization at storage in the tub.

In the case of a wrapper margarine, a certain minimum consistency is required for proper operation of the packing machine. To enable a proper wrapping, the product has to be a deformable solid at the instant of entering the packing machine. This is realized by means of an unstirred crystallizer or resting tube at the end of the line (B in Figure 7–3). Because there is hardly any flow or deformation in those resting tubes, the structure formed as a consequence of crystallization is not broken down immediately, as it is in a stirred crystallizer or eventually a heat exchanger.

Fat crystals are usually platelets or needle-shaped (Juriaanse & Heertje, 1988) β' crystals (Birker, de Jong, Roijers, & van Soest, 1991). Figure 7–4 shows some typical examples of fat crystals in margarine.

A margarine can develop consistency only if the crystallites that have been formed during processing agglomerate in larger flocs. Agglomeration has to take place within the contact time of the crystallites, which is of the order of the reciprocal shear rate $1/\dot{\gamma}$. Agglomeration occurs if the crystallites grow together within that time. If the crystallites do not grow together, the hydrodynamic forces will separate those crystallites again. Later in the agglomeration process, small flocs will take over the role of the crystallites. Results by Vreeker, Hoekstra, den Boer, and Agterof (1992) and by Kloek, van Vliet, and Walstra (1997) demonstrate that the fat agglomerates are fractal and have a relatively open structure (see Figure 7–5). These results suggest that the agglomerates are formed by a process in which crystallites (or small flocs) immediately stick together upon collision. Also, the fact that the agglomerates remain relatively open during further processing implies that the connections are so strong by the end of the contact period during a collision that the aggregates do not separate or rearrange to form denser fractal structures under the influence of the hydrodynamic forces.

Johansson and Bergenståhl (1995) were able to show that the connections between the crystallites are of the same crystal morphology as the crystals themselves (i.e., β crystals are connected only by a fat that tends to crystallize in the β modification, and β' crystals are connected only by a fat that crystallizes in the β' modification; interestingly, α crystals do not agglomerate).

The agglomerate structure of a margarine resides primarily in the higher melting and first-crystallizing triglycerides, and thus in the structure of the flocs forming during the initial stages of crystallization. This follows from temperature-cycling experiments, which demonstrate that an almost complete melting of the structure (leaving only 1–2% of the fat in the crystalline state) followed by recrystallization to the original solids content has very little effect on the consistency. However, full melting destroys the floc structure completely, in which case the consistency does not recover upon lowering of the temperature. Apparently, the fractal agglomerate structure is intimately related to processing, and more specifically to the hydrodynamic conditions at the onset of crystallization.

Low-Fat Spreads

Low-fat spreads have appeared on the market during the last decade. The first product in this range was halvarine, which contains typically 40% fat and 60% water.

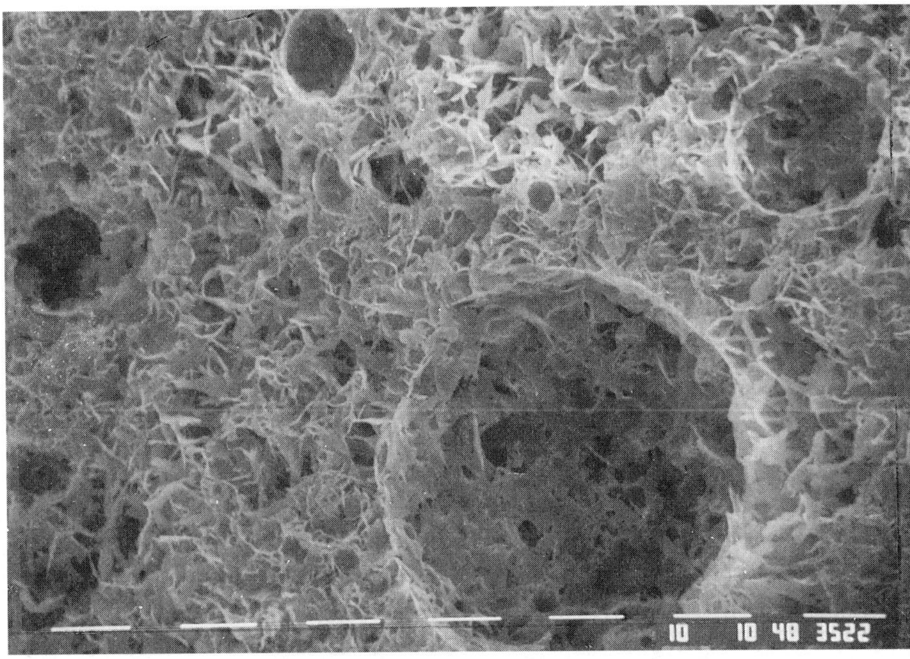

Figure 7–4 Electron Microscopy Photograph of Fat Crystal Structure in a Commercial Margarine

The presence of so much water in the product formulation requires a modified production process. Instead of starting with separate flows of water and molten fat, an o/w premix is prepared in the supply tank. This emulsion is cooled in a scraped heat exchanger, and further crystallization of the fat and subsequent inversion of the emulsion occur in a pin stirrer operated at high stirring speeds. The details of the inversion process are still under debate, but it appears that the increase of the effective volume fraction of the oil droplets through the inclusion of small water droplets as a result of intense stirring (i.e. w/o/w duplex emulsion formation) forms an aspect of the mechanism. Eventually, the effective volume fraction of the oil droplets becomes so high that droplet breakup as a result of stirring cannot compensate for the coalescence of the oil droplets anymore, and a w/o emulsion is formed (Groeneweg, Agterof, Jaeger, Janssen, Wieringa, & Klahn, 1998).

The inverted phase is stabilized by fat crystals. The crystals formed during the cooling and crystallization steps tend to stick to the droplet. Since the crystals are wetted better by oil than by water, they protrude more into the oil than into the water phase. By a mechanism similar to that of molecular emulsifiers, the crystals promote dispersion of the phase in which they do not dissolve (Bancroft rule), thus destabilizing o/w emulsions and stabilizing the w/o emulsions, provided the fat crystal concentration is high enough. W/o emulsions at solids concentrations above 1% or 2% are so stable that coalescence does not occur within the lifetime of the product. This stabilization mechanism for w/o emulsions is called the *Pickering mechanism.*

Johansson, Bergenståhl, and Lundgren (1995) found that w/o emulsions can become

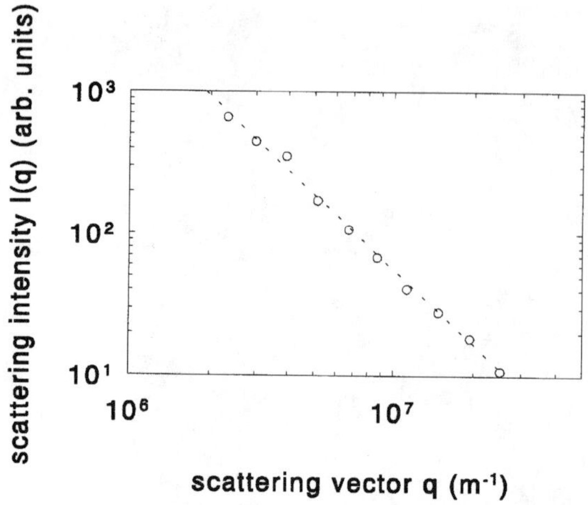

Figure 7–5 Scattering Angle Dependence of the (Static) Light-Scattering Intensity from Glycerol Tristearate Aggregates Formed under Quiescent Conditions in Olive Oil. The data satisfy a power-law relation $I(q) = q^{-d_f}$, with q the scattering vector and d_f the fractal dimension of the aggregate. The fractal dimension is found to be 1.7 during the initial stages of crystallization (as in the graph above) and increases to 2.0 over the next few days due to recrystallization.

unstable at very–low-fat crystal concentration. This was attributed to bridging of water droplets by fat crystals, which may occur for fat crystal concentrations less than those corresponding to "monolayer" surface coverage of all emulsion droplets. This is expected to occur especially if the fat crystals are wetted by water.

Further reduction of fat, to levels of 20% and lower, leads to products so lean that usually fat continuity cannot be maintained anymore. These products are water continuous; their microstructure is similar to that of low-fat dressings. The rheology of the water phase determines the product consistency more than that of the fat phase.

These products are water continuous during all stages of processing. This has technological advantages because the dosage of ingredients and handling of rework are easier than in the case of high-fat spreads.

Zero-Fat Spreads

Actually, a "zero-fat" claim in products does not necessarily imply a fat level of exactly 0%. U.S. legislation defines *zero fat* as a certain negligible quantity per "serving." The exact percentage of fat permitted in a product thus depends on the size of the "serving" and turns out to be different for different products. For spreads, "zero fat" comes down to somewhat less than 3%.

There are not many zero-fat spreads on the market yet. Two methods of structuring these aqueous products are used in practice: structuring with monoglyceride crystals and structuring with biopolymers.

Structuring with Monoglycerides. Monoglycerides are known to form lamellar liquid crystalline phases upon dissolution in water. When cooled down, these lamellar phases transform into a crystalline bilayer structure of α crystals, the so-called α-gel phase. The strength of this structure is rather low. This is analogous to α crystals formed in oil, which do not develop any consistency either. The α phase transforms slowly into a "coagel phase" of β crystals. The strength of these crystal networks is similar to that of an ordinary fat crystal network crystallized in oil. When viewed under a microscope, these monoglyceride structures are quite similar to ordinary fat crystal structures in oil.

Monoglyceride crystal networks are not strong enough for adequate structuring of a spread at a level of 3%. One actually uses mixtures of monoglycerides and biopolymers to arrive at spreadlike consistencies.

Structuring with Biopolymers. Spreadlike consistencies can also be obtained by using precipitated biopolymers. Such biopolymers are water soluble at high temperatures and gel or precipitate upon cooling. Cooling of pure biopolymers would lead to continuous gel networks with the macroscopic rheological properties of a rubber or a pudding (Bot, van Amerongen, Groot, Hoekstra, & Agterof, 1996). That continuous gel structure is in practice disrupted by adequate working during gel formation. The microstructures obtained are in fact dispersions of gel particles.

An effective route of making dispersions of gel particles is based on phase separation. Different macromolecules can be "incompatible." If that is the case, each macromolecular species will form a rather pure aqueous solution that does not mix with the equally pure aqueous solutions of the other species. Tolstoguzov (1991) was the first to investigate this phenomenon extensively for biopolymers.

At sufficiently high temperatures, when all biopolymers are still dissolved, one expects that w/w emulsions will be formed in the case of binary mixtures. The dispersed phase is an aqueous solution of one of the biopolymers, and the continuous phase is an aqueous solution of the other biopolymer (Kasapis, Morris, Norton, & Brown, 1993). These emulsions will be broken up by the hydrodynamic flow in the cooling unit, a process that will finally facilitate the formation of a sufficiently fine dispersion of gelled particles.

Partially hydrolyzed starches are commonly used in zero-fat spreads. Gelatin is often used to impart "oral melting."

Chocolate

Chocolate is essentially composed of cocoa powder, sugar, milk solids (in the case of milk chocolate), vanillin, and lecithin, suspended in a crystalline matrix of cocoa butter. About one third of the composition is fat (cocoa butter). Cocoa butter is very stable against oxidation (Morgan, 1994). It is composed predominately of 2-monounsaturated triglycerides such as POP, POS, and SOS. The most important characteristic is that it melts completely at body temperature, whereas it is almost completely solid at, for example, 20°C.

Fat bloom (which should be distinguished from sugar bloom) is a recrystallization process that causes the surface of a chocolate bar to become dull and apparently covered by a grey film. However, Jewell (1972) found not only that bloom formation is a surface phenomenon but that it permeates throughout the bulk of the chocolate.

To prevent the recrystallization process, a very specific temperature program is applied to the chocolate during production. This process is called *tempering*. The liquid chocolate is cooled from 50 to 32°C under constant agitation; then the temperature is lowered to 28°C to allow the formation of stable cocoa butter seed crystals, after which the temperature is increased again to 30–32°C to promote growth of the crystals (Manning & Dimick, 1985).

The polymorphic crystal modification of tempered chocolate fat is thought to be β', whereas that of bloomed fat is thought to be in the more stable β modification (Giddey & Clerc, 1961). Fats rich in POS and SOS, like cocoa butter, tend to bloom (Steiner & Bonar, 1961). Also, the use of additives such as milk fats and vegetable oils has been investigated (Campbell, Anderson, & Keeney, 1969; Hicklin, Jewell, & Heathcock, 1985).

Rheology

Fat networks are space-filling structures of fractal aggregates of fat crystals. These aggregates are brittle, fragile, and highly compressible. Nevertheless, margarine is relatively incompressible, shows little tendency to fracture, and usually deforms smoothly like a paste. The reason for this apparent contradiction lies in the fact that the aggregates are immersed in a liquid oil phase. The fat crystals form a structure with tiny pores that does not allow rapid draining of the liquid oil from the pores. The relative immobility of the oil phase renders the margarine virtually incompressible. The hardness—that is, the resistance offered by the material against a change in shape—is a consequence of the frictional forces between the crystals. That resistance is therefore in the first instance independent of the degree and rate of the deformation.

A complicating factor in representing or modeling the rheological behavior of foods, and of margarine in particular, is that the microstructure changes permanently and irreversibly during deformation. For that reason, it is inadequate to compare margarine straightforwardly with materials that have a constant microstructure (such as elastic, viscous, or viscoelastic materials). This section is intended to address this complex behavior in order to achieve better insight into the relevant phenomena taking place during relatively fast large-deformation processes such as spreading and eating, as well as the slow consolidation processes occurring during storage.

First, the plastic rheological behavior of margarine will be addressed, both in the case where the structure can be considered as incompressible and in the case where it is fully compressible. Then a criterion will be derived that determines the time and length scales at which margarine becomes compressible. Finally, a number of rheological test methods will be discussed that are particularly suited for margarine. In addition, some standard rheological characterizations will be given. The advantage of the latter is that they enable a straightforward comparison of margarine with other foods.

Plasticity of Compressible and Incompressible Materials

The fat phase of margarine is a mixture of solid fat crystals and liquid oil. The "hardness" or yield stress of the products is predominately determined by the solid fat content ϕ. The hardness for a series of arbitrary margarines at temperatures between 10 and 35°C shows an amazingly good correlation with ϕ^2 (Haighton, 1976):

Equation 1

$$\text{yield stress} \propto \text{hardness} = c\phi^2$$

The variation in magnitude of the proportionality constant c represents the combined effect of all factors contributing to the consistency other than the solid-fat content. Surprisingly, the exact chemical composition of the fat in terms of different triglycerides is relatively unimportant at a fixed solid-fat content. The constant c varies only marginally between different products or different triglyceride compositions. The variation of c between different margarine compositions is in practice only about a factor of 2. The factor c for butter is about half of that for margarines, indicating that "butter triglycerides" are marginally less effective structurants than "margarine triglycerides."

The behavior of margarine is similar in first approximation to that of the so-called ideal plastic solid. Ideal plastic behavior implies that the resistance against deformation of the sample depends neither on the rate nor on the degree of deformation and that the material remains isotropic during deformation. The constant stress required to deform the sample, the yield stress, is the only characteristic rheological parameter for ideal plastic materials.

The rheology of real margarine and fat dispersions differs from ideal plasticity in two aspects only, as is indicated schematically in Figure 7–6.

1. Fats and margarines show a small elastic region where the stress increases with deformation.
2. The stress does not remain perfectly constant beyond the elastic regime but

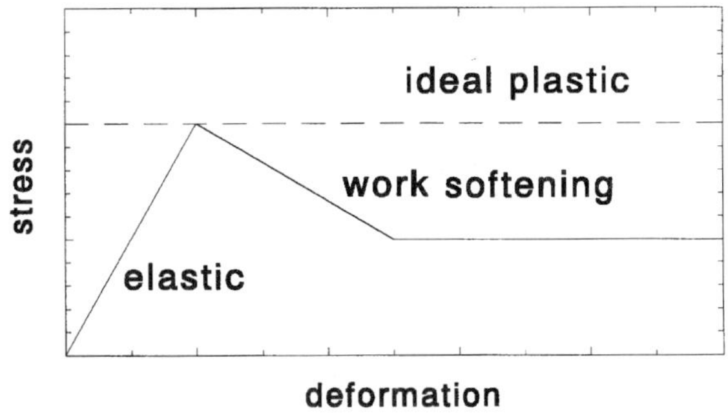

Figure 7–6 Schematic Representation of the Stress-Strain Curve for an Ideal Plastic Material and an Elastoplastic Work-Softening Material.

> decays as one continues to deform. That phenomenon is known as structural breakdown or "work softening."

Therefore, the description "work-softening elastoplastic material" represents the rheological behavior of margarine in more exact terms.

Yield stress and hardness are in principle equivalent, as was already indicated in Equation 1. However, the term *yield stress* emphasizes the tensorial character of the material property. The yield stress τ_y as measured in shear has to be generalized to yield stresses in other types of deformation, such as tension and compression.

The stress tensor is symmetric and contains six components in a general loading situation. The yield criterion is a relation between those six components, which has to be invariant against rotations in the case of isotropic plasticity. This implies that we may transform the stress tensor to its principal axes, where it has three principal components only. A criterion for isotropic plasticity may thus be formulated as a relation between the three principal stress components.

That relation between the three principal components has to be insensitive to an interchange of the components. Instead of using the principal components themselves, one often uses symmetric functions of these principal components for the formulation of a yield criterion. It can be shown that there are three of these independent symmetric functions, also called the invariants of the stress tensor (Freudenthal & Geiringer, 1958).* The first of these invariants represents the pressure, $p = (\Sigma_i \sigma_{ii})/3$. The pressure is not relevant if it is assumed for the moment that margarine is incompressible—that is, none of the material properties, including the yield criterion, depend on the pressure. A yield condition for incompressible materials thus is a relation between the second and third invariant, J_2' and J_3'.

The yield condition can be represented graphically as a two-dimensional plot of the (J_2', J_3') plane. However, meaningful visualizations of yield conditions are easier in stress space. This is a Cartesian coordinate system with the magnitude of the principal stresses along the axes. The (1, 1, 1) direction represents the axis along which the hydrostatic pressure on the sample increases, and stress vectors perpendicular to the hydrostatic axis represent deformational (deviatoric) stresses as shear, compressive, or tensile stresses. A yield condition for incompressible materials can now be represented as some surface enclosing the (1, 1, 1) direction of stress space. The cross sections of that surface at different pressure will be identical for incompressible materials.

Actually, the yield conditions of incompressible materials can range only between the so-called von Mises condition and the Tresca condition. The von Mises condition is represented in stress space by a circular cylinder around the (1, 1, 1) direction, and the Tresca condition by a hexagonal prism

*For the sake of completeness the expressions for the invariant are $J_2' = (\Sigma_i\Sigma_j\ \sigma_{ij}s_{ji})/2-(\Sigma_i\ \sigma_{ii})^2/6$ and $J_3' = (\Sigma_i\Sigma_j\Sigma_k\ \sigma_{ij}\sigma_{jk}\sigma_{ki})/3-2/3((\Sigma_i\ \sigma_{ii})(\Sigma_i\Sigma_j\ \sigma_{ij}\sigma_{ji})/2-(\Sigma_i\ \sigma_{ii})^3/9)$.

around that direction. That limitation is a consequence of the fact that yield surfaces cannot be concave in stress space (Prager & Hodge, 1951). In the case of the Tresca condition, yielding occurs at a critical difference between the greatest and the smallest principal stress, the intermediate principal stress being of no importance here. In the case of the von Mises condition, yielding occurs at a critical value of the second invariant ($J_2' = \tau_y^2$, with τ_y being the yield stress in shear), which represents the (square of the) deviatoric stress vector. The intermediate principal stress is of some importance in this case.

The projection of the cylinder and the hexagonal prism are drawn in Figure 7–7. The principal-stress axes make angles of 120° in this projection. The von Mises condition is represented by a circle, and the Tresca condition by a hexagon. The difference between the two conditions is not great; the largest difference occurs when extension or compression is compared with shear. Evaluating the ratio of the yield stress in compression σ_y and the yield stress in shear τ_y, one obtains $\sigma_y/\tau_y =$ 3 for the von Mises condition and $\sigma_y/\tau_y = 2 =$ 4 for the Tresca condition. This implies that yielding occurs within narrow limits and that it is unnecessary to choose between the two criteria. For the remainder of this chapter, we will assume that yielding is represented accurately enough by the Tresca condition.

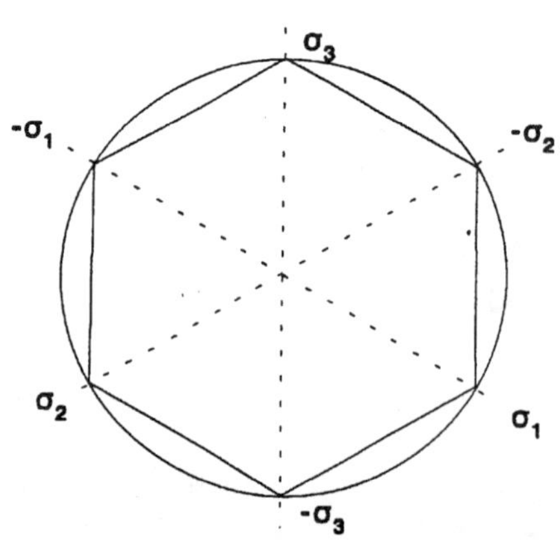

Figure 7–7 Intersection of the Yield Surface with an Isopressure Plane in Stress Space. The circle represents the von Mises criterion; the hexagon represents the Tresca criterion. The dotted lines represent the projection of the principle axes on the isopressure plane.

The above discussion assumed that margarine was an incompressible material. However, margarine is not incompressible if the external comprehensive load is applied to a small sample for a long period of time because the liquid oil may then be squeezed out of the sample. The yield condition becomes more complex in that case and becomes pressure dependent. The so-called *Mohr-Coulomb equation* offers a possibility to account for the pressure dependence (Coulomb, 1773). That equation has proved its value in the areas of soil mechanics and powder technology.

Equation 2

$$\tau_y = f(T + p)$$

Here τ_y is the yield stress in shear, T the tensile stress at which the coherence is lost (i.e., rupture occurs), p the pressure acting on the particles as a consequence of external loading forces, and f the friction coefficient. The Mohr-Coulomb equation is a three-dimensional analogue of the one-dimensional law of friction between sliding surfaces. Equation 2 is represented graphically in Figure 7–8.

The effect of Equation 2 on the yield condition is best visualized in stress space. The von Mises and Tresca conditions for incompressible materials are represented by a cylinder and a hexagonal prism with their axis along the (1, 1, 1) direction, respectively. The Mohr-Coulomb condition for compressible materials modifies, for example, the prism to a hexagonal pyramid, with its top at a pressure of $-T$ and with a cross section (perpendicular to the

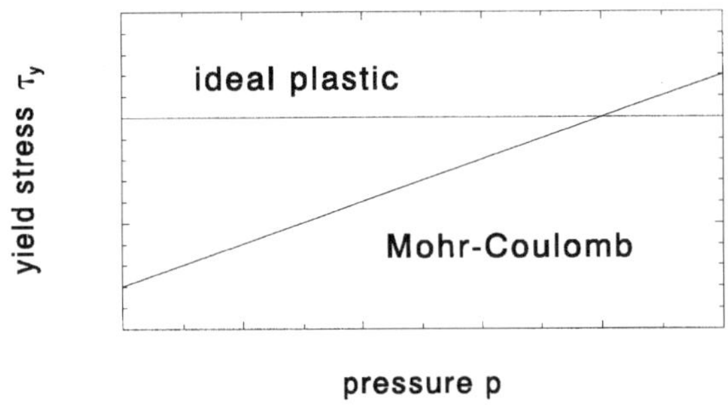

Figure 7–8 Dependence of the Yield Stress on the Hydrostatic Pressure p for an Incompressible Ideal Plastic Solid and for a Mohr-Coulomb Material

pressure axis) increasing with the pressure p. The tensile yield stress T is associated with attractive forces between particles and is usually negligibly small. The friction factor f depends on microstructural parameters, such as particle or agglomerate hardness, surface roughness, and shape, and ranges typically between 0.1 and 0.3 for most powders. Our preliminary experiments suggest friction factors of 0.18 and 0.17 for margarine and butter, respectively. These values are probably an underestimate of reality because they were obtained by means of equipment designed to measure friction coefficients in soils. Soils drain much more easily and quickly than margarine or butter. Drainage times had to be longer than a week to achieve adequate consolidation in the case of fats. Recrystallization certainly occurs in that period, which probably reduces the friction factor.

The Mohr-Coulomb criterion is regularly used to characterize powders and soils (Atkinson & Bransby, 1978). A prerequisite for the applicability is that compression forces are indeed acting on the particles and not on the pore liquid. The pore liquid should be given enough time to drain. This draining process is called *consolidation*. Powders and soils consolidate relatively easily because air or water can readily be removed from the pores upon pressurizing. For the characterization of soils, one uses a so-called *drained triaxial compression test*. A sample is placed on a permeable sieve plate and loaded on the top and from the sides by being immersed in a pressurized (external) liquid. The cylindrical soil sample, which is higher than it is wide, is protected from the pressuring liquid by being covered by a rubber sleeve. The pressure on the sample can be adjusted by pressurizing the external liquid. For soils, the sample size in the test can be quite large, typically a few centimeters. Proper drainage of these sample sizes requires less than an hour for most soils (and more than a week for fats!).

Powders are essentially dispersions in air. The low viscosity of air makes powders "drain" or consolidate almost instantaneously: application of an external pressure leads to an immediate consolidation and subsequent increase of the yield stress. A simple Jenicke shear cell can then be used to determine the increase in shear strength as a function of applied pressure.

Whether margarine should be treated as a compressible or as an incompressible plastic material depends on the time scale of the experiment and the size of the sample studied. This question requires a study of the compression of networks immersed in a liquid. In this process, an external load results in an immediate increase of the pore liquid pressure. Since the network does not deform as long as the liquid has not drained from the sample, there will be no pressure acting on the network at the start of the consolidation process. The excess pore liquid pressure is slowly relaxed by exudation. Consolidation can be described adequately by the so-called *consolidation equation* (Terzaghi, 1960):

Equation 3

$$\frac{\partial p}{\partial t} = c_v \frac{\partial^2 p}{\partial x^2}$$

Here, p is the excess hydrostatic pressure resulting from the imposed compressive load on the sample, t the time elapsed since loading started, x a spatial coordinate, and c_v the consolidation coefficient. The excess hydrostatic pressure is a function of both t and x. An example of a consolidation problem is the drainage of a slab of material between parallel plates that are permeable for the pore liquid and impermeable for the particles. The temporal evolution of p in this problem is sketched in Figure 7–9.

The transport coefficient that determines the rate of pressure relaxation in the sample, the consolidation coefficient c_v, is the quotient of the Darcy constant k_D and the compressibility of the network m_v.

Equation 4

$$c_v = \frac{k_D}{m_v}$$

The compressibility of a granular material is governed by deformability of the particles at the contact points with their neighbors, which is determined by the particle compressibility (or the inverse particle modulus). In the case

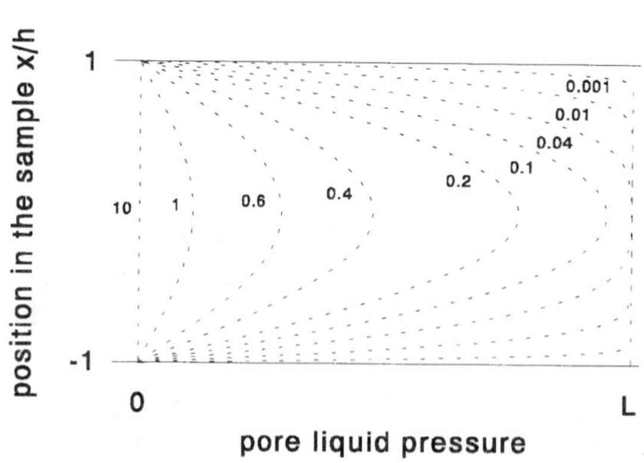

Figure 7–9 Decay of the Pore Liquid Pressure in a Consolidated Slab of Material, Drained Both at the Top and at the Bottom Surface (–1 x/h 1). Numbers in the figure represent the dimensionless drainage time $Fo^{-1} = c_v t/h^2$. Pressure on the network and pore liquid pressure add up to the load L.

of a network of fractal fat agglomerates, the macroscopic compressibility will be related to the compressibility of those aggregates rather than to the compressibility of individual particles. A typical value for m_v in a fat network is 10^{-4} Pa^{-1}.

The Darcy constant characterizes the permeability of the particulate network for the pore liquid. It is inversely proportional to the viscosity of the liquid and depends strongly on the pore distribution and pore connectivity. However, for randomly packed monodisperse spherical particles, a relatively simple relation between k_D and the volume fraction ϕ of particles in the immersed network has been found, the so-called *Kozeny-Carman relation*:

Equation 5

$$k_D = \frac{1}{5\eta S^2} \bullet \left(\frac{1-\phi}{\phi}\right)^2$$

where η is the viscosity of the pore liquid, and S the specific surface area of the particulate material per unit of volume (i.e., the ratio of external surface and volume of a particle). Since the specific surface area is inversely proportional to the particle diameter, Equation 5 shows that fine particles immersed in viscous pore liquids result in low permeabilities. Typical values for fat crystal networks are $\eta = 0.07$ Pa • s and $S = 5 \bullet 10^7$ m^{-1} (de Jager, van den Tempel, & de Bruijne, 1963).

Equations 3 through 5, together with the numerical values obtained by de Jager et al. (1963), allow a qualitative assessment of the value of the consolidation coefficient. Taking ϕ to equal 0.20 (though the actual value will depend strongly on storage temperature), a value of $2 \bullet 10^{-10}$ m^2s^{-1} is found for the consolidation coefficient c_v.

As in heat conduction problems, a Fourier number F_O can be defined that compares (the squares of) the sample thickness h and the penetration depth of the draining profile:

Equation 6

$$F_O = \frac{h^2}{c_v t}$$

Fo >> 1 implies that the sample is too thick to show appreciable draining after loading the sample for a time *t*. In this case, the material may be considered to be incompressible during the experiment. Fo << 1 means that the sample consolidates significantly during the experiment.

The analysis above allows an assessment of the risk of oil exudation during storage in a warehouse as a result of the high pressures on the lowest margarine packs caused by the improper stacking of products. Assuming a typical sample thickness of 0.05 m, a Fourier number near unity is found after approximately 10^7 s ≈100 days. This means that it takes about 100 days to consolidate a pack of margarine completely; the first signs of oily exudation will become visible sooner. Better storage conditions (especially lower temperatures, because this will increase ϕ) will delay the consolidation process. In materials in which ϕ is very high (i.e., materials without a liquid phase, such as chocolate), the consolidation proccss docs not occur.

The volume of oil exuded per unit surface area after a certain time *M*(*t*) can thus be calculated by means of penetration theory as long as the Fourier number is greater than 1:

Equation 7

$$M(t) = 2\left(\frac{k_D m_v t}{\pi}\right)^{1/2} L$$

with *L* being the load on the sample. Since the material constants k_D and m_v are known, the above equation enables a quantitative evaluation of exudation risks under practical circumstances.

The physics of drainage is analogous to diffusion of heat out of a slab of warm material immersed in a cold bath, for which analytical solutions exist (Carslaw & Jaeger, 1959; Crank, 1975). Parameters of the consolidation problem such as pressure, rate of exudation, and amount of exudation are fully equivalent to the heat conduction parameters of temperature, heat flow, and heat content. The material constants permeability (Darcy constant), compressibility, and consolidation coefficient are the full analogues of thermal conductivity, heat capacity, and thermal diffusivity, respectively.

The frictional nature of the margarine, and very likely of many other foods, implies that tensile stresses on a sample are supported not by the fat network but by the liquid oil. The external tensile stresses are balanced by an equal underpressure in the oil. Due to the incompressibility of the oil, however, an extension of the margarine sample in one direction has to be accompanied by a contraction in perpendicular directions. This contraction is opposed by the fat network, however, because it is able to support compressive forces. The consequence is that the network actually experiences compressive stresses only, even in the case of exclusively tensile external forces. An analogous line of reasoning can be applied to shear loading, where two of the principal stresses on the sample are equal and of opposite sign while the third principal stress is zero. The network will experience zero stress in the direction of principal tension, maximum compressive stress in the direction of principal compression, and intermediate compressive stress in the direction of zero stress on the sample.

The underpressure in the pore liquid as a consequence of tensile external forces can be observed at the surface of the samples. Here, the pore liquid at the free surface may be drawn inward. As a result, the fat crystals will protrude into the air, and the surface will become rough. Indeed, the surface of a margarine becomes dull upon deformation. The phenomenon is also called *dilatancy*. The curvature of the oil-air interface between the fat crystals may be interpreted in terms of a difference in pressure in the pore liquid p_{pore} relative to atmospheric pressure p. The underpressure $p - p_{pore}$ will result in a radius of curvature of the interface of $2\gamma/(p - p_{pore})$, where γ is the surface tension of the oil-air interface.

The underpressure in the pore liquid during deformation plays an important role in fracture. The interface is not stable anymore if the underpressure in the pore liquid becomes so high that the radius of curvature of the interface becomes smaller than the size of the pores. The liquid interface will be drawn inward, and air will enter the structure. Since the tensile stresses on the sample are no longer supported as the pore liquid is replaced by air, the network can be torn apart further, and a crack will form. Thus, fracture is expected to be controlled by the tensile strength of the network, the consolidation coefficient of the network, its pore size distribution, and the rate of the deformation. However, systematic studies of fracture of margarine in relation to these parameters have not been done.

Experimental Techniques To Study Plastic Materials

The mechanical behavior of a plastic material such as margarine can be characterized in different ways. On the one hand, margarine may be characterized in terms of the prime rheological parameters for a plastic material, such as yield stress and degree of work softening. Such tests are meaningful for practical applications and are product specific but are not very suitable for comparing margarine with completely different foods. On the other hand, more standard rheological parameters such the dynamic moduli and shear thinning behavior may be determined. These parameters are easy to obtain by means of sophisticated and easy-to-operate standard rheological equipment but are not always very meaningful as a characterization of the behavior in practical situations.

In this section, examples of both types of techniques will be described. Large-deformation rheology and penetration tests will be discussed as examples of the techniques that are especially appropriate for plastic materials. The dynamic small-deformation test will be discussed as an example of a test that is available for a wide range of foodstuffs.

We have already discussed the deviations of margarine rheology from ideal plastic behavior: margarine shows a small region of elastic deformation and a certain degree of work softening. The extent to which margarine deviates from ideal behavior can be determined by measuring the stress-strain curve. However, meaningful measurements at large deformations are difficult to perform because intrinsically work-softening plastics tend to deform inhomogeneously under most experimental conditions. Inhomogeneous deformation makes a proper recording of the stress-strain curve impossible.

These inhomogeneities can be suppressed by a proper choice of the measuring geometry. An example is compression of samples with a low aspect ratio (i.e., height smaller than width) under well-lubricated conditions. The principle is shown in Figure 7–10. Samples having a high aspect ratio develop a slip plane at an angle of $\pi/4$, with the direction of compression. As this plane develops, the macroscopic deformation becomes meaningless. However, when the height is smaller than the width of the sample, slip planes still develop at the same angle with the direction of compression, but there is no opportunity to accommodate a significant displacement along those planes because sliding will be obstructed by the plates of the compression geometry. Because new slip planes are formed continuously during compression, the time and space average of the deformation of the sample becomes approximately homogeneous.

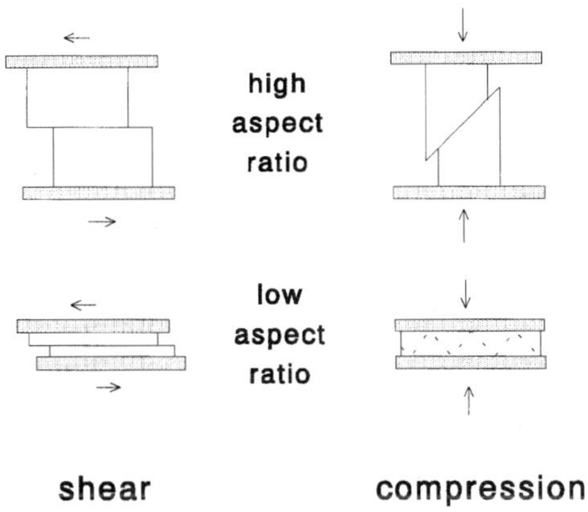

Figure 7–10 The Work-Softening Character of Margarine Makes Slip Planes Develop upon Shear Deformation. Slip planes do develop in compression also, but their effect is ironed out in the case of a low–aspect ratio sample.

The average deformation of the sample will be a uniform compression only if the lubrication is good enough. This was verified by means of internal markers inside the sample. That lubrication requirement is fulfilled more or less automatically in the case of margarines because the product adjacent to the plates is very much softer than the bulk as a consequence of work softening. A very thin layer of work-softened product lubricates the bulk of the product.

The disadvantage of the low aspect ratio of the samples is that friction due to this lubrication layer has a relatively great effect on the measured force. Therefore, the measured force has to be corrected for the friction with the plates. The most convenient procedure for this is to compare samples of different height and extrapolate the measured force for different sample heights to the value for infinite height. The relation between the measured force F per unit area A and the actual compression stress σ is given by

Equation 8

$$\frac{F}{A} = \sigma + \left(\frac{2\tau_f R}{3h_0}\right) \frac{h_0}{h}$$

where τ_f is the friction due to the shear forces in the lubricating layer, R the radius of the compressing plates, h_0 the initial height of the sample, and h the actual height during compression. The most straightforward method to extract σ is by plotting, for each value of the deformation h/h_0, the measured stresses σ in the different samples versus the inverse initial sample height $1/h_0$. The method requires that samples of different height experience identical deformation histories at equal degrees of deformation. This requirement is fulfilled by compressing the sample, with cross-head speeds being proportional to the sample height. The intercept with the stress axis gives the stress for infinite sample height, which is in fact the stress at that deformation. The friction τ_f with the plates is obtained from the slope. The friction with the plates has to be smaller than the stress σ (e.g., less than 1/4 of σ); otherwise, the sample deformation would deviate too much from a uniform compression.

Experimental results are shown in Figure 7–11 for margarine and butter. The degree of work softening is much smaller for butter than for margarine. The data also show that the maximum stress is already reached at quite small deformations and that margarines and butters are therefore hardly elastic.

The compression test provides information about work softening of a margarine at moderate degrees of deformation (typically of the order of unity). To obtain such information at larger deformations, it is more appropriate to perform extrusion experiments. A popular test is to extrude a product through a sieve plate of known dimensions and to take the relative loss in hardness as a measure for the degree of work softening. Such measurements confirm that work softening is indeed more pronounced for margarine than for butter. The experience shows that the smoothness of the deformation under practical conditions (e.g., kneading, calendering, or

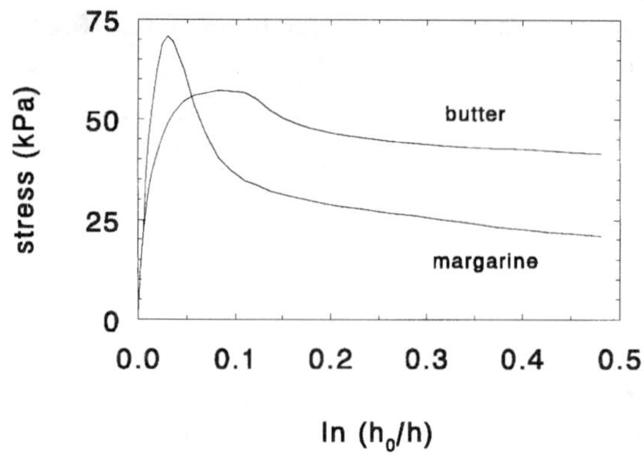

Figure 7–11 Stress-Strain Curves for Butter and a Margarine, as Obtained from a Well-Lubricated Compression Test. The drop in stress level between the maximum of the curve and the value at large deformations is called work softening, and is considered to be a measure for the "butter-likeness" of the margarine.

spreading) correlates well with the so-obtained degree of work softening.

Penetration tests are widely used as a simple and adequate way to determine the yield stress. In this method, a rod with diameter d is made to penetrate into the product over a short distance that is not much longer than the diameter of the rod. The penetration distance can best be standardized. The force required to penetrate the sample may be used as a product characteristic as long as the consistencies of rather similar products are compared. However, when the rod diameter is changed in order to characterize margarines over a broader range of consistencies, it is necessary to derive a geometry-independent parameter from the experiments. For this, the yield stress (τ_y) of the material is best suited.

The problem of penetration of a rod into a plastic material is a special case of the more general problem of compression of a truncated cone. The force F to compress such a conical sample depends on the cone angle θ, and is in accordance with the plasticity theory given by

Equation 9

$$\frac{F}{\pi d^2/4} = \sigma_y (1 + \theta)$$

A cone angle $\theta = 0$ represents compression of a cylinder of material having the diameter of the probe. Cone angles $\theta = \pi/2$ and π are relevant to indentation of a material by means of a rod. A cone angle $\theta = \pi/2$ refers to the situation of a rod just touching the surface. A rod penetrating into the product corresponds to $\theta = \pi$. The latter is the condition of practical interest. The relation between yield stress and penetration force in the case of the Tresca yielding condition ($\sigma_y = 2\tau_y$) is given by:

Equation 10

$$\tau_y = \frac{2F}{(1 + \pi) \cdot \pi d^2}$$

Sometimes a "collar" is formed as the rod penetrates the sample. The degree of collar formation is a qualitative measure for the degree of work softening of the product. This can be understood as follows. The material adjacent to the penetrating rod will deform more than the material far away from the rod; therefore, the material near the rod will become much softer than the material far away as a result of work softening. If these differences in consistency remain small enough, the material adjacent to the rod will not flow alongside the rod and will not form a collar. This behavior is observed for butter or smoothly deforming margarines. If, however, the same experiment is done with a typical margarine, some material underneath the probe will be pushed alongside the probe and will form a collar. This behavior is illustrated in Figure 7–12. This qualitative measure for work softening correlates very well with the more quantitative measure that may be derived from an extrusion test or a lubricated compression test.

Finally, a number of small-deformation rheological characterizations are included in Figures 7–13, 7–14, and 7–15. The results in the first two figures show that the storage

A B

Figure 7–12 During the Pencil Penetration Test, a Collar Is Formed in Margarine (B) and Not in Butter (A)

modulus and the loss modulus are, in general, frequency independent and start to depend on the frequency only as the percentage of solids goes to zero. This shows that these dilute dispersions become viscous at these very low concentrations. The results in Figure 7–15 show that the linear elastic region extends to strains of only 10^{-3}.

Low-Fat and Zero-Fat Spreads

Fat reduction in spreads implies an increase in the water content of the product. Above a certain level, the rheological properties of reduced-fat spreads are significantly affected by the presence of water, in contrast to margarine, for which only the properties of the fat phase matter. A major problem with halvarines (i.e., reduced-fat spreads with 60% water) is the occurrence of loose moisture at spreading. This defect is remedied by thickening the aqueous phase. At low temperatures, the fat phase is solid and dominates the rheology. At higher temperatures, however, the fat has melted, and the contribution of water is no longer negligible. At those temperatures, the consistency or yield stress of the 60% w/o emulsion is related to the drop size, as for mayonnaise.

Breakdown

The breakdown characteristic is an essential property of margarine, both during production and during end use. During production, the microstructure of margarine is strongly influenced by the fierce action of scraped heat exchangers and pin stirrers. The extent to which the fat aggregates are broken down during processing affects the product consistency at packing. During end use, the

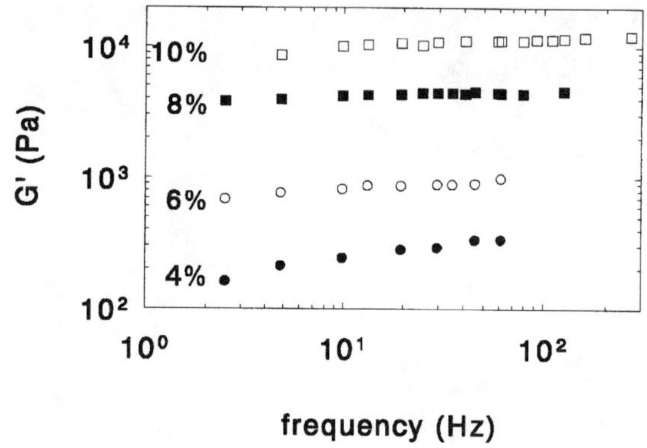

Figure 7–13 The Storage Modulus *G*' of Tristearin Solid-Fat Dispersions in Paraffin Is Almost Independent of Frequency. However, there is a clear increase with increasing solid content ϕ.

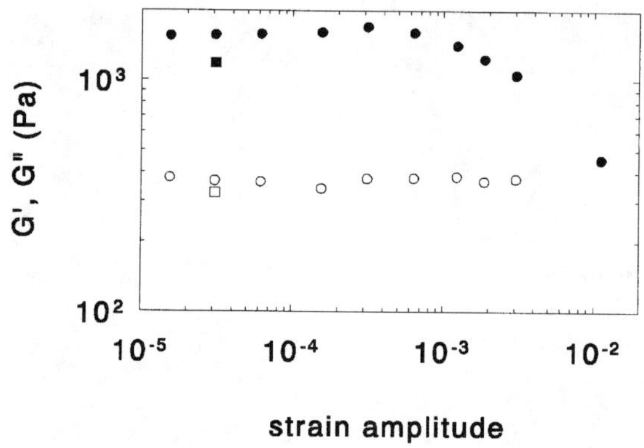

Figure 7–15 Effect of Strain Amplitude of an Oscillatory Deformation on the Dynamic Properties of 8 wt% Tristearin Fat Crystals in Paraffin Oil at a Frequency of 20 Hz: ● G'; ○ G". The squares refer to a deformation cycling test, in which the sample is first deformed up to a strain amplitude of $3 \cdot 10^{-3}$ (i.e., smaller than the yield strain) and subsequently measured at a strain amplitude $3 \cdot 10^{-5}$ (■ G'; □ G").

appreciation of the consumer depends strongly on the ease of spreading of the product and the response to mastication and dispersion in saliva. The end-use properties are far less processing dependent.

During production of a margarine, the flow of the product through the processing line and the continuous working on the product will prevent the formation of a percolating fat crystal network. Instead, the growth of the aggregates will be limited to a certain size, which depends on the intensity of the flow fields around the aggregates. Since the growth of the aggregates is much slower than the breakdown, the products will remain fluid during processing.

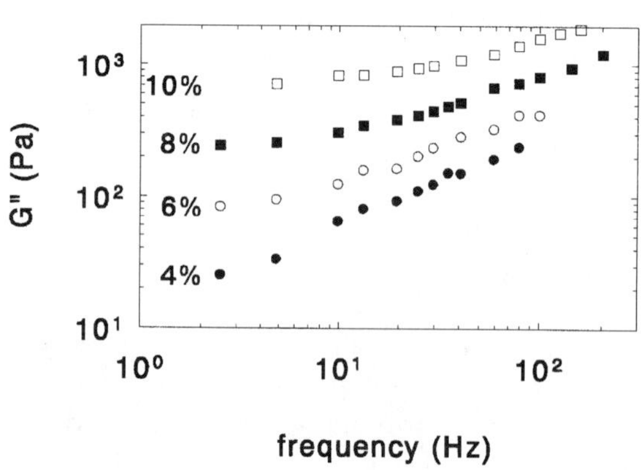

Figure 7–14 The Loss Modulus *G*" of Tristearin Solid Fat Dispersions in Paraffin Increases Somewhat with Frequency. In addition, there is a clear increase with increasing solid content ϕ.

This behavior can be demonstrated in margarine samples or fat dispersions after processing also. In Figure 7–16, the response of a fat dispersion is shown during a shear rate sweep cycle—that is, a linearly increasing shear rate followed by a linearly decreasing shear rate. It can be seen in the up-curve that the material shows a yield stress at small shear rates. When the yield stress has been overcome, the material starts to flow.

The decreasing shear rate branch shows that the stress decreases linearly with decreasing shear rate: the fat crystal dispersion behaves as a perfect Newtonian fluid. Reversal at higher or lower values of the shear rate

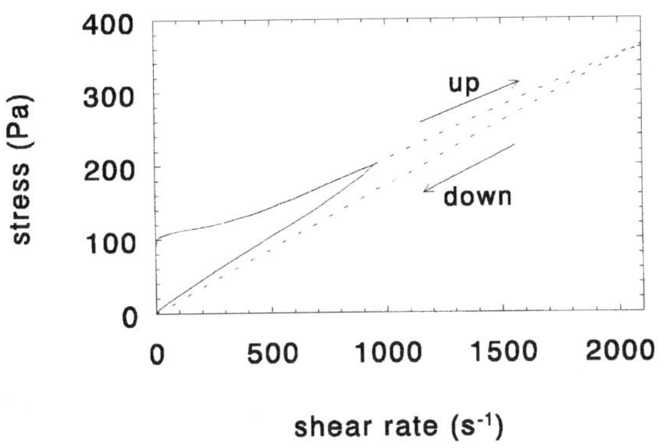

Figure 7–16 The Aggregate Structure of a 4% Fat Dispersion at 5°C Is Broken Down to a Level Depending on the Shear Rate Applied to the System. Unsheared margarine behaves in a non-Newtonian manner with increasing shear rate, but like a perfectly Newtonian fluid with decreasing shear rate. This implies that the viscosity of the fat dispersion is determined by the highest stress that the sample has experienced.

would have given Newtonian behavior in the decreasing branch also. The viscosity in the downgoing branch would have been different in that case and appears to be determined exclusively by the maximum applied stress.

Immediate repetition of the experiment would show perfectly Newtonian behavior at both increasing and decreasing shear rates because the growth process for the aggregates is much slower than the breakdown process. Repeating the experiment after a day of resting time would show a result similar to that of a fresh, nonsheared sample.

The almost unlimited ability of fat crystal structures to recover from damage caused by working is a consequence of the ability to recrystallize. Two recrystallization phenomena may contribute to this. The first process is recrystallization due to Ostwald ripening, in which local dissolution of crystalline material is promoted because the smoothing of the sharp edges of the fractured sintered crystals will reduce the radius of curvature (Knoester, de Bruijne, & van den Tempel, 1968; Walstra, 1987).

The second process is dissolution of material from the core of a crystal after formation of a fresh fracture plane through that crystal. This process occurs because a crystal will be a more or less layered structure having high-melting components in its core and lower melting fats at its surface as a consequence of the temperature evolution in the fluid oil during the crystallization process. Only the outermost layer of a crystal will actually be in equilibrium with the liquid phase of the fat blend. The core of the crystals will not be at equilibrium with that liquid but will not dissolve into that liquid as a consequence of the encapsulation with the outer layers. The core will dissolve completely as soon as the outer layers are damaged or interrupted as a consequence of working. The dissolved triglycerides will crystallize elsewhere in proportions compatible with the equilibrium composition at that temperature.

During spreading on bread or toast, the time of deformation is so short that neither recrystallization nor consolidation processes are of any importance. Under those conditions, margarine can be considered to be an incompressible ideal plastic material for which the yield stress is proportional to the square of the solid fraction ϕ. This causes the ease of spreading to depend on temperature because ϕ depends on temperature. Fat blends for margarines are composed in such a way that the spreadability of the product is ensured. Soft margarines are even spreadable at refrigerator temperatures (in contrast to butter). This is illustrated in Figure 7–17.

Spreading induces the rupture of the fatty films separating the water droplets in the product. This will lead to coalescence of the water drops. For low-fat products having high water levels, such as halvarines, coales-

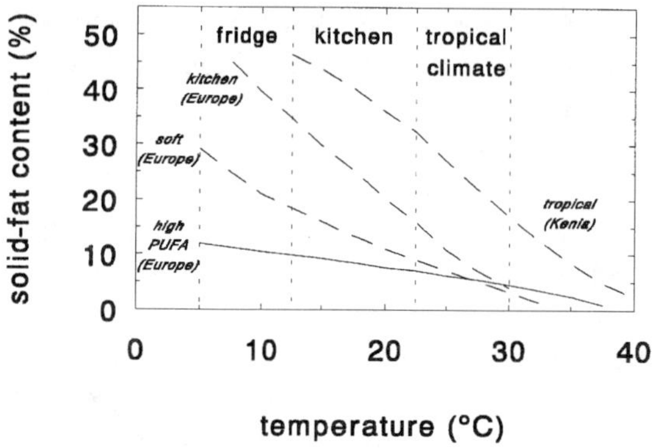

Figure 7–17 The Spreadability of Margarines Is Controlled by the Amount of Solid Triglycerides at the Temperature of Use. The solids content can be adjusted by an optimal selection of the fat blend. PUFA = polyunsaturated fatty acids.

cence may become so frequent that products tend to lose moisture at spreading. This is considered to be a defect that can be cured by adequate structuring of the internal water phase with biopolymers. The biopolymers will not prevent the rupture of the fatty films between the drops during spreading but their thickening action will help to avoid coalescence of the droplets to large pools of water.

Breakdown processes occurring in the mouth are decisive for the mouth-feel of the products. The characteristics of the mouth as a food processor and the behavior of different types of w/o emulsion in the mouth are discussed below.

Foods, and spreads in particular, disintegrate during mastication. Several stages can be distinguished in that process:

1. Comminution of the food by the action of the teeth; the food is cut to pieces having a typical size of only a few millimeters.
2. Softening as a consequence of the temperature rise; fat in the pieces melts, resulting in weaker and softer products
3. Breakup of these pieces in the saliva flow in the mouth when these pieces have become soft enough

A number of important physical properties for the effectiveness of the breakup process can be identified: the rheological properties of the food material, the stresses exerted by the saliva flow in the mouth, and the interfacial tension between food material and saliva. To understand the role of food in this complex physical process, some knowledge of the hydrodynamic flow conditions of saliva in the oral cavity is necessary.

A straightforward approach is to study the in-mouth breakup properties of model emulsions with a known yield stress. It is found that emulsions with a yield stress exceeding 50 Pa cannot be broken up in the mouth, while those with lower yield stresses can. This suggests that the (intermittent) maximal stress level in the oral cavity is approximately 50 Pa (de Bruijne, Hendrickx, Alderliesten, & de Looff, 1993). This stress level refers to an oral cavity with saliva only. The stress levels during mastication will certainly be higher than 50 Pa if other food components thicker than saliva will be present in the mouth.

Apart from the stress levels in the mouth, the type of flow is important also. Two major types can be distinguished: shear and elongation. The dependence of the breakup efficiency on the viscosity of the oil droplets is completely different in the two cases. Breakup in shear flow is strongly dependent on viscosity of the oil droplets, whereas breakup in elongation is almost independent of the droplet viscosity. The behavior for shear flow is explained by the fact that shear deformation contains a component of rotation, which causes the axis of maximal elongation to rotate away during deformation. This makes a breakup process in shear less effective, particularly in the case that the drop viscosity becomes higher than that of the external phase.

The experimental results for the breakup of droplets in the mouth are quantified in Figure 7–18. The critical capillary number for breakup, a dimensionless drop size, is lower in elongational than in shear flow. The experimental data show that the average droplet size after breakup is independent of the viscosity of the droplets, and therefore it can be concluded that flow in the oral cavity that is responsible for breakup is elongational (de Bruijne et al., 1993). In addition, the absolute values for the critical capillary number in the mouth clearly coincide with the values for breakup in elongational flow in better defined geometries if the hydrodynamical stresses in the mouth are taken as 50 Pa.

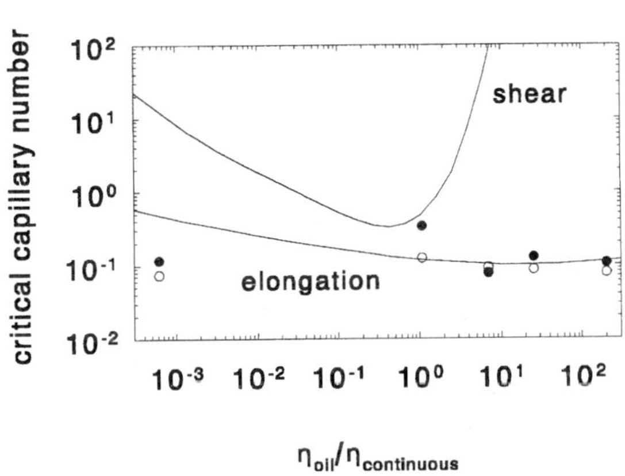

Figure 7–18 Critical Capillary Number for Breakup of Drops of Newtonian Liquids in Steady Elongational and Shear Flow. Fully drawn lines show results according to Grace (1982) and Bentley and Leal (1986). The capillary number Ω is defined as $\Omega = \sigma_c d/\gamma$, with σ_c being the external stress in the continuous phase, d being the diameter of the oil droplet, and γ being the interfacial tension of the droplet. Also plotted is the critical capillary number for oil droplets broken up in the mouth during mastication, assuming a constant deformation rate in the mouth (using $\eta_c = 50$ mPa • s, $\sigma_c = 50$ Pa). The data indicate that breakup in the mouth occurs in elongational flow.

Margarine can be broken down in the mouth only if its yield stress is less than 50 Pa, as the stress level in the mouth does not exceed this value. Margarines with a higher yield stress will not release their taste and flavor components when tasted in the presence of saliva only. Since this yield stress is determined by the solids content ϕ only, the degree of breakup of a margarine in the mouth is thus determined by the amount of solids at body temperature. This is illustrated in Figure 7–19, which shows a number of expectorated margarine samples having an increasing amount of solids at mouth temperature. It can be seen that the hardest product still shows rather large lumps after mastication, whereas the softest product is broken up in pieces more than 10 times smaller. If the fragments in masticated food are large, the consumer will experience the product as sticky, fatty, or waxy (de Bruijne et al., 1993). In addition, the release of salt and flavors is impaired from products that do not break up to a large extent. Indeed, the product that shows the largest lumps in Figure 7–19 tastes and smells the least.

Reduced Fat Spreads

For halvarines at oral temperature, there is a considerable contribution by the dispersed water droplets to the yield stress of the product. Thus, the melting of solid fat is not sufficient to attain the critical value of 50 Pa. The magnitude of this contribution depends primarily on the average size of the drops: small drops lead to a relatively high yield stress compared with large drops. The mechanical properties of the aqueous phase of the droplets contribute far less to the consistency of the spread than the size of the droplets. As a result, the yield stress of a halvarine can be reduced

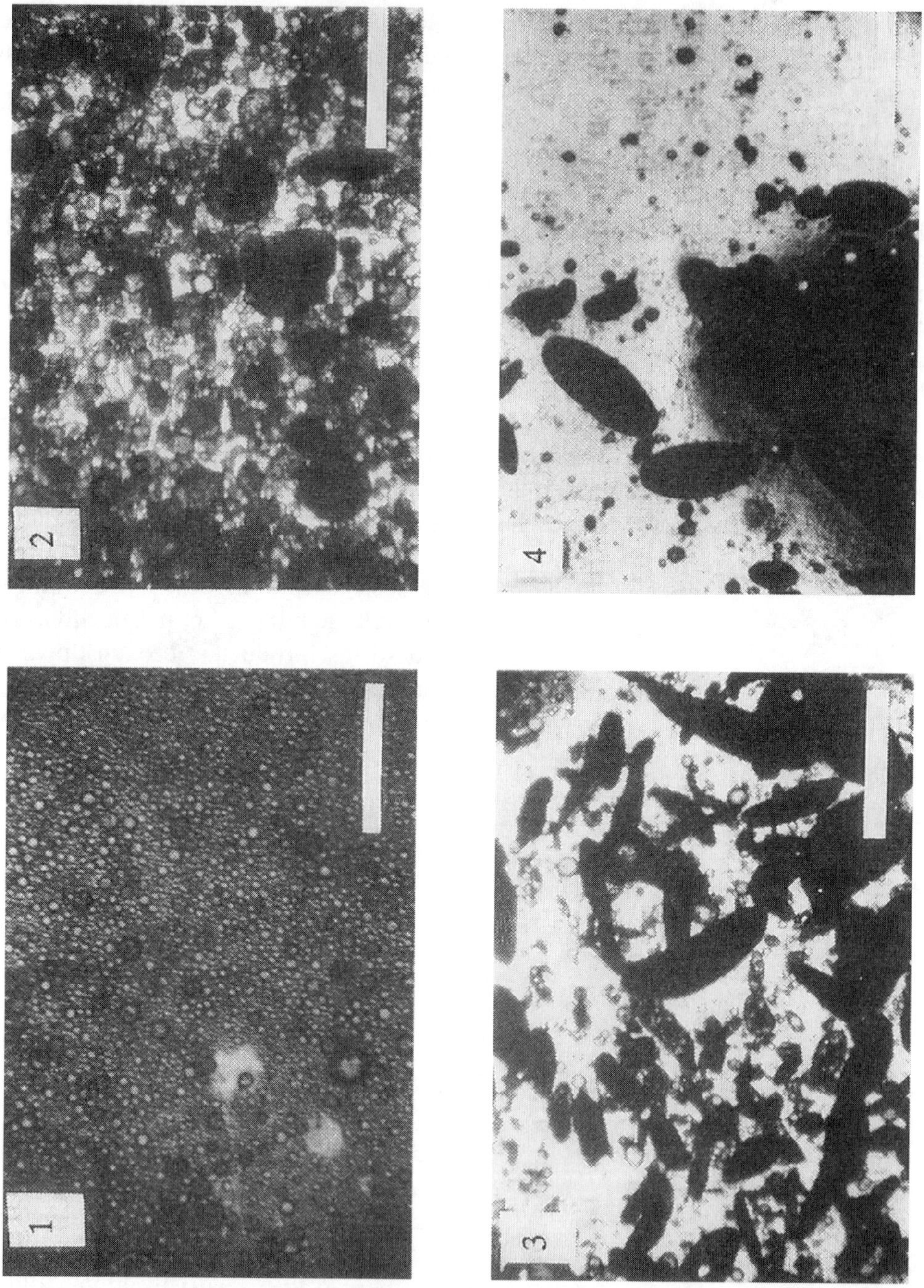

Figure 7–19 Expectorated Spread Samples with Increasing Consistency at Body Temperature (1 > 2 > 3 > 4). The bar length represents 1 mm in all photographs. *Source:* Courtesy of the Royal Society of Chemistry, *Food Colloids and Polymers: Stability and Mechanical Properties*, E. Dickinson and P. Walstna, eds., p. 206, April 1992, Proceedings of a conference held at Lunteren, The Netherlands.

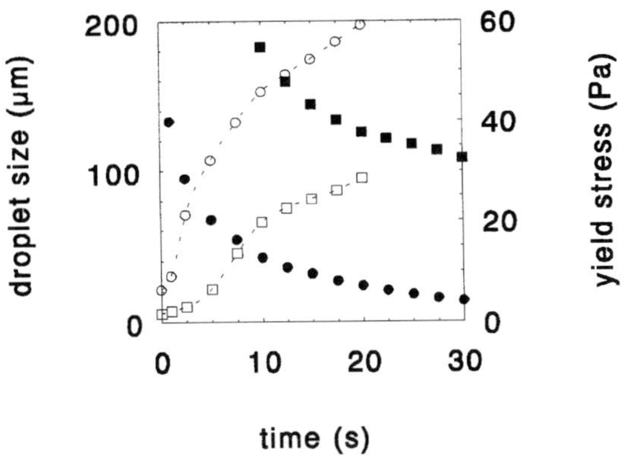

Figure 7–20 Coalescence of Internal Droplets and Decay of the Yield Stress for Two Different Halvarines That Were Shock Heated to Mouth Temperature. □, ○ droplet size as obtained from turbidity measurements; ■, ● yield stress. Squares represent a halvarine with a slowly coalescing internal water phase; circles represent a rapidly coalescing internal phase.

sufficiently for proper oral dispersion only by coalescence of the water droplets.

Figure 7–20 shows coalescence of droplets in a very thin sample that is shock-heated to mouth temperature suddenly (within 1 s). The water droplets start to coalesce, and the droplet size increases. Simultaneously, the yield stress (actually the stress at a very small shear rate of 1 s^{-1}) is found to decrease. The experiments indicate a perfect correlation between drop size and yield stress after shock-heating the samples to body temperature.

The difference between the halvarines in Figure 7–20 stems from differences in intrinsic stability of the water droplets in the w/o emulsion. This is determined by factors such as the type of emulsifiers used, the residual amount of solid fat at mouth temperature (Pickering stabilization), and the viscosity of the water phase in the droplets.

Zero-Fat Spreads

Premium margarine as well as butter shows a drop in consistency or yield stress typically of three orders of magnitude as the temperature rises from 20 to 35°C. This allows both convenient spreading at room temperature and acceptable oral breakdown at mouth temperature. Since melting does not occur in most zero-fat spread formulations, it is difficult to combine an adequately high consistency at spreading (several kPa) with a sufficiently low consistency at eating (50 Pa).

An adequate mouth-feel and an adequate release of taste and flavor components from a zero-fat spread again require a drop of the in-mouth yield value of the product to approximately 50 Pa. For products containing gelatin as a structuring agent, the necessary drop in consistency can be achieved partly by the same mechanism as for full-fat products: melting. The typical time t needed to heat up a piece of product after breakdown to particles of a characteristic dimension R due to comminution is

Equation 11

$$t \approx \frac{R^2}{(\lambda/\rho c_p)}$$

where λ is the thermal conductivity, ρ the density, and c_p the specific heat. For water, $\lambda/\rho c_p \approx 10^{-7}$ m²/s. If the dimension of the sample is 0.3 mm, a period of 1 s is required to heat up the fractured product. This period is similar to the residence time of food in the mouth, allowing sufficient time for melting of the gelatin during mastication of the product.

In nonmelting polymer systems, another mechanism could be responsible for the reduction of the yield stress to approximately 50 Pa. The most likely mechanism for this process is saliva uptake as a consequence of the stresses occurring during mastication. The tensile forces acting on the product during mastication will lead to an underpressure in the pore liquid between the structural elements of the zero-fat spread. This will induce an inward flow of saliva because the pore liquid and saliva are miscible. (Note that for full-fat products, such an underpressure does not lead to an inward flow of the saliva because the oil-saliva interface at the surface of the food particle cannot be drawn inward.) The saliva flow may contribute considerably to the required consistency drop.

The value of the consolidation coefficient for zero-fat spreads has not been determined. The critical value can be estimated on the basis of the assumption that the teeth comminute food to an average particle size $R \approx 0.3$ mm within a typical residence time $t \approx 1$s. Substituting these values in a Fourier-type equation

Equation 12

$$c_v = \frac{R^2}{t}$$

renders a critical value $c_v \approx 10^{-7}$ m²/s. The mechanism of saliva uptake may contribute to the oral dispersion of the product if c_v exceeds this critical value.

Sensory Properties

The structural, rheological, and breakdown properties of margarine should culminate in a product with desirable sensory attributes. This is the *only* relevant product property as far as the consumer is concerned. Consumers do not use rheometers or study expectorated products under microscopes. Such laboratory techniques are important for diagnosing insufficiencies in sensory performance.

The sensory performance itself has to be evaluated by panels of assessors. The use of humans for evaluation of product properties introduces many problems such as variability in judgment of different assessors, long-term drift in the appraisal scales, and even the influence of psychological factors. Although many of these restrictions can be circumvented by using techniques such as quantitative descriptive analysis (QDA), such approaches are cumbersome and unsuited for routine analysis.

For this reason, it is usually best in research to take an intermediate position between the convenient instrumental approach and the desirable human evaluation. This can be done by identifying and understanding the role of the product in the process relevant for the consumer.

Complementary to a description in which the mouth is viewed mainly as a food processor, it is also possible to highlight especially its sensing capabilities. The mouth contains specialized areas that can be considered as consistency, temperature, and taste sensors. In addition, there is a connection to the nasal cavity that contains sensors for a variety of flavors.

Food consistency is sensed by comparing the deformability of the food with that of the mouth, and in particular the tongue tissue (Peleg, 1980, 1983). Hard foods deform the tongue, and soft foods do not. It is actually the tissue deformation that is sensed. Typically, particles smaller than 10 to 20 μm cannot be perceived, but this depends somewhat on the hardness of the particles (Tyle, 1993). The tongue is also the location of the tastebuds, responsible for the sensation of salt, sweet, bitter, sour, and umami taste (Bloom & Fawcett, 1968; Kawamura, Kare, & Tannenbaum, 1987). The nose contains the olfactory epithelium, in which some 10^7 flavor receptors cells are located.

The specific balance between the responses from the different sensors determines the satisfaction of a consumer with a product. This balance is rather critical, and deviations due to defects may induce a consumer to reject the product. Two important attributes for margarine are the absence of inhomogeneities in a product (smoothness) and a desirable taste and flavor.

A margarine should be perceived by a consumer as a homogeneous deformable material. The hardness of a margarine can be assessed easily and, after some training, rather accurately by means of fingers only. Work softening can be judged from the change of feeling while kneading the product by hand. However, the determination of those characteristics by means of simple instrumental tests is in the end more reliable because humans suffer more from drifting standards than instruments do. Feeling of the product during kneading by hand is indispensable for the detection of inhomogeneities in the structure. Such invisible defects are usually characterized as lumpiness. Lumpiness is not tasted but has a negative effect on spreading performance; it hampers the formation of a smooth surface of the spread film.

Margarine should not contain inhomogeneities; even larger fat crystals or agglomerates are considered defects. This implies that fat crystals and fat crystal agglomerates in spreads should be so small that they cannot be detected individually in the mouth. The formation of large crystals or agglomerates occurs during storage and is a result of recrystallization processes. Two types of defects can be distinguished: sandiness and graininess.

Sandiness relates to crystals that are so hard and strong that they fail to break or deform in the mouth. Sandiness often occurs in blends of oils from a single source that are hardened to different temperatures so as to obtain a desired melting behavior. These blends are usually rich in tri-elaidate (elaidic acid = 9-trans-octadecenoic acid), which crystallizes separately during recrystallization.

Graininess relates to crystal structures that are only marginally stronger than their environment. The impression on eating is sometimes described as mealy. Graininess is often caused by a compound crystal of PPO and POP, unsymmetric and symmetric di-palmitate mono-oleate triglycerides. Mixtures of palm oil and interesterified palm oil are prone to have this defect.

Even more important than smoothness is the flavor cocktail and the flavor release profile of a margarine—that is, the mixture of perceived flavors and its change in intensity during mastication (Overbosch, Agterof, & Haring, 1991). The flavor release rate is determined by three subsequent steps: the release of the flavor from the product to the saliva, the transport from the saliva in the air in the oral cavity, and the flow of flavor-loaded air from the oral cavity to the nasal cavity. The last two processes depend on the physiological characteristics of the consumer (e.g., the time required for mastication, the amount of saliva produced, and the breathing pattern during mastication). Only the first step in the sequence depends on the properties of the product. Therefore, attention may be focused on the equilibration of the flavor between food particle and saliva.

The equilibration time depends strongly on the degree to which the product has been mixed into the saliva. The concepts of modern mixing theories enable an excellent visualization of the process of mixing food and saliva. These theories consider a mixing process as a sequence of stretching and folding operations (Ottino, 1989). Each stretching step leads to a reduction of the dimensions of the phases to be mixed perpendicular to the direction of stretching, while the folding steps are necessary to keep the overall external dimensions of the mixture constant. In the context of mastication, both food and saliva lumps are permanently being stretched, but the dimensions of the bolus of food and saliva remain about constant. The equilibration time of the flavor, which was initially present only in the food, will depend on the characteristic dimensions of both the food phase and the saliva phase. If different foods are assessed for their flavor release—in, for example, a tasting panel—this will be done by comparing equal amounts of product. The ratio of the volumes of the food phase and the saliva phase will then be equal for all products, and therefore only one independent dimension remains. In addition, the diffusivities in the food are usually lower than those in saliva and thus the dimensions of the food phase control the rate of flavor equilibration. The time to equilibrate will then in the first instance be determined by the food properties and not by the saliva properties.

Equation 13

$$t \approx \frac{R^2}{D}$$

R is the characteristic dimension of the food phase—that is, the average radius of the particle—and D is the diffusion coefficient of the flavor. A reasonable order-of-magnitude estimate for D in a water-continuous system is $10^{-9}\ m^2 \cdot s^{-1}$. Since the viscosity of edible oils is about 50 times greater than that of water, D is typically 50 times lower for a fat-continuous system. The mastication process takes only a few seconds. If the consumer is to perceive all the flavor present in the food, aqueous foods should be reduced to particle sizes of less than 70 μm, and fatty foods typically below 10 μm. The time it takes to reach these particle sizes will clearly determine the perceived flavor-release profiles. The rate at which these particle sizes are attained depends mainly on the properties of the food product but will depend somewhat on the chewing habits of the consumer as well. These critical particle dimensions are actually reached as soon as the food consistency or yield stress drops below the critical value of about 50 Pa.

The release of water-soluble taste components such as salt is primarily controlled by the stability against coalescence of the internal water phase. Those components can become released only if the internal water drops coalesce with the external saliva phase. However, coalescence of the internal aqueous phase is equally important for a reduction of the yield stress below the critical value of 50 Pa. The release patterns of both water-soluble taste components and oil-soluble flavor components are finally controlled by the same mechanism: coalescence of the internal aqueous phase.

The flavor-release profile is relatively difficult to measure experimentally. Sensory release profiles or time-intensity graphs are the assessor's manual records of the perceived intensity with time. Instrumental release profiles can be obtained by, for example, a mass spectrometer or another flavor-sensitive sensor measuring the flavor concentrations in the exhaled air.

Reduced Fat Spreads

The rate at which water droplets in halvarines coalesce affects the consumer's appreciation for the product: halvarines in which coalescence is relatively fast are experienced as more pleasant than products for which this coalescence is relatively slow. The slowly coalescing product in Figure 7–20 was perceived as more sticky, fatty, and waxy than the readily coalescing product. It is therefore not surprising that halvarines are always perceived as more waxy and more fatty than margarines of exactly the same fat phase, notwithstanding the lower fat content of the halvarines. The effect of rate of coalescence on fatty impression is illustrated in Figure 7–21 also. An increasing yield stress of the shock-heated product will coincide with a reduced rate of coalescence, resulting in an enhanced fatty impression of the product.

Zero-Fat Spreads

The majority of flavor substances are soluble in fat and relatively insoluble in water. This makes fatty foods excellent flavor carriers: small amounts of flavored fat are enough to add flavor to a large amount of fat-

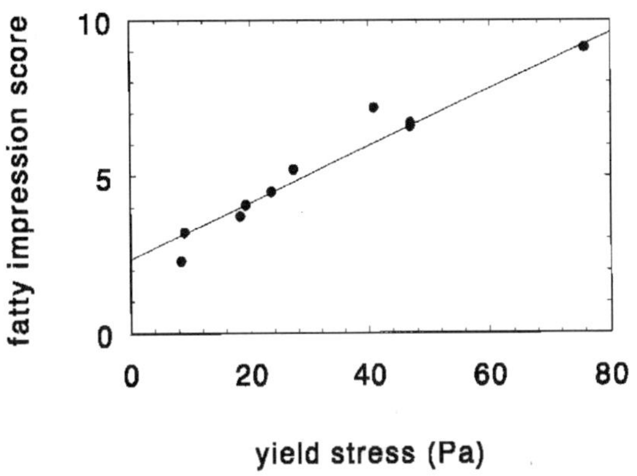

Figure 7–21 Relation between Thickness or Fatty Impression Panel Score and the Yield Measured 10 s after Shock-Heating a Number of Halvarines

free foods. Fat-free foods cannot act as carriers for oil-soluble flavors. Zero-fat spreads fall short with respect to this property in particular.

The flavor-carrying capability is determined by the partition coefficient of the flavor of interest—that is, the ratio of the equilibrium flavor concentrations in oil and water respectively. That quantity is an adequate measure for the degree to which one may dilute the product without affecting the flavor perception.

Similar effects cause the difference in "aftertaste" between full- and zero-fat spreads. Minute remainders of a full-fat spread in the oral cavity—for example, between the teeth or on the tongue—are enough to flavor adequately all the saliva present in the oral cavity and give the assessor the impression of an aftertaste. This is true in particular for flavors with a high partition coefficient. Zero-fat spreads do not have that "resistance" to dilution and usually fail to produce an aftertaste.

Chocolate

Chocolate is produced by mixing cocoa butter and ground cocoa beans. The addition of cocoa butter is necessary because otherwise the product will remain coarse and less workable (Hoskin, 1994). The solid materials in chocolate should be present as very small particles to allow the creamy texture and appealing melting properties that consumers associate with chocolate (Morgan, 1994). Premium chocolate blocks may have particle sizes as small as 10 to 12 μm, whereas chocolate used as a coating in bars typically has particle sizes of 20 to 40 μm, and other forms of chocolate may contain particles with sizes up to 40 to 80 μm (Morgan, 1994). The minimum size of particles that can be detected by the tongue is 20 to 30 μm (Hoskin, 1994).

Cocoa butter is composed of a number of very similar fats. This causes it to melt over a much narrower temperature interval than, for example, fats used in margarines. In combination with the fact that the melting points of these fats are slightly lower than mouth temperature, melting of the chocolate is experienced as "cooling" (McGee, 1984).

For chocolate, it is possible to correlate sensory scores on "bite firmness," "melting properties," "smoothness," and "stickiness" with instrumental texture studies (Tscheuschner & Markov, 1986; Markov & Tscheuschner, 1989). The instrumental techniques are the breaking stress in compression for the bite firmness; the rate of melting of a solid chocolate block placed on a plate thermostated at 37°C as measured by the remaining height of the block during gradual melting, for the melting properties; and the maximum size of the dispersed solid components, as measured by a precision film thickness-measuring machine, for the smoothness.

O/W EMULSIONS: MAYONNAISE AND DRESSINGS

The origin of mayonnaise cannot be traced in as much detail as that of margarine. Mayonnaise appears to be the outcome of a gradual evolution of strongly tasting sauces to modern, more subtly flavored sauces. It is thought to have been introduced into French cuisine after Richelieu conquered the isle of Menorca in 1756, and it derives its name from the harbor of Mahon on that island (McGee, 1984).

Mayonnaise is an o/w emulsion, containing typically 80% oil. Interestingly, its properties are determined mainly by the dispersed droplet phase; these products derive their consistency from the fact that the oil droplets form a space-filling structure. With regard to the importance of the dispersed phase, mayonnaise can be considered to be the exact opposite of margarine, for which only the properties of the continuous phase matter. This qualitative difference between mayonnaise and margarine justifies treating margarine (and other w/o emulsions) and mayonnaise (and other o/w emulsions) separately in this chapter.

Other products in the o/w emulsion category are salad dressings and reduced- or zero-fat dressings. Salad dressings are usually 40% o/w emulsions, which renders the volume fraction of oil too low to structure the product. Additional structure or product strength has to come from added biopolymers or from enhanced sticking forces between the drops (i.e., relatively high contact angles). At still lower fat levels, the amount of oil is no longer important for the structure, but only for the flavor. The structure or thickness of the product has to come from the added proteins or polysaccharides.

Structure

Mayonnaise is an emulsion containing typically 80% liquid vegetable triglyceride oil, such as sunflower oil. The size of the droplets is usually between 1 and 5 μm. The polydispersity of the droplets determines the amount of oil that can be emulsified: if a wide range of droplet sizes is present in the emulsion (as is usually the case for homemade mayonnaise), somewhat more oil can be dispersed than when a narrow range of droplet sizes is produced.

It is therefore important to control the emulsification process. This is relatively difficult, as anyone who failed once to make mayonnaise at home is likely to agree. On an industrial scale, however, the process can be controlled adequately by using specialized equipment. In such a process, homogenization usually occurs in two steps. In the first step, oil is dispersed coarsely to 20- to 100-μm droplets in a slow mixing process in a stirred tank. In the second step, this preemulsion is dispersed more finely in the vigorous mixing process—for example, in a colloid mill.

In a colloid mill, the emulsion is forced through a narrowing gap between a stator and a rotor. In this gap, very high shear rates (10^4 to 10^5 s^{-1}) break up the droplets very efficiently. The breakup process is so efficient because the high shear rates in the gap allow the droplet to stretch to a very long filament before breakup occurs as a result of Rayleigh instabilities. The resulting fragments are in this case smaller than in the case of binary breakup (Wieringa, van Dieren, Janssen, & Agterof, 1996). A binary breakup process, such as that occurring at much lower shear rates, would require many more breakup steps.

The presence of emulsifiers such as lecithin reduces the interfacial tension γ to only a few mPa • m instead of 30 mPa • m at clean o/w interfaces. A lower interfacial tension helps to obtain smaller droplets in the homogenization process. The proportionality of droplet size and interfacial tension was already implied by the definition of the critical capillary number for binary breakup, as given in the caption for Figure 7–18.

The resulting droplet size of 1 to 5 μm is sufficient for mayonnaise, but a further size reduction significantly below 1 μm is possible if more vigorous mixing equipment is used. The drop size is very important for the mechanical properties of the mayonnaise. New homogenization techniques for smaller droplet sizes may be considered if stiffer mayonnaises are desired.

In a concentrated mayonnaise emulsion, droplets are densely packed. The presence of neighbors deforms the droplet. The exact type of interaction of the droplet with its neighbors determines the details of the deformation. An external pressure acting on the drops (e.g., gravity, stirring) will result in a smooth transition from the flat surfaces at the contact between two droplets to the curved free surfaces of the droplets, whereas short-range physicochemical interaction forces, such as van der Waals forces, will result in a sudden change from flat lamella at the droplet contact to the curved surfaces away from the lamella. The sudden transition between free droplet surface and lamella is characterized by a contact angle.

The situation for two droplets with short-range physicochemical interactions is drawn schematically in Figure 7–22. The radius *a* of the (circular) contact area between the droplets and the impression *h* of the droplet at the contact are related to the contact angle ϑ:

Equation 14

$$a = R \sin \vartheta$$
$$h = R (1 - \cos \vartheta)$$

with *R* being the radius of the undeformed droplet. Contact angles are usually very difficult to establish by microscopic observations because they tend to be small in practice, of the order of a few degrees.

Concentrated emulsions require effective emulsifiers for their stability. Three processes are important for the stability of emulsions: first, the rate of drainage of the lamella between two droplets as a result of an external force; second, the depletion of the emulsifier and the extent to which this process is counteracted by diffusion of the emulsifier back into the lamella; and third, the attractive or repulsive nature of the short-range physicochemical interaction forces between two droplets when the lamella between the droplets has thinned extremely. These three processes will be discussed in some detail below.

The interdroplet forces in the emulsion, which are a result of stirring, sedimentation, or creaming processes, determine the size of the contact area. Drainage of the contact films is a first step toward the eventual coalescence of the drops. The drainage depends strongly on the size of the contact film. Greater interdroplet forces produce larger contact films, which drain more slowly. Equivalently, an increase of the (draining) forces between the drops slows down the drainage of the separating films and reduces the coalescence frequency.

The drainage is determined by the mobility of the interface between film and drop. In the absence of emulsifiers, the viscosity of the drop phase is the parameter controlling the interfacial velocity. Drainage rates and coa-

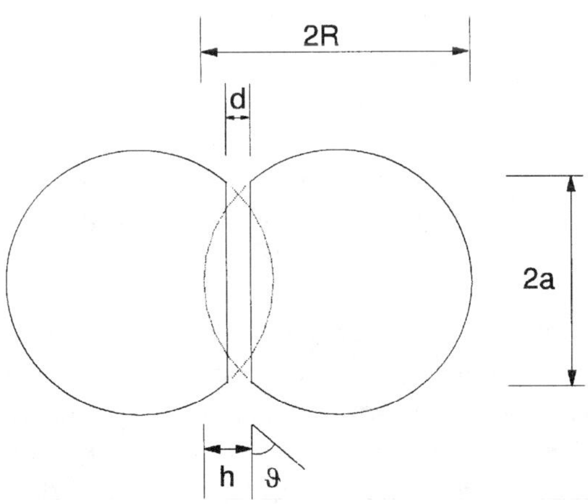

Figure 7–22 Schematic Representation of a Deformed Droplet. *R*, radius of the undeformed droplet; *a*, radius of the contact film; *h*, impression of the droplet; *d*, distance between contact films; ϑ, contact angle.

lescence frequencies are determined by the drop viscosity and not by the viscosity of the film between those drops.

The presence of emulsifiers may decelerate the drainage further. An important consequence of the interfacial motion is that the emulsifiers are depleted from the approaching films. This leads to a relative increase in the interfacial tension in the film, which immobilizes the interfaces (Lucassen-Reynders, 1996). This is called the *Gibbs-Marangoni effect*. Diffusion counteracts the emulsifier depletion. That process is relatively fast for emulsifiers dissolving preferentially in the oil phase, and slow for water-soluble emulsifiers. This is caused by the fact that oil-soluble emulsifiers can be readily supplied from the drops, whereas water-soluble emulsifiers have to diffuse "upstream" through the narrow film to reach the depleted interface.

If the films reach a certain critical thickness, short-range physicochemical forces between the interfaces may stop the drainage process. This involves a balance between the external forces on the droplets, the attractive van der Waals forces, and the repulsive double-layer or steric forces.

If the physicochemical interaction force between the droplets is large enough to prevent coalescence, the droplets may still flocculate. The degree of flocculation may be characterized by the contact angle between the film interface and the free-drop interface.This contact angle relates to the interaction energies rather than the forces. To understand that relation, consider two free interfaces that approach each other from initial infinite separation to an equilibrium film thickness d. The work done by the interaction forces between the interfaces during the approach will equal the difference in surface energy between film interfaces and free interfaces. According to Young's law, this equals $2\gamma(1 - \cos \vartheta)$. Qualitatively, this means that contact angles relate rather uniquely to the equilibrium film thickness: thinner equilibrium film thickness (i.e., more work done) leads to greater contact angles ϑ. In practice, large contact angles can be obtained by using emulsifiers with a small hydrophilic part or by using a combination of ionic emulsifiers in a high–ionic strength environment.

The contact angle is representative of the degree to which drops stick to each other. Apart from depending on the size of the emulsifier, contact angles are affected by such phenomena as bridging flocculation and depletion. These effects are often related to specific ingredients in the food emulsion.

Food proteins are surface active and stabilize o/w emulsions by adsorbing at the droplet interfaces. Some proteins, particularly globular proteins, are able to adsorb at two interfaces simultaneously. Protein adsorption at the film interfaces becomes higher than at the free-drop interfaces in that case, and the contact angles may become quite large. The term *bridging flocculation* is used to describe this sticking mechanism between drops.

Polysaccharides are seldom surface active and do not adsorb at the droplet interface. However, when the radius of gyration of the nonadsorbing polysaccharide is larger than the equilibrium film thickness, the polysaccharide will be present everywhere in the continuous aqueous phase but will be excluded from the film between the drops. The absence of the polysaccharide in the film will cause an osmotic pressure difference between the film and the rest of the continuous phase, which can be compensated for only by pulling the droplets nearer to each other. As a consequence, the contact angle increases. This process is called *depletion flocculation*.

The stability of o/w food emulsions is affected negatively by the presence of a fat crystal network in the oil droplets. The network reduces the deformability of the oil phase, resulting in a smaller contact area between the droplets. Smaller contact areas lead to higher pressures and pressure gradients in the lamella, and this leads to more rapid drainage of the lamella.

The effect of the reduced deformability of the drop phase on the emulsion stability can be understood more quantitatively by considering the contact between two drops of radius R, pressed together by a force F (e.g., gravity, stirring). The force will deform the droplets, which will develop flat circular contacts with their neighbors. The radius a of these lamella is determined by a balance between the force and the forces opposing drop deformation: the yield stress τ_y of the fat crystal network in the droplet and the interfacial tension γ:

Equation 15

$$\frac{F}{\pi a^2} = (1 + \pi) \bullet 2\tau_y + \frac{2\gamma}{R}$$

The left-hand side of Equation 15 represents the mean overpressure (relative to ambient pressure) in the separating film. The right-hand side represents the sum of the plastic deformation force on the fat crystal network and the Laplace pressure. The force required for plastic deformation of the network can be derived from the problem of subsidence of a truncated cone with cone angle $\pi/2$ (see Equation 9). The second term represents the Laplace pressure, which is typically 10^3 Pa. Equation 15 illustrates that the radius a of the lamella decreases with increasing yield stress τ_y, particularly if the yield stress exceeds 10^3 Pa. This is the case at a typical fat solids level of about 3% in the oil phase. Most o/w food emulsions contain a crystal-free oil phase to avoid stability problems (in contrast to w/o food emulsions).

The order of magnitude of the pressure in the film is F/a^2, implying a pressure gradient of the order F/a^3. To balance this pressure gradient in order to stabilize the emulsion against drainage, a gradient in interfacial tension is required of magnitude hF/a^3. Using Equation 15, it can be shown that

Equation 16

$$\frac{d\gamma}{dr} \propto \{(1 + \pi) \bullet 2\tau_y + \frac{2\gamma}{R}\}^{3/2}$$

This shows that greater gradients or better emulsifiers are required to stabilize emulsions of less deformable drops (i.e., greater yield stresses or greater Laplace pressure).

The churning process is a practical illustration of the consequences of drop deformability. The milk proteins in the cream are capable of stabilizing the o/w emulsion as long as fat crystals are absent in the oil drops. Stabilization fails as soon as crystallization sets in and structure building in the oil phase starts. Coalescence sometimes occurs only "partly"—that is, the separating film at the contact disappears, but the colliding drops fail to form a single larger spherical drop. Partial coalescence occurs in the case that the yield stress is greater than the Laplace pressure. The yield stress prevents the surface tension from making the new droplet perfectly spherical. Since coalescence does not occur in the case of yield stresses smaller than the Laplace pressure, coalescence is always expected to be partial in churning processes.

Rheology

The viscosity of an emulsion is determined predominately by the volume fraction ϕ of the dispersed drops. At low volume fractions, the viscosity will increase with the amount of dispersed phase according to the Einstein equation:

Equation 17

$$\eta_{\text{emulsion}} = \eta_{\text{continuous}} (1 + 2.5\phi)$$

At intermediate volume fractions, the relation between viscosity and ϕ will become nonlinear. Under such conditions, it is more appropriate to use the so-called Krieger-Dougherty relation,

Equation 18

$$\eta_{\text{emulsion}} = \eta_{\text{continuous}} \left(1 - \frac{\phi}{\phi_0}\right)^{-K\phi_0}$$

with ϕ_0 the maximum packing fraction of (undeformed) droplets in the emulsion. The Krieger-Dougherty relation reduces to the Einstein equation in the limit of vanishingly low volume fractions if $K = 2.5$ (Krieger & Dougherty, 1959). In principle, the deformability of the emulsion droplets gives a viscoelastic character to the emulsion. However, this effect is negligible as long as the deformation of the droplets is negligible, which is the case as long as the stresses applied are below those of the homogenization process. In practice, this means that an emulsion at intermediate volume fractions will behave as purely viscous during handling or eating because the imposed stresses remain well below the stresses of the homogenization process during production of the emulsion.

At high volume fractions, $\phi > \phi_0$, the emulsions show plastic rheology rather than viscous behavior. This is caused by the fact that each drop is encaged between neighboring drops. The system of drops is completely locked. Flow, which is to be associated with motion of drops relative to neighboring drops, can be realized only if some of the drops deform sufficiently, such that these can jump to a new, lesser stressed position. Flow will thus occur only at stresses high enough to deform the drops. This implies that concentrated emulsions exhibit a yield stress. This yield stress and the deformation at which this stress is reached are the prime rheological characteristics for these products.

Mayonnaise is an emulsion containing a high volume fraction of oil droplets. From the above considerations, it is now possible to predict the main features of the stress-strain curve of mayonnaise in terms of the microstructure of the emulsion. To do this, consider the droplet microstructure as depicted in Figure 7–22.

When two droplets of radius R are brought together, the impression h of a droplet at the point of contact (or the radius of the contact area a) is determined by a combination of the pressure acting on the drops and the attractive forces between them. In the case of zero pressure and finite attractive forces, the drops remain perfectly spherical except for the film between the drops. The sharp transition between the spherical surface and the film will have a characteristic finite contact angle. In the case of a finite pressure and vanishing attractive forces, the free surface of the droplet curves continuously toward the film. There is no contact angle, but a gradual transition between the free surface of the droplet and the film.

First, consider the situation where a densely packed system of (many) droplets is deformed, neglecting any attractive forces between the droplets. Two regimes can be distinguished under these conditions. When moderate stresses are applied to the system, the deformation of the droplets will be small. The configuration of droplets will not be disturbed, and the response of the system will be elastic. The macroscopic deformations can be accommodated by local compression and extension of the droplets at their contact points, and the number of contacts of each drop with its neighbors will remain constant. The elastic properties of the deforming droplet will be determined via the interfacial tension by the Laplace pressure of the droplet. This quantity can be taken as measure for the elastic modulus of a single oil drop, and automatically as the modulus of the complete emulsion.

Equation 19

$$G \propto \frac{\gamma}{R}$$

Greater macroscopic deformations cannot be accommodated by the deformation of the contacts between the droplets only. Under these conditions, the systems starts to flow, and the configuration of the droplets will change: drops will move or jump with respect to each other in order to accommodate the macroscopic deformation. This will lead macroscopically to plastic flow of the emulsion.

The crossover from elastic deformation to plastic flow may be identified with the onset of breakage of contacts between drops. The critical deformation γ_c at which the contacts between drops will break is typically of the order h/R, with h being the impression of the droplets at a given external pressure. This ratio determines the linear elastic region of deformation.

Equation 20

$$\gamma_c \propto \frac{h}{R}$$

The consistency or yield stress is the product of the elastic modulus and the extent of the elastic regime, γ^h/R^2.

This behavior is illustrated in the schematic stress-strain curve for mayonnaise that is plotted in Figure 7–23. At vanishingly small deformations, the stress increases proportionally to the applied deformation, and deformation is elastic. In this deformation regime, the contact area between the droplets in the emulsion will increase in the case of compression and decrease in the case of extension. At larger deformations, the stress becomes constant, and flow is plastic. Typical values for mayonnaise are $\gamma/R = 10^3$ Pa, and $h/R = 0.05$.

Equations 19 and 20 imply that the shape of the stress-strain curve of an emulsion can be manipulated in principle by two physical

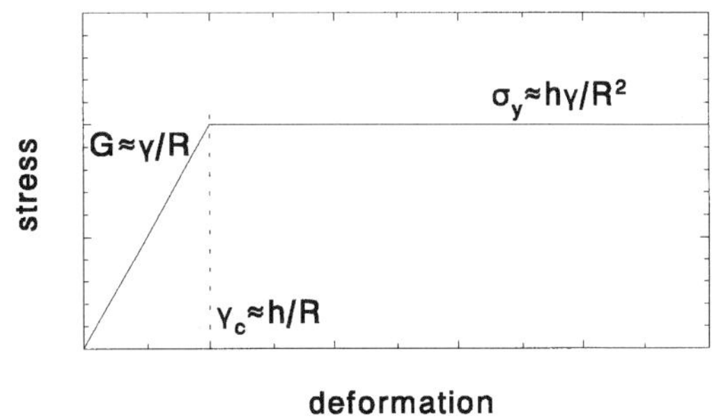

Figure 7–23 Schematic Representation of the Stress-Strain Curve of Mayonnaise, Showing the Essence of the Relation with the Microstructure

properties, the Laplace pressure and the degree of contact between the droplets. The Laplace pressure in the emulsion droplets can be varied by means of the homogenization process: further homogenization leads to smaller droplets and thus to stiffer emulsions. The contact between the droplets can be changed by adding emulsifiers: since $h/R = 1 - \cos \vartheta \approx 1/2\ \vartheta^2$, a strong dependence of rheological properties on contact angles is expected. In addition to reducing the drop size, homogenization may increase the contact angle through bridging flocculation. These effects are quite pronounced in the case of proteins and partial coverage of the drops (Dickinson & Stainsby, 1988).

The validity of Equation 19 has been demonstrated in both theoretical and experimental work. The proportionality of the modulus of the emulsion to γ/R was confirmed by Princen (1979) and Mason, Bibette, and Weitz (1995). However, there is some disagreement on the dependence of the modulus on the volume fraction of the dispersed phase ϕ. Princen and Kiss (1986) found the following dependence for the shear modulus from a generalization to three dimensions of the behavior of two-dimensional foams in the absence of attractive forces:

Equation 21

$$G = \frac{A\gamma\,\phi^{1/3}}{R}(\phi - \phi_0)$$

where A is a dimensionless constant of about unity, ϕ is the volume fraction of the dispersed phase, ϕ_0 is the maximal packing fraction of the dispersed phase, and R is the radius of the emulsion droplet. Mason et al. (1995) reported a similar result for perfectly monodisperse emulsions, using moduli scaling with ϕ instead of $\phi^{1/3}$. Experimental data cannot distinguish between the theories: satisfactory fits can be obtained for both approaches (Pons, Taylor, & Tadros, 1997; Taylor & Pons, 1998).

As argued above, the critical strain is expected to increase proportionally to the square of the contact angle. For more concentrated emulsions, that dependence will be weaker. In a highly concentrated emulsion, where all flattening of the contacts between the drops is a consequence of crowding, one expects no dependence on the contact angle or the type of emulsifier anymore. The importance of the contact angle is confirmed by the data of Taylor and Pons (1998). They showed that the critical strain depends also on the nature of the emulsifier used for preparing the emulsion. In addition, they found an increase of the critical strain with the volume fraction ϕ. However, there are still insufficient data to develop a general theory incorporating the effect of volume fraction and contact angle on the rheological properties of concentrated emulsions.

From the considerations above, it is clear that the density of packing of the oil droplets determines the consistency of the mayonnaise. Minor reductions in fat level (e.g., below 70%, as in low-fat dressings) would lead to a considerable loss in consistency. This is illustrated in Figures 7–24 and 7–25. In practice, the problem is solved by replacing the oil droplets with biopolymer precipitates.

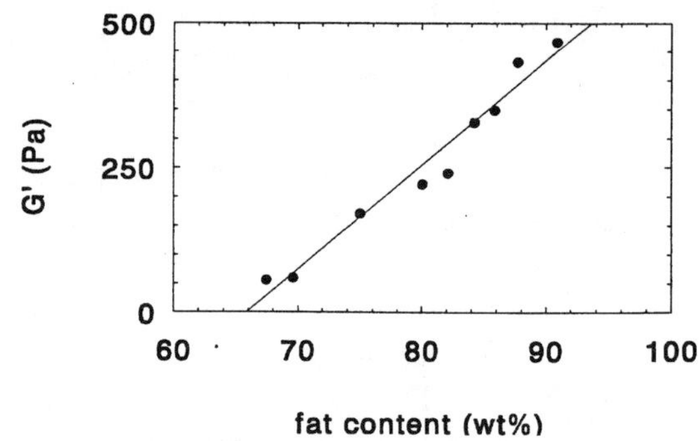

Figure 7–24 The Storage Modulus G' (at a strain of 0.05) for Dressings Increases Linearly as a Function of Fat Content

The use of precipitates will ensure that the critical strain of the low-fat product remains in the range from 0.05 to 0.10. A continuous biopolymer matrix would have a much larger critical strain (Bot et al., 1996), and the product would be perceived as a gelly pudding rather than as a dressing.

The problem of reproducing the mechanical properties of full-fat mayonnaises by using such biopolymer dispersions comes down to matching both the impression at the droplet contact (Equation 20) and the modulus at the contact (Equation 19) to the values of these quantities in a mayonnaise. The primary elastic contacts between the biopolymeric par-

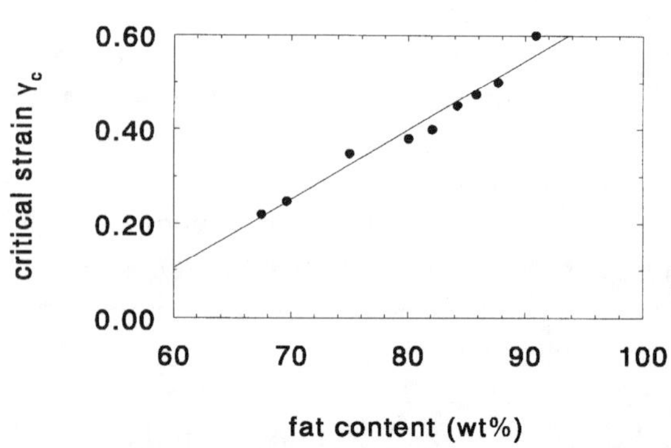

Figure 7–25 The Critical Strain γ_c for Dressings Increases Linearly as a Function of Fat Content Also

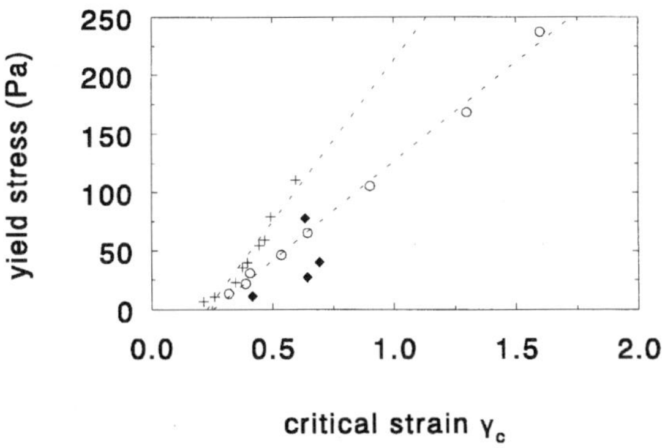

Figure 7–26 Yield Stress versus Critical Strain for Various Dressings. +, Full-fat mayonnaise; ○, model dressing prepared from Colflo; ◆, model dressing prepared from Simplesse.

ticles behave as Herzian contacts (Landau & Lifshitz, 1960), which implies that the impression at the contact varies with the applied force as $F^{2/3}$. Therefore, F varies as $(h/R)^{3/2}$ for a biopolymer precipitate, in contrast to h/R for an emulsion. The consequence of this behavior is that the proportionality factor between stress and deformation h/R becomes $E(h/R)^{1/2}$, with E as Young's modulus, instead of $G = \gamma/R$ as for an emulsion. This quantity replaces the Laplace pressure if we are to obtain rheological behavior similar to that of a mayonnaise. The particle concentration must also be adjusted to match the yield stress and the critical strain.

Figure 7–26 illustrates that the yield stress is a function of the critical strain for mayonnaise having different fat levels and a few fat-free dispersions of biopolymer precipitates: modified starch (Colflo) and aggregated whey protein (Simplesse). The considerations above about Herzian contacts imply $\tau_y \propto \gamma_c^{3/2}$. The data are in qualitative agreement with this expectation.

Figure 7–26 clearly shows that the critical strain (and thus the yield stress) of these fat-free dressings deviates from mayonnaise behavior as a function of the concentration. At equal yield stress, the model dressings containing biopolymer precipitates show a larger critical strain than mayonnaise. In other words, the biopolymeric precipitates used in Figure 7–26 are not dense enough to meet the modulus requirements of a mayonnaise. Special processing routes are often necessary to make the precipitate dense enough for this purpose. A mayonnaiselike rheology can be obtained by mixing different biopolymeric precipitates of different modulus.

Finally, a small-deformation rheological characterization is shown in Figure 7–27. The results show that the storage modulus decreases with increasing strain, whereas the loss modulus is more or less independent of applied strain. G' and G'' cross at a strain of approximately 0.4.

Breakdown

The nature of mayonnaise, an emulsion with relatively little interaction between the droplets, does not allow for much breakdown as a consequence of working. This is illustrated in the flow-curve shown in Figure 7–28. The graph shows the stress during a shear rate sweep cycle. It appears that the difference between the stress during the increasing shear rate sweep and the stress during the decreasing shear rate sweep is only marginal. Breakdown may occur during eating as a consequence of dilution by saliva.

Emulsion droplets cannot break up during the relatively slow deformation processes of handling and consumption but will break up only at flow rates similar to those occurring during the homogenization process. Any changes in behavior of the emulsion should find their origin in effects on the interactions between the droplets. The main interaction is the pressure that the droplets exert on each other. This is determined by the volume fraction of the oil droplets, which can change only if an additional water phase is present

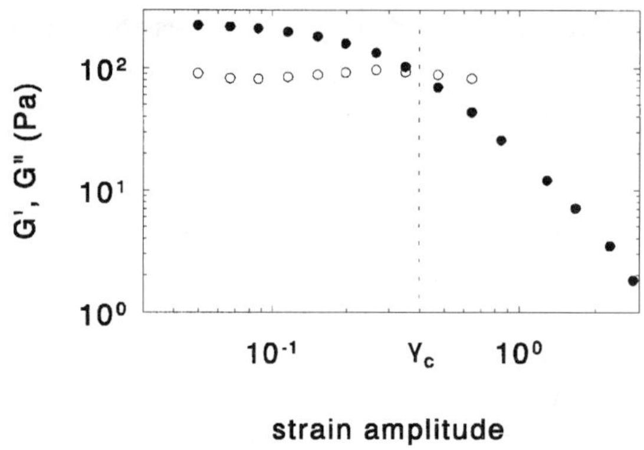

Figure 7–27 Dynamic Measurements as a Function of Deformation for Full-Fat Mayonnaise. ●, storage modulus G'; ○, loss modulus G''. The strain at the intersection of G' and G'' is taken as a measure for the critical strain γ_c.

(e.g., saliva). Without such a water phase present, the contact angle between the drops is the only microstructural parameter that may change. A decrease of this angle as a consequence of flow leads to a reduction of the yield stress of the mayonnaise.

Gentle stirring may sometimes lead to the opposite of structural breakdown—namely, shear thickening. In such cases, the yield stress increases somewhat as a consequence of the randomization of the droplet configuration, which results in a less efficient packing. The so-called *critical volume fraction* at which the emulsion is just closely packed decreases somewhat as a consequence of randomization of the configuration. The effect of randomization disappears slowly at rest, when the drops gradually move into a position where they fit somewhat better between their neighbors. The critical volume fraction increases, and the yield stress of the emulsion returns to its initial, slightly lower value.

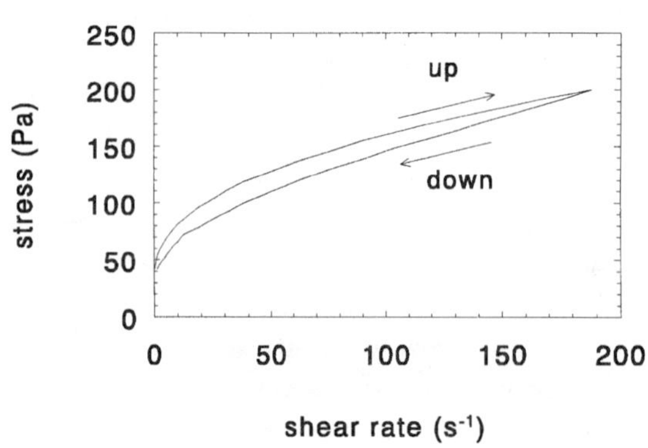

Figure 7–28 Response to a Shear Rate Cycle of a Full-Fat Mayonnaise (80% oil) at 20°C. From the approximate coincidence of the increasing and decreasing branches, it can be concluded that no significant structural breakdown occurs.

If a water phase is present, however, breakdown of the emulsion occurs more easily. The most relevant situation in which this occurs is in the mouth under the influence of saliva. The mechanism by which this occurs is quite similar to the mechanism proposed earlier for zero-fat spreads.

The tensile stresses exerted on the mayonnaise lumps during oral dispersion will induce an inward flow of water because the emulsion droplet network cannot support these stresses. The emulsion will become more dilute as a consequence of external stresses, and the yield stress will decrease. The interaction with water or saliva could be a factor of importance for the oral dispersion of dressings with a yield stress considerably greater than 50 Pa. However, the yield stress of full-fat mayonnaise products is often low enough to allow adequate oral dispersion even without dilution by saliva.

Other breakdown processes occurring in the mouth are rare. For stability reasons, few o/w emulsions contain melting fat crystallite structures, as was already discussed above. An example is sour cream, a dairy product containing partially crystallized butter fat that will melt as the temperature rises to mouth temperature.

Sensory Properties

The main function of dressings is to add taste and flavor and to improve the visual impression of other foods. The role of the dressing is essentially a combination of a decorative material and a flavor carrier. This role requires an optimum combination of mechanical, oral dispersive, and mouth-feel characteristics. These characteristics will be discussed in more detail below.

The mechanical properties of a dressing are mainly related to the decorative function for other foods. To this end, dressings have to exhibit a yield stress because otherwise the decorative pattern would drain away from the food. The yield stress of traditional full-fat mayonnaise is typically about 50 Pa. This yield stress should withstand the hydrostatic pressure of a decorative whorl. A full-fat mayonnaise can support a decoration 5 mm in height, which is high enough to allow shaping of a decoration. However, a higher decoration will start to flow. It should also be noted that the requirements for decoration and oral dispersion are conflicting. Decoration becomes easier at yield stresses higher than 50 Pa, whereas oral dispersion requires yield stresses less than 50 Pa. Thus, the yield stress of mayonnaise is a compromise.

In addition, dressings can be used to apply a coating to a food product. It can be demonstrated easily that the maximum thickness of a plastic film clinging to a vertical wall equals $\tau_y/\rho g$, where τ_y is the yield stress, ρ the density, and g the acceleration of freefall. For dressings with a yield stress of 50 Pa, this implies a film thickness of 5 mm. Dressings with a lower yield stress can still coat a food product adequately. Therefore, the yield stress of a dressing should not be regarded as critical for its coating performance. However, it is important that the yield stress remains finite because it affects the flavor-carrying performance of the dressing as well. Rather than dripping through a bowl of salad, the product should form an adequate coating on salad and vegetables to distribute the added taste and flavor components uniformly throughout a food.

The quantity of other foods that can be flavored adequately by the dressing depends to a large extent on the amount of oil and oil-soluble flavors in the product: the most desirable flavors in foods are fatty flavors, which are better soluble in oil than in water. Actually, the flavor-carrying capability will be directly proportional to the oil/water partition coefficient of the flavor and the oil content of the product (de Roos & Wolswinkel, 1994; Overbosch et al., 1991). This implies that the minimum amount of oil in the dressing will be determined by the nature of the flavor that has to be delivered by the dressing. Experiments by Bakker and Mela (1996) have shown that o/w and w/o emulsions release their taste and flavor similarly. Thus, the microstructural details of the emulsion are not important provided that the (w/o) emulsion is thin enough to be broken up in the mouth.

The sensory properties just discussed are related to and consequences of a limited number of structural, mechanical, or breakdown characteristics of the food. These relations are quite important, since they open the possibility to adjust the sensory performance of a product by manipulating its structural, mechanical, and breakdown characteristics.

However, such direct relations between structural or mechanical characteristics and the sensory properties of the emulsion can be identified only for a limited number of attributes. Often, attributes cannot be determined through an objective instrumental approach, and human assessors are required to quantify the sensory performance. The standardized technique for this is QDA. The sensory performance is in essence represented by means of appraisal scores of a large number of descriptors such as thickness, smoothness, and creaminess (Szczesniak, 1963; Kramer & Szczesniak, 1973). A large portion of the sensory studies aim at establishing correlations among the various descriptors and correlations between the descriptors and the results of simple instrumental techniques.

In the QDA perspective, the texture of food is considered to be the combined sensory experiences from the moment at which it is placed in the mouth until swallowing. Afterfeel effects, such as astringency and mouth coating, are often considered to be a separate experience. Texture perception is a highly dynamic process because the physical properties of foods change continuously in the mouth (Guinard & Mazzucchelli, 1996).

The perception of fat in foods depends largely on the structure felt in the mouth, and to a lesser degree on olfaction (Drenowski, 1993). For semisolid and liquid products such as dressings, texture is expressed in terms of attributes such as thickness, smoothness, or creaminess. These sensory attributes have been related to objective physical quantities.

Thickness has been identified as a quantity that is correlated to the viscosity, in particular the viscosity at high shear rates (Kokini, 1987; Stanley & Taylor, 1993).

Smoothness has been correlated to friction forces, which relate to the yield stress at low shear rates. It has been suggested that smoothness is decreased by increasing the average size of the oil droplets and by increasing their polydispersity (Guinard & Mazzucchelli, 1996; Richardson & Booth, 1993). Smoothness is regarded as a highly desirable property of dressings.

Creaminess is an attribute that has viscous, flavor, and taste aspects (Stanley & Taylor, 1993). The flavor component depends strongly on the presence and release of certain typical dairy flavors. If the influence of flavor on the panel scores is minimized by using nose clips, creaminess is found to correlate very well with viscosity (Mela, 1988). For mixtures of dairy cream and xanthan gum, creaminess was correlated to the degree of non-Newtonian rheological behavior. For a power law fluid, creaminess appeared to be optimal for mixtures with a power law index of 0.2 (Daget, Joerg, & Bourne, 1991). Similar results were obtained by Kokini and Cussler (1983), who related creaminess with thickness and smoothness. The general idea that creaminess relates to non-Newtonian flow behavior is supported by earlier studies by Wood (1974). Viscosity is not the only mechanical factor of importance, since creaminess is found to correlate with the size of the fat globules for dairy emulsions of equal viscosity (Richardson, Booth, & Stanley, 1993).

CONCLUSION

In this chapter, the structure, rheology, and perception of fabricated fat-based foods were reviewed. It was shown that the structure of a product is intimately related to its perception. For o/w emulsions, the main structural characteristic is the fat crystal network, with its melting properties and recrystallization capabilities. For w/o emulsions, the main characteristic is the dense packing of droplets. In both cases, these features can be translated into terms of the observed rheological behavior, and ultimately into many of the sensory properties of the product.

REFERENCES

Atkinson, J.H., & Bransby, P.L. (1978). *The mechanics of soils: An introduction to critical state soil mechanics.* New York: McGraw-Hill.

Bakker, J., & Mela, D.J. (1996). Effect of emulsion structure on flavor release and perception, in R.J. McGorrin & J.V. Leland (Eds.), *Flavour food inter-*

actions (ACS Symposium Series 633), pp. 36–47, Washington: American Chemical Society.

Batchelor, G.K. (1983). *An introduction to fluid mechanics.* Cambridge, UK: Cambridge University Press.

Bentley, B.J., & Leal, L.G. (1986). A computer controlled four roll mill for investigation of particle and drop dynamics in two-dimensional shear flow. *Journal of Fluid Mechanics, 167*, 219–240.

Birker, P.J.M.W.L., de Jong, S., Roijers, E.C., & van Soest, T.C. (1991). Structural investigations of β' triacylglycerols: An x-ray diffraction and microscopic study of twinned β' crystals. *Journal of the American Oil Chemists Society, 68*, 895–906.

Bloom, W., & Fawcett, D.W. (1968). *A textbook of histology*. Philadelphia: W.B. Saunders.

Bot, A., van Amerongen, I.A., Groot, R.D., Hoekstra, L.L., & Agterof, W.G.M. (1996). Large deformation rheology of gelatin gels. *Polymer Gels and Networks, 4*, 189–227.

Campbell, L.B., Anderson, D.A., & Keeney, P.G. (1969). Hydrogenated milk fat as an inhibitor of the fat bloom defect in chocolate. *Journal of Dairy Science, 52*, 976–979.

Carslaw, H.S., & Jaeger, J.C. (1959). *Heat conduction in solids*. Oxford, UK: Clarendon.

Chrysam, M.M. (1996). In I.H. Hui (Ed.), *Bailey's industrial oil and fat products* (Vol. 3, pp. 41–126). New York: John Wiley.

Coulomb, C.A. (1773). Essai sur une application des règles de maximis et minimis a quelques problémes de statique. *Memoires de Mathematique de l'Academie Royale des Sciences à Paris, 7*, 343–382.

Crank, J. (1975). *The mathematics of diffusion.* Oxford, UK: Clarendon.

Daget, N., Joerg, M., & Bourne, M. (1991). Creamy perception I: In model dessert creams. *Journal of Texture Studies, 18*, 367–388.

de Bruijne. D.W., Hendrickx, H.C.A.M., Alderliesten, L., & de Looff, J. (1993). Mouthfeel of foods. In E. Dickinson & P. Walstra (Eds.), *Food colloids and polymers: Stability and mechanical properties* (pp. 204–213). Cambridge, UK: Royal Society of Chemistry.

de Jager, E.M., van den Tempel, M., & de Bruijne, P. (1963). Permeability of plastic disperse systems. *Proceedings of the Koninklijke Nederlandse Academie van Wetenschappen, 66*, 17–35.

de Roos, K.B., & Wolswinkel, K. (1994). Non-equilibrium partition model for predicting flavour release in the mouth. In H. Maarse & D.G. van der Heij (Eds.), *Trends in flavour research* (pp. 15–32). Amsterdam: Elsevier.

Dickinson, E., & Stainsby, G. (1988). Emulsion stability. In E. Dickinson & G. Stainsby (Eds.), *Advances in food emulsions and foams* (pp. 1–44). London: Elsevier Applied Science.

Drenowski, A. (1993). Individual differences in sensory preferences for fat in model sweet dairy products. *Acta Psychologica, 84*, 103–110.

Freudenthal, A.M., & Geiringer, H. (1958). The mathematical theories of the inelastic continuum. In: S. Flügge (Ed.), *Encyclopedia of Physics, Vol. 6, Elasticity and plasticity* (pp. 229–433). Berlin: Springer.

Giddey, C., & Clerc, E. (1961). Polymorphism of cocoa butter and its importance in the chocolate industry. *Revue Internationale de la Chocolaterie, 16*, 548–554.

Grace, H.P. (1982). Dispersion phenomena in high viscosity immiscible fluid systems and application of static mixtures as dispersion devices in such systems. *Chemical Engineering Communications, 14*, 225–277.

Groeneweg, F., Agterof, W.G.M., Jaeger, P.H., Janssen, J.J.M., Wieringa, J.A., & Klahn, J.K. (1998). On the mechanism of the inversion of emulsions. *Chemical Engineering Research and Design Transactions of the Institution of Chemical Engineers Part A), 76*, 55–63.

Guinard, J.X., & Mazzucchelli, R. (1996). The sensory perception of texture and mouthfeel. *Trends in Food Science and Technology, 7*, 213–219.

Haighton, A.J. (1976). Blending, chilling, and tempering of margarines and shortenings. *Journal of the American Oil Chemists Society, 53*, 397–399.

Hicklin, J.D., Jewell, G.G., & Heathcock, J.F. (1985). Combining microscopy and physical techniques in the study of cocoa butter polymorphs and vegetable fats blends. *Food Microstructure, 4*, 241–248.

Hoffmann, G. (1989). *The chemistry and technology of edible oils and fats and their high fat products*. London: Academic Press.

Hoskin, J.C. (1994). Sensory properties of chocolate and their development, *American Journal of Clinical Nutrition, 60*, 1068S–1070S.

Jensen, L.H., & Mabis, A.J. (1963). Crystal structure of β-tricaprin. *Nature, 197*, 681–682.

Jewell, G.G. (1972). Some observations on bloom on chocolate. *Revue Internationale de la Chocolaterie, 27*, 161–162.

Johansson, D., & Bergenståhl, B. (1995). Sintering of fat crystal networks in oil during postcrystallization processes. *Journal of the American Oil Chemists Society, 72*, 911–920.

Johansson, D., Bergenståhl, B., & Lundgren, E. (1995). Water-in-glyceride oil emulsions: Effect of fat crystals on stability. *Journal of the American Oil Chemists Society, 72*, 939–950.

Juriaanse, A.C., & Heertje, I. (1988). Microstructure of shortenings, margarine and butter: A review. *Food Microstructure, 7*, 181–188.

Kasapis, S., Morris, E.R., Norton, I.T., & Brown, C.R.T. (1993). Phase equilibria and gelation in gelatin/maltodextrin systems. Part III: Phase separation in mixed gels. *Carbohydrate Polymers, 21*, 261–268.

Kawamura, Y., Kare, M.R., & Tannenbaum, S.R. (1987). *Umami, a basic taste: Physiology, biochemistry, nutrition, food science*. New York: Marcel Dekker.

Kellens, M., & Reynaers, H. (1992). Study of the polymorphism of saturated monoacid triglycerides: 2. Polymorphic behavior of a 50/50 mixture of tripalmitin and tristearin. *Fat Science and Technology, 94*, 286–293.

Kloek, W., van Vliet, T., & Walstra, P. (1997). Structure of fat crystal aggregates. In E. Dickinson & B. Bergenståhl (Eds.), *Food colloids: Proteins, lipids and polysaccharides* (pp. 168–181). Cambridge, UK: Royal Society of Chemistry.

Knoester, M., de Bruijne, P., & van den Tempel, M. (1968). Crystallisation of triglycerides at low supercooling. *Journal of Crystal Growth, 3/4*, 776–780.

Kokini, J.L. (1987), The physical basis of liquid food texture and texture-taste interactions. *Journal of Food Engineering, 6*, 51–81.

Kokini, J.L., & Cussler, E.L. (1983). Predicting the texture of liquid and melting semi-solid food. *Journal of Food Science, 48*, 1221–1225.

Kramer, A., & Szczesniak, A.S. (1973). *Texture measurements of foods*. Dordrecht: Reidel.

Krieger, I.M., & Dougherty, T.J. (1959). A mechanism of non-Newtonian flow in suspensions of rigid spheres. *Transactions of the Society of Rheology, 3*, 137–152.

Landau, L.D., & Lifshitz, E.M. (1960). *Theory of elasticity*. Oxford, UK: Pergamon.

Larsson, K. (1964). The structure of the β-form of trilaurin. *Arkiv för Kemi, 23*, 1–15.

Larsson, K. (1994). *Lipids: Molecular organization, physical functions and technical applications*. Dundee, UK: Oily Press.

Lucassen-Reynders, E.H. (1996). Dynamic interfacial properties in emulsification. In P. Becher (Ed.), *Encyclopedia of emulsion technology*, (Vol. 4, pp. 63–91). New York: Marcel Dekker.

Manning, D.M., & Dimick, P.S. (1985). Crystal morphology of cocoa butter. *Food Microstructure, 4*, 249–265.

Markov, E., & Tscheuschner, H.D. (1989). Instrumental texture studies on chocolate IV: Comparison between instrumental and sensory texture studies. *Journal of Texture Studies, 20*, 151–160.

Mason, T.G., Bibette, J., & Weitz, D.A. (1995). Elasticity of compressed emulsions. *Physical Review Letters, 75*, 2051–2054.

McGee, H. (1984). *On food and cooking: The science and lore of the kitchen*. New York: Charles Scribner.

Mela, D.J. (1988). Sensory assessment of fat content in fluid dairy products. *Appetite, 10*, 37–44.

Morgan, J. (1994). Chocolate: A flavor and texture unlike any other. *American Journal of Clinical Nutrition, 60*, 1065S–1067S.

Ottino, J.M. (1989). *The kinematics of mixing: Stretching, chaos and transport*. Cambridge, UK: Cambridge University Press.

Overbosch, P., Agterof, W.G.M., & Haring, P.G.M. (1991). Flavour release in the mouth. *Food Reviews International, 7*, 137–184.

Papenhuijzen, J.M.P. (1971). Superimposed steady and oscillatory shear in dispersed systems. I. Experimental and some results. *Rheologica Acta, 10*, 493–502.

Peleg, M. (1980). A note on the sensitivity of fingers, tongue and jaws as mechanical testing instruments. *Journal of Texture Studies, 10*, 245–251.

Peleg, M. (1983). Some theoretical rheological characteristics of the mechanical signals in sensory evaluation of texture. *Journal of Food Science, 48*, 1187–1191.

Pons, R., Taylor, P., & Tadros, T.F. (1997). Investigation of the interactions in emulsions stabilized by a polymeric surfactant and its mixtures with an anionic surfactant. *Colloid and Polymer Science, 275*, 769–776.

Prager, W., & Hodge, P.G. (1951). *Theory of perfectly plastic solids*. New York: John Wiley.

Princen, H.M. (1979). Highly concentrated emulsions. I. Cylindrical systems. *Journal of Colloid and Interface Science, 71*, 55–66.

Princen, H.M., & Kiss, A.D. (1986). Rheology of foams and highly concentrated emulsions. Part 3: Static shear modulus. *Journal of Colloid and Interface Science, 112*, 427–437.

Richardson, N.J., & Booth, D.A. (1993). Multiple physical patterns in judgements of the creamy texture of milks and creams. *Acta Psychologica, 84*, 93–101.

Richardson, N.J., Booth, D.A., & Stanley, N.L. (1993). Effect of homogenization and fat content on oral perception of low and high viscosity model creams. *Journal of Sensory Studies, 8*, 133–144.

Sato, K., Arishima, T., Wang, Z.H., Ojima, K., Sagi, N., & Mori, H. (1989). Polymorphism of POP and SOS. I. Occurrence and polymorphic transformation. *Journal of the American Oil Chemists Society, 66*, 664–674.

Skoda, W., Hoekstra, L.L., van Soest, T.C., Bennema, P., & van den Tempel, M. (1967). Structure and morphology of β-crystals of glyceryl tristearate. *Kolloid-Zeitschrift und Zeitschrift für Polymere, 219*, 149–156.

Stanley, N.L., & Taylor, L.J. (1993). Rheological basis of oral characteristics of fluid and semi-solid foods: A review. *Acta Psychologica, 84*, 79–92.

Steiner, E.H., & Bonar, A.R. (1961). Separation of some glycerides of cocoa butter by paper chromatography. *Journal of the Science of Food and Agriculture, 12*, 247–250.

Szczesniak, A.S. (1963). Classification of textural characteristics. *Journal of Food Science, 28*, 385–389.

Taylor, P., & Pons, R. (1998). The low shear rheology of concentrated o/w emulsions. In R.H. Ottewill & A.R. Rennie (Eds.), Modern aspects of colloidal dispersions (pp. 225–234). Dordrect: Kluwer.

Terzaghi, K. (1960). *Theoretical soil mechanics* (2nd ed.). New York: John Wiley.

Tolstoguzov, V.B. (1991). Functional properties of food proteins and the role of protein-polysaccharide interactions. *Food Hydrocolloids, 4*, 429–468.

Tscheuschner, H.D., & Markov, E. (1986). Instrumental texture studies on chocolate I: Methods of measurement and texture characteristics. *Journal of Texture Studies, 17*, 37–50.

Tyle, P. (1993). Effect of size, shape and hardness of particles in suspension on oral texture and palatability. *Acta Psychologica, 84*, 111–118.

Vreeker, R., Hoekstra, L.L., den Boer, D.C., & Agterof, W.G.M. (1992). The fractal nature of fat crystal networks. *Colloids and Surfaces, 65*, 185–189.

Walstra, P. (1987). Fat crystallisation. In J.M.V. Blanshard & P. Lillford (Eds.), *Food structure and behaviour* (pp. 67–85). London: Academic Press.

Wieringa, J.A., van Dieren, F., Janssen, J.J.M., & Agterof, W.G.M. (1996). Droplet breakup mechanisms during emulsification in colloid mills at high dispersed phase volume fraction. *Chemical Engineering Research and Design Transactions of the Institution of Chemical Engineers, Part A), 74*, 554–562.

Wood, F.W. (1974). An approach to understanding creaminess. *Die Stärke, 26*, 127–130.

CHAPTER 8

Flesh Foods and Their Analogues

Marion L. Greaser and Albert M. Pearson

Texture in flesh foods is usually referred to as *tenderness*, which is the most important quality factor in meat, poultry, and fish products (Chrystall, 1994). Thus, *tenderness* will be used to mean *texture*, and the two terms will be used interchangeably. Lack of tenderness is a much greater problem with beef and lamb than with the other meat-producing species. Consumers are able to identify beef steaks that are more tender and are willing to pay a premium for improved tenderness (Boleman et al., 1997).

In some cases, the flesh foods become too soft or mushy; this will simply be referred to as the lack of texture or simply as being soft and mushy. Lack of texture does not normally occur in whole-muscle red meats or poultry, although it may be a problem in some fish products or in minced meats and some sausage products. The current chapter will review the structure of muscle (flesh), the changes it undergoes during contraction and development of rigor mortis, and the effects of postmortem changes in muscle upon tenderness. This will be followed by a discussion of the influence of pre- and postslaughter treatments on texture. Finally, some unique methods of measuring texture in flesh foods will be reviewed.

STRUCTURE, COMPOSITION, AND FUNCTION OF MUSCLE

Muscle Structure

Details about the structural arrangement of muscle have been determined by light and electron microscopy, and a diagram showing the levels of organization is shown in Figure 8–1 (see Craig, 1994, and Squire, 1997, for recent more extensive reviews). A whole muscle is encased in a layer of connective tissue called the *epimysium*. Groups of long, cylindrical muscle cells (also referred to as muscle *fibers* or *myofibers*) are surrounded by a thinner connective tissue layer called the *perimysium*. This layer is visible to the naked eye and gives a textured appearance to muscle when viewed in cross section. A third, very thin endomysium layer surrounds each individual muscle cell (see Bailey & Light, 1989, and Rowe, 1989, for reviews of connective tissue structure and composition).

Muscle cells are often several centimeters in length and may extend the whole length of the muscle. Their diameter is usually between 10 and 100 μm. The cells are multinucleate (many nuclei), and the nuclei are located near the periphery of the cell. Several other cell

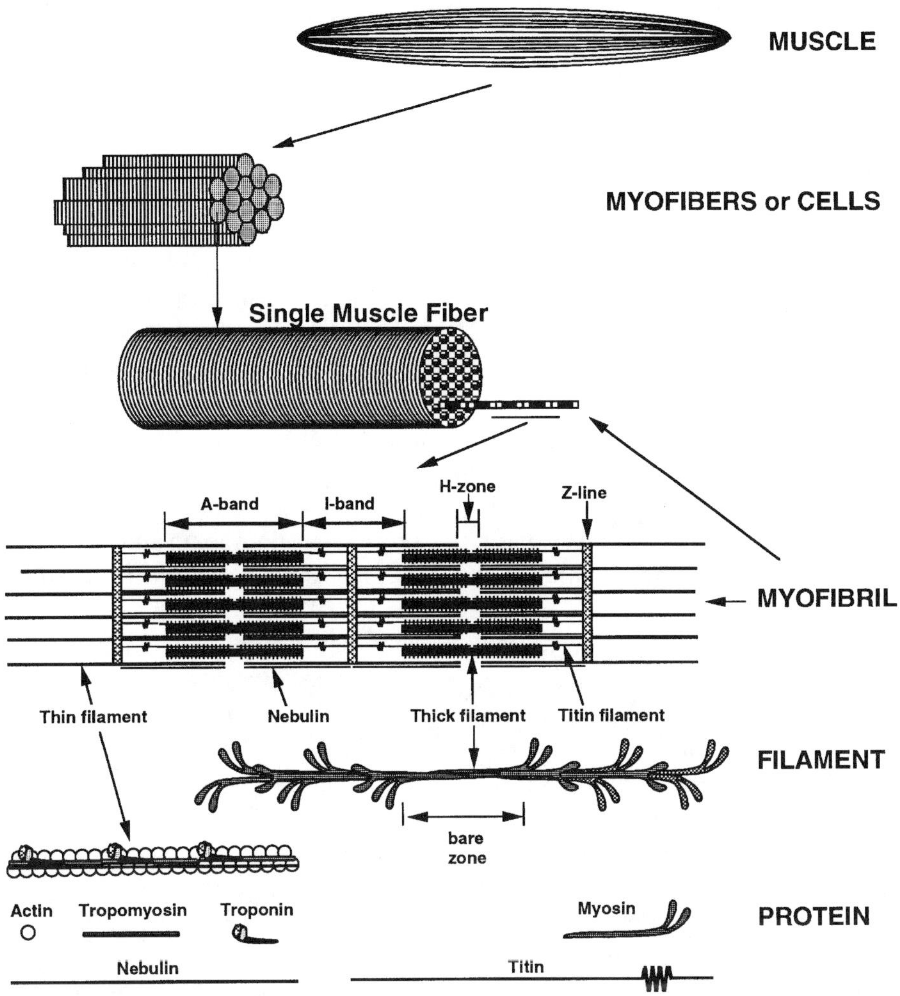

Figure 8–1 Diagram Showing the Different Levels of Organization of Skeletal Muscle

types are typically found in muscle tissue. Adipose (fat) cells are clustered in the perimysium and outside the epimysium regions. The clusters of fat cells are referred to as *marbling*. Axon branches from nerves extend to each muscle fiber and terminate at the motor endplates. Fibroblasts also reside in the connective tissue layers. In addition, there are cells that provide the linings of the blood vessels and capillaries.

A striking difference between muscle cells and other tissue cell types is the inclusion of unique organelles called *myofibrils*. These are the sites of force production in the cell. They are not bounded by a membrane but exist as an insoluble structure in the living cell. Myofibrils are oriented parallel to the long axis of the myofiber and may run its complete length. Their diameters are about 1 to 2 μm. They occupy about 80% of the cell volume in skeletal muscle. There are typically 500 to 1,000 myofibrils in the cross section of a mature skeletal muscle fiber. Myofibrils often branch, and this may be a mechanism for increase in myofibril number during muscle growth. The myofibril consists of groups of thick and thin myofilaments whose long axes run parallel to the long axis of the cell (Figure 8–1). The myofibrils have alternating light and dark bands when viewed by phase-contrast microscopy. The light and dark bands are aligned in adjacent myofibrils. The origin of these bands is related to the location of the thick and thin filaments. The A bands extend the full length of the thick filaments. The I bands contain thin filaments and the Z line. The less dense region in the center of the A band (where only thick filaments are located) is called the *H zone*. The distal portion of the A band appears more dense because of the presence of both thick and thin filaments. In the very center of the A band is a dark line called the *M line* that consists of cross-connections between adjacent thick filaments. Over most of the length of the thick filaments, there are perpendicular projections (or cross-bridges) extending toward the thin filaments. In a very narrow zone on either side of the M line, there are no cross-bridges, and this region is referred to as the *pseudo–H zone* or *bare zone*. The smallest contractile unit is called a *sarcomere*, and it extends from one Z line to the next. Note that the two halves of the sarcomere have opposite polarities. Thick filaments are about 1.5 to 1.6 μm long (in mammals, birds, and fish) and 12 to 15 nm in diameter. Thin filaments are about 1.0 to 1.3 μm in length and 6 to 8 nm in diameter. Each thin filament terminates at the Z line, and the polarity is reversed on each side of this structure.

The filament pattern varies with location of the slice through the myofibril. Near the Z lines, the thin filaments are arranged in a square pattern. In the I zone, the thin filaments appear randomly spaced. The overlap region in the A band shows a hexagonal array, with each thin filament surrounded by three thick filaments and each thick filament surrounded by six thin filaments. The thick filament pattern in the H zone also is hexagonal, and cross-connections are seen at the M line. The thick filament consists predominately of the protein myosin. The heads of the myosin compose the cross-bridges, and these are arranged in a helical fashion throughout the majority of the filament's length (absent only in a bare zone at the center). A set of bridges occurs every 14.3 nm, and the helix turns every 42.9 nm. The number of cross-bridges at each 14.3-nm location is three in mammalian muscle. Thin filaments contain a double-stranded helix of actin along with the proteins tropomyosin and troponin (Figure 8–1). There are 13 actin monomers per half-twist of the helix (6.5 per chain), and since each actin monomer extends approximately 5.5 nm in length, the actin structure repeats every 35.8 nm. Lying in the groove on either side of the actin is tropomyosin. It extends along the length of seven actins. Each tropomyosin binds one troponin complex. The length of the tropomyosin-troponin repeat is thus 38.5 nm. Note that neither of these repeat distances coincides with the thick-filament repeat.

There is now fairly strong evidence to suggest that there are filaments in the myofibril in addition to the thick and thin filaments. When living muscle fibers are stretched to sarcomere lengths of 4 to 5 μm, they spontaneously return to normal sarcomere lengths (2.2–2.5 μm). Since thick filaments are about 1.5 to 1.6 μm and thin filaments are about 1.0 micron long, the sarcomere should theoretically fall apart if stretched beyond 3.6 μm. Electron micrographs of muscle fibers stretched to lengths greater than 3.6 μm reveal superthin filaments (2–6 nm in diameter) in the gap between the ends of the thick and thin filaments (termed *gap filaments* by Sjostrand, 1962). Some fibers have been stretched to sarcomere lengths as long as 12 to 14 μm. These gap filaments are thought to be elastic and normally may function to protect the myofibril from overstretch. They may also prevent nonuniform dislocation of the A band in the sarcomere or may contribute to the restoration of the muscle to rest length after shortening. They may provide a template to control thick-filament length.

Muscle Proteins

Protein makes up 19% to 20% of the fresh weight of skeletal muscle. This protein is roughly classified into three groups on the basis of solubility (Helander, 1957; Scopes, 1970). Homogenizing muscle tissue in 5 to 10 volumes of a low–ionic strength buffer (such as 20 mM potassium phosphate, pH 7.0) and then subjecting the slurry to centrifugation results in a supernatant fraction containing the sarcoplasmic proteins and a pellet. Subsequent homogenization of this pellet in a high–ionic strength buffer (such as 1.1 M KI–20 mM potassium phosphate, pH 7.0) and centrifugation separate the myofibrillar proteins (supernatant) and the connective tissue proteins (pellet). The protein composition and properties of the different proteins associated with each class will be discussed below.

Connective Tissue Proteins

The major proteins of the extracellular matrix are listed in Table 8–1. These proteins make up about 20 to 30 mg per gram of muscle (Scopes, 1970). Collagen is the most prominent protein of this fraction and constitutes over 40% of the dry matter of this protein class (Bailey & Light, 1989). It is a triple-stranded protein with subunits approximately 100,000 Da in size. The monomers undergo extensive postranslational modifications that include removal of peptide fragments from both ends of the synthesized protein, hydroxylation of proline and lysine side chains, addition of sugar residues, oxidation of several lysine and hydroxyllysine side chains, and formation of both heat-labile and relatively heat-stable intermolecular covalent bonds (see Kalder, 1994; McCormick, 1994; Prockop & Kivirikko, 1995, for reviews). Collagen type I is the most abundant type found in skeletal muscle, followed by collagen type III. Both are fibril-forming collagens: that is, they associate in a side-to-side manner to form long, interconnected strands. Type IV collagen, in contrast, forms "chicken wire" network-type complexes. The different collagen types are unequally distributed in the different muscle connective tissue layers (Bailey & Sims, 1977; see Table 8–1).

A second protein found in significant concentrations in muscle connective tissue is elastin. As the name implies, this protein has the capacity to elongate when stretched. Most elastin is located in the epimysium and perimysium layers. The elastin content varies considerably between different muscles—from 37% of the connective tissue protein in bovine semitendinosus to 2% in the longissimus (Bendall, 1967).

A number of proteins in connective tissue function in the linking of the muscle cells to the collagen matrix. These include fibronectin, laminin, and decorin. All are glycosylated and are referred to collectively as *proteoglycans*. A significant proportion of the extracellular matrix is composed of these highly hydrated polysaccharide-containing molecules (Bailey & Light, 1989). The postmortem changes in the connective tissue constituents have been recently reviewed (Greaser, 1997).

Myofibrillar Proteins

Since the myofibril dominates the structure of the muscle cell, it would be expected that

Table 8–1 Major Proteins of the Muscle Extracellular Matrix

Name	Properties	Location		
		Epimysium	*Perimysium*	*Endomysium*
Collagens				
Type I	Fibrous	***	*	*
Type III	Fibrous	*	***	*
Type IV	Network	-	-	***
Type V	Fibrous	*	*	**
Type VI	Fibrous	*	**	*
Type VII	Anchoring	-	*	*
Elastin	Elastic	**	**	*
Laminin	Heterotrimer	-	-	***
Fibronectin	Homodimer	*	*	*
Decorin	Leu-rich proteoglycan	-	**	**

Source: Adapted from M. Greaser, *50th Annual Reciprocal Meat Conference*, Vol. 50, pp. 53–59, © 1997, American Meat Science Association.

these organelles would provide a major proportion of the total muscle protein. The myofibrillar protein (also sometimes referred to as *salt soluble*) concentration is approximately 115 mg per gram of muscle (Scopes, 1970). Table 8–2 lists the major constituents of this fraction. Myosin makes up the largest proportion of this fraction and nearly half of the total. Myosin is an exceptionally large protein with a molecular weight of 521,000 Da. It has a long tail and two pear shaped heads. Myosin is a hexamer consisting of two 223,000 Da chains and four 15,000 to 20,000 Da chains. Each of the larger chains (called *heavy chains*) extends through the full length of the tail and one of the heads. Two light chains are associated with each head. The molecular weights of the light chains vary with muscle type. The three-dimensional structure of the head portion of chicken skeletal muscle (a proteolytic fragment called *subfragment-1*) has now been reported (Rayment et al., 1993). It shows a long (85 amino acids) alpha-helical neck connecting the head to the myosin tail. The light chains wrap around this helix near its junction with the tail. The ATP-binding site is found in a large cleft; the actin-binding site is on the opposite side of the head from the ATP site. Myosin's tails provide the core of the thick filaments, and the heads constitute the cross-bridges between the thick and thin filaments that can be seen in electron micrographs of muscle.

Actin (see Sheterline, Clayton, & Sparrow, 1996, for a review) makes up about 22% of the myofibrillar protein and serves as the core of the thin filaments. The monomer has a molecular weight of 42,000 Da and is referred to as *G-actin* (globular). Actin polymerizes into a double-stranded helix upon addition of physiological concentrations of neutral salts. The polymer is called *F-actin* (fibrous). G-actin binds one mole of ATP and one mole of divalent ion (Mg^{++} in the thin filament) per mole. During polymerization, a phosphate is liberated from the ATP. The nucleotide and metal ion are readily exchangeable on G-actin but are very firmly bound in the F form. The X-ray diffraction

Table 8–2 Major Myofibrillar Proteins

Protein	*% (weight/weight)*
Myosin	43
Actin	22
Titin	8
Tropomyosin	5
Troponin	5
Nebulin	3
C-protein	2
M-protein	2
α-actinin	2

Source: Adapted with permission from L. Yates and M. Greaser, *Journal of Molecular Biology*, Vol. 168, pp. 123–141, © 1983, Academic Press Ltd.

pattern of actin has been determined (Kabsch, Mannherz, Suck, Pai, & Holmes, 1990). G-actin has four subdomains: 16,400 Da, 4,000 Da, 11,500 Da, and 10,000 Da. The ADP and divalent metal ion are located in a cleft between the four domains, the metal being complexed with the phosphates of the ADP. The polypeptide chain is wrapped so that the N and C terminus are located in subdomain 1; the myosin head is believed to bind in this domain during thc contraction cycle. All four subdomains are involved in interactions with adjacent actin monomers in the polymerized state. The contact regions appear to be greater with subunits along a strand than with those on adjacent strands. It is the interaction of the myosin heads with actin that results in force production and muscle shortening. Actin also provides a backbone for binding of tropomyosin and troponin, the two proteins that regulate the actin-myosin interaction.

Tropomyosin is a long rod-shaped molecule with a molecular weight of 65,000 Da. It contains two polypeptide chains, each 32,500 Da, which are completely α-helical and associated with each other in a side-to-side fashion. Tropomyosin has an interesting amino acid sequence, with nonpolar (N) and polar (P) amino acid side-chain patterns of (NPPNPPP) repeated 41 times. There are thus two nonpolar side-chains in every seven residues, a pattern that is close to the typical alpha helix pitch of 3.6 amino acids per turn. This amino acid repeat pattern allows the hydrophobic (nonpolar) side chains to be positioned in a stripe along one side of the helix. The two hydrophobic nonpolar stripes on each chain interact to hold the two chains together. Tropomyosin is positioned in the two grooves on either side of the thin filament (Figure 8–1). Each tropomyosin overlaps with the next one by about eight or nine amino acids. The position of the tropomyosin in the actin groove varies with the calcium state. At high calcium, the tropomyosin moves deeper into the groove, presumably exposing the myosin-binding site on actin. At low calcium, tropomyosin may sterically block one of the binding sites on actin. Calcium binding to troponin triggers this movement. The position change also occurs when myosin heads bind to actin.

Troponin consists of three subunits (TnC, TnI, TnT), whose total molecular weight is 70,000 Da (see Farah & Reinach, 1995, for a review). The shape of the complex is somewhat elongated (Figure 8–1). TnC (MWt 18,000 Da) has two high-affinity sites that bind either Ca^{++} or Mg^{++} (carboxyl end) and two lower affinity sites that are Ca^{++} specific (amino terminal end). The exchange rate for the high-affinity sites is much slower than for the low-affinity sites. The high-affinity sites are probably occupied by Mg^{++} at rest and remain so during contraction. The Ca^{++}-specific sites are believed to be the location where contraction is triggered. The TnC undergoes large conformational changes in the transformation between Ca^{++} states. The TnC has been found to be dumbbell shaped, with the two halves connected by a long α-helix (Satyshur, Pyzalska, Greaser, Rao, & Sundaralingam, 1993). There are nine alpha helices: one near the N terminus and four pairs (A-B, C-D, E-F, and G-H) that flank the metal binding loops. The D and E helices are fused into a long helix that connects the two ends of the molecule. Following Ca^{++} binding, the B and C helices move as a unit away from the A and D helices. This may involve a straightening of a bend in the B helix. TnI (MWt 20,900 Da) strongly inhibits the actomyosin ATPase activity in the presence of tropomyosin. This inhibition is calcium insensitive. TnI is believed to bind to actin in the absence of Ca^{++} and to be released when Ca^{++} binds to TnC. TnT (MWt 30,500 Da) functions to bind the troponin complex to tropomyosin. This subunit is highly elongated and extends from near residue 190 on the tropomyosin past the carboxyl terminus to the amino terminus of the next tropomyosin in series, a distance of approximately 15 nm. It seems likely that the association between TnT and tropomyosin involves a three-stranded coiled coil (two strands from tropomyosin, the other from TnT). Antibody studies have shown that the troponin complex is located every 38.5 nm along the thin filament. Troponin is necessary for calcium regulation of contraction and serves to move tropomyosin to different positions when it binds or releases calcium.

Several other proteins are believed to play primarily a cytoskeletal role in the myofibril. Titin, also called *connectin*, contributes nearly 8% to 10% of the myofibrillar protein fraction. This protein has been postulated to be responsible for the passive elasticity. The subunit size of titin is more than 3 million Da (Labeit & Kolmerer, 1995). Thus, titin is the largest known protein. Isolated titin molecules are long and stringlike with dimensions of 2 to 4 nm × 1 μm. Two sets of titin molecules are believed to be aligned in parallel in each half-sarcomere, with each molecule extending from near the M line to the Z line (Figure 8–1). The protein is believed to be a major constituent of the gap filaments. The portion in the I band stretches with increase in sarcomere length, while that in the A band remains anchored to the thick filaments. The full titin message has been sequenced by Labeit and Kolmerer (1995), and a complex domain structure has been found. The protein contains a large number of 100 amino acid repeating motifs that have some sequence similarities to regions of fibronectin type 3 (FN3) and immunoglobulins (Ig) domains. The former have been referred to as *type I* motifs, the latter as *type II*. Both types of motifs are folded into seven-stranded beta sheets with very similar three-dimensional structure. A large portion of the A-band region contains an 11-domain super-repeat with a pattern of (II - I - I - I - II - I - I - I - II - I - I). Since one domain is approximately 4 nm long, the 11-domain length of about 44 nm corresponds very closely with the 42.9-nm myosin repeat. There is also a kinase domain in the A-band region of the titin. All of the I-band and Z-line domains are of the Ig type. Near the middle of the half–I-band region is an unusual sequence that is rich in proline (P), glutamic acid (E), valine (V), and lysine (K) residues and has been termed the *PEVK region*. The length of this region varies among muscle types and seems to correlate with the passive tension characteristics of the muscle (skeletal muscle soleus < skeletal muscle psoas < cardiac). The elasticity of titin has been recently demonstrated by experiments where the stretch resistance of single titin molecules has been measured (Kellermayer, Smith, Granzier, & Bustamante, 1997; Tskhovrebova, Trinick, Sleep, & Simmons, 1997).

Another cytoskeletal protein is nebulin. It has a subunit weight of 600,000 to 800,000 Da. Nebulin is associated with the thin filaments over most of their length and may in fact provide a length template for thin-filament assembly (Figure 8–1). The size of nebulin appears to be related to the thin-filament length when several different muscle types are compared. It too has a repeating amino acid sequence (35-residue repeat) with a SDxxYK sequence near the center of each repeat.

Several other proteins are located at the Z line. The protein α-actinin contains two 100,000-Da chains arranged in an antiparallel fashion. The amino terminus binds to actin, and it has recently been shown that the carboxyl terminus binds to titin. Desmin is an intermediate filament protein with a subunit size of 55,000 Da. It is found only around the periphery of the Z line and is believed to function in linking adjacent myofibrils. Dystrophin is a very large protein (427,000 Da) found associated with the cell membrane. The protein is defective in individuals with muscular dystrophy. In spite of its importance for muscle function, the protein is found in very small quantities in muscle (0.01% of total muscle protein).

Table 8–3 Major Glycolytic and Related Enzymes in Skeletal Muscle

Enzyme	*Amount (mg/g)*
Phosphorylase	2.0
Amylo-1, 6-glucosidase	0.1
Phosphoglucomutase	0.6
Phosphoglucose isomerase	0.8
Phosphofructokinase	0.35
Aldolase	6.5
Triose phosphate isomerase	2.0
Glyceraldehyde 3-phosphate dehydrogenase	0.3
Glyceraldehyde phosphate dehydrogenase	11.0
Phosphoglycerate kinase	1.4
Phosphoglycerate mutase	0.8
Enolase	2.4
Pyruvate kinase	3.2
Lactic dehydrogenase	3.2
Creatine kinase	5.0
AMP kinase	0.4
Total protein	~40

Source: Reprinted with permission from *Biochemical Journal*, Vol. 134, p. 201, © 1973, The Biochemical Society and Portland Press.

Sarcoplasmic Proteins

Muscle tissue typically has about 55 mg of soluble protein per gram of fresh weight (Scopes, 1970). The major constituents include the enzymes of the glycolytic cycle (Table 8–3). Most of the rest consists of other enzymes needed for normal cell metabolism. Myoglobin, the oxygen-carrying component of the muscle cell and the substance that gives red muscles their color, is also found in this fraction. The sarcoplasmic proteins are believed to have limited importance in affecting meat texture, so their description will not be detailed in this chapter.

Muscle Contraction Mechanism

It is now clear that muscles contract by a relative sliding of the thick and thin filaments past each other. This is accomplished by the asynchronous attachment and detachment of cross-bridges in a cyclical pattern to cause the sarcomere to shorten. The lengths of the individual myofilaments do not change significantly (i.e., less than 0.4%), only their degree of overlap. The contractile system is activated by calcium ions, which are released from the sarcoplasmic reticulum in response to a nerve impulse. Acetylcholine is released at the motor endplate, where it causes the outer cell membrane to depolarize. The depolarization wave passes along the surface of the muscle cell and then to the interior via the T-tubule system. The calcium released from the sarcoplasmic reticulum diffuses to the myofibrils, where it binds to the TnC subunit of troponin. This causes a shape change in the protein that is transmitted through TnI and TnT to cause a movement of the tropomyosin deeper into the actin groove. The myosin heads can then bind the actin, rotate, and thus slide the two filaments a small distance past each other. The asynchronous rotation of the myosin heads, which were previously energized by the splitting of ATP, results in force development and/or muscle shortening (see Amos & Cross, 1997, for a recent discussion of molecular motors).

RIGOR MORTIS

Physical and Chemical Changes

Rigor mortis is the term that describes the stiffness that occurs in muscle tissue within the first few hours after death. Living muscle and muscle immediately postmortem are extensible, meaning that the length will increase when the tissue is subjected to stretch. However, this ability to be stretched declines postmortem to a level about 1/20th to 1/40th of that at the time of death. The time to rigor mortis can be determined experimentally by use of an apparatus that periodically (8-min cycle) puts a weight (50 g) on a 1-cm^2 strip of muscle and then removes the weight (Bendall, 1960). The displacement of the bottom end of the muscle strip is then converted to a tracing on a strip chart recorder. An example of several such records is shown in Figure 8–2. The rigor process is divided into three different phases. The first is called the *delay* phase. During this time, the muscle extensibility remains constant. The second phase is referred to as *rapid*, and during this period the extensibility declines fairly quickly. The final phase is called *completion*, and the extensibility remains constant throughout, but with a much reduced range. Rigor mortis time course varies considerably between species. Completion of rigor requires 12 to 24 hr in beef, 8 to 12 hr in lamb, 3 to 6 hr in pigs, 1 to 4 hr in poultry, and 5 to 24 hr in fish.

The antemortem muscle activity, the amount of muscular activity at the time of death, and the temperature at which the muscle is held postmortem all affect the time and pattern of rigor development. Figure 8–2 demonstrates several of these situations with rabbit muscle. Figure 8–2A shows the extensibility pattern of a muscle strip held at 17°C for an animal that was anesthetized before sacrifice. Note that the extensibility remains essentially constant for 7 to 8 hr (the delay phase). However, the extent of the stretch becomes increasingly smaller (the rapid phase) until there is barely any displacement during the loading cycle (completion, at about 11 hr). This lack of extensibility would continue indefinitely if bacterial spoilage of the muscle sample was prevented. Figure 8–2B shows the pattern for a muscle strip held at 37°C instead of 17°C. Note that the delay phase is much reduced (a 2.5-fold decline) and that the completion occurs at about 5 hr. In addition, the upward displacement of the top of the trace shows that the muscle has shortened during the rapid phase (approximately 15% of the rest length). A trace from Figure 8–2C muscle held at 17°C (rabbit not anesthetized) has virtually no delay phase, a completion by 6 hr, and a small amount of shortening. A muscle strip held at 17°C from another rabbit whose glycogen had been depleted by antemortem insulin injection had no delay phase and completion prior to 1 hr postmortem.

It is now understood that the loss of extensibility is related to the loss of ATP in the muscle cells (Newbold, 1966). ATP is required to dissociate the myosin heads from actin during the contraction cycle in living muscle, and the heads remain detached dur-

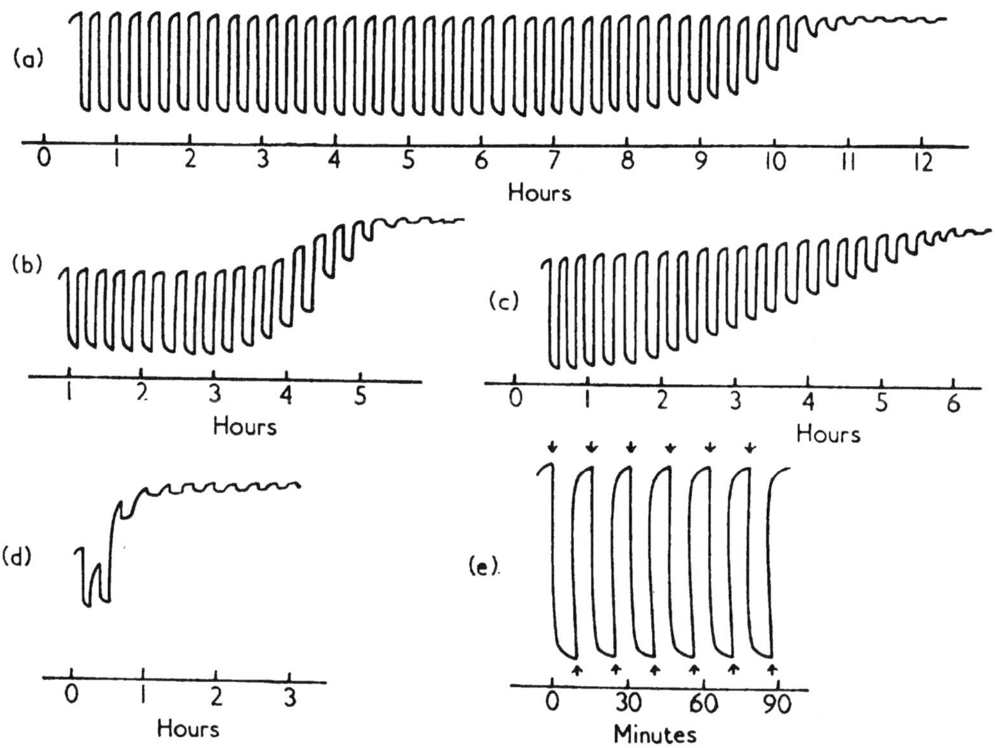

Figure 8–2 Effects of Different Antemortem and Postmortem Treatments on Development of Rigor Mortis in Rabbit Psoas Muscle. Muscle samples were loaded and unloaded every 8 min and the length of the strip recorded. (a) Well-fed rabbit, immobilized before death, 17°C. (b) As in (a), but muscle held at 37°C. (c) Well-fed rabbit, killed without immobilization, 17°C. (d) Glycogen depleted antemortem by insulin injection, 17°C. (e) Enlarged tracing from the delay phase of rigor. ↓ means load on; ↑ means load off. *Source:* Reprinted with permission from J.R. Bendall, *Structure and Function of Muscle*, Vol. 3, p. 232, © 1960, Academic Press, Inc.

ing muscle rest. However, when ATP is absent, myosin heads form rigid bonds with actin that would require considerable force to break. This is the explanation for the inextensibility that develops postmortem. Thus, a biochemical measure of rigor mortis would be the ATP level in the tissue. It has been found that if the average muscle ATP concentration drops below 2 mM, muscle inextensibility will begin to develop (Bendall, 1960). This concentration is somewhat dependent on the size of the muscle piece being examined. Modern biophysical experiments using individual skinned muscle cells show that extensibility still remains with ATP levels below 0.1 mM. The difference between single cells and larger strips probably reflects the fact that larger strips have heterogeneous patterns of ATP loss: that is, part of the cells will have ATP concentrations below 0.1 mM, while others may still have 3 to 4 mM; the extensibility may be affected when only a part of the cells in a larger strip are in rigor.

The tissue remains in its prerigor, readily extensible state (delay phase) for as long as the enzymatic depletion of ATP is balanced by ATP resynthesis from glycolysis and from the rephosphorylation of ADP from the limited reserve of creatine phosphate, CP. The glycolysis pathway converts glycogen to lactic acid with the generation of ATP. When the CP supply is exhausted, however, glycolysis alone cannot maintain the ATP level, and cross-bridging occurs in rough proportion to the ATP decline. Except in rare situations where glycogen reserves are exceptionally low, the tissue's pH declines to about 5.3 to 5.7 (the "ultimate pH") in full rigor, due to the accumulation of lactic acid from anaerobic glycolysis. Muscles that will have an extended rigor mortis time course will have a high pH (near 7.0) immediately after death and an ultimate pH near 5.5. If the pH is about 6.5 immediately postmortem, then rigor will develop more rapidly. A high initial pH (>6.5) and a high ultimate pH (>6.0) will be associated with a very rapid rigor mortis time course. In reality, the rigor mortis pattern depends on the death levels of creatine phosphate (CP), ATP, and glycogen. Creatine phosphate serves as a backup supply for high-energy bonds and is created by the creatine phosphokinase–catalyzed reaction:

$$C + ATP \leftrightarrow CP + ADP$$

In times of high-energy bond needs, the reaction operates in the reverse direction. Postmortem ATP levels thus remain constant as long as CP remains. Once the CP is gone, ATP begins to decline. However, the ADP levels never rise significantly. A possible explanation is that the ADP generated drives the reaction catalyzed by adenylate kinase shown below:

$$2\ ADP \leftrightarrow ATP + AMP$$

Although some ATP is restored, the resulting AMP is rapidly converted to IMP plus ammonia (essentially only in the forward direction), and the net result is a reduction of the total adenine nucleotide pool. The inosine nucleotides are poorly phosphorylated and thus do not function to prevent myosin-actin rigor bond formation.

The shortening associated with rigor mortis at high temperature is sometimes mistakenly considered to be a required part of the rigor mortis process (Bendall, 1951). However, histological examination of muscle held at 17°C reveals that only a small portion of the fibers show loss of striation patterns indicative of contraction (Dransfield, Lockyer, & Prabhakaran, 1986). Even rigor at 37°C with the whole-muscle strip shortening by 40% results in many fibers with clear banding patterns (Bendall, 1960). In addition, the forces developed by muscle strips as they go into rigor are small; a weight equal to 15% of the maximum force developed by contracting living muscle will prevent the shortening associated with rigor mortis.

The inextensibility of rigor mortis is accompanied by a resistance to bending at the time of rigor completion. However, within a few hours, the muscle becomes more flaccid again. This was originally termed the "resolution of rigor mortis." However, since once the ATP has been depleted and the actin-myosin bonds are set there are no new sources of ATP, there must be another explanation of this postmortem tissue softening. Dransfield et al. (1986) have shown that the extensibility increases again after rigor mortis completion (Figure 8–3). This additional extensibility was not accompanied by significant sarcomere length changes but rather by periodic breaks in the sarcomere pattern. However, the additional length from these breaks could explain only a portion of the elongation during stretch, and the authors suggested that there might be some extracellular matrix slippage as well (Dransfield et al., 1986).

Abnormal Rigor Mortis Patterns

The normal rigor process is perturbed by a number of conditions. First, if muscle is frozen during the early postmortem period and then later thawed rapidly, it undergoes what

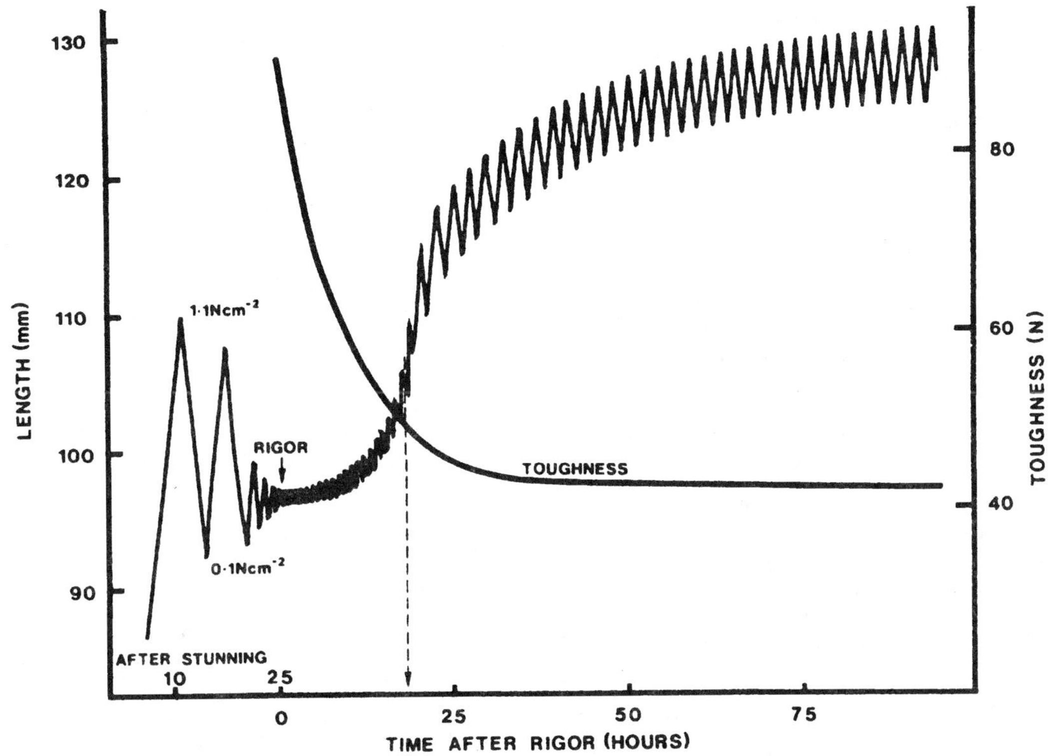

Figure 8–3 Changes in Extensibility of Raw Beef Muscle. A strip of bovine *M. pectoralis profundus* was subjected to alternating stresses of 1.1 N/cm^2 and 0.1 N/cm^2 and the muscle length plotted against time after stunning (upper numbers) or time after rigor (lower numbers). The extensibility declines from soon after death until rigor mortis but then increases again with further time after rigor. In addition the muscle length increases significantly in response to the cycling stress conditions. The timing of the increase in muscle length and extensibility corresponds to the timing of the decline of toughness that occurs in parallel samples that have been cooked. *Source:* Reprinted from *Meat Science*, Vol. 16, E. Dransfield et al., Page No's. 127–142, Copyright (1996), with kind permission from Elsevier Science Ltd., The Boulevard, Langford Lane, Kidlington OX5 1GB, UK.

has been termed *thaw rigor* (Perry, 1950). The muscle shortens quickly (within 1 to 2 min, depending on the diameter and the thawing rate) to a length as short as 20% of the frozen state. Actually, this initial shortening is not rigor mortis, since application of a weight to the muscle strip will reextend the muscle. However, the strip will proceed to rigor completion within the next 30 min (Kushmerick & Davies, 1968). The mechanism of this contraction is believed to be the ice crystal–induced rupture of the sarcoplasmic reticulum and the muscle cell plasmalemma, which releases large amounts of calcium to activate the myofibrils. Muscle strips that had been previously loaded with ^{45}Ca were shown to release larger amounts into the medium after a freeze-thaw cycle (Kushmerick & Davies, 1968).

The rigor mortis pattern is much shorter than normal in certain bovine animals. This appears to be due to an antemortem stress that largely depletes the muscle glycogen stores. The resulting muscles have a brief time to rigor mortis and an elevated ultimate pH. The muscle pH may be as high as 6.5 to

6.7, and the meat is very dark in appearance. This condition is referred to as "dark cutting" and occurs more often in bulls, particularly if animals from separate feeding locations are mixed prior to slaughter (Tarrant, 1981).

A similar condition can occur in pigs, where it is termed *DFD* for "dark, firm, and dry." It too is caused by antemortem stress. Animals displaying this condition usually are carriers of a genetic defect in the ryanodine receptor, the protein that normally releases calcium during a muscle contraction. Animals with the defective protein, however, leak calcium without a nerve impulse and thus deplete their glycogen stores.

The same genetic defect also causes pigs with an adequate glycogen level at death to undergo an extremely rapid postmortem pH decline. The muscle reaches its ultimate pH and rigor mortis completion as early as 15 min postmortem in some cases. This low pH while the muscle temperature is still high results in a denaturation of the myosin and an excessive fluid loss from the muscle surface. The condition is referred to as *PSE*, since the meat is pale in color, soft in texture, and exudative (loses large amounts of fluid) (see Briskey, 1964, and Greaser, 1986, for more extensive reviews). Swatland (1997) has recently shown that the loss of extensibility is correlated with the postmortem changes in capacitance and resistance properties of muscle.

INFLUENCE OF PRESLAUGHTER TREATMENTS

Both pre- and postslaughter treatments can affect texture of meat, poultry, and fish products. The influence of several factors affecting texture will be discussed in the sections below.

Preslaughter Treatments and Muscle Properties

Dransfield (1994) has reviewed the effects of some preslaughter treatments on texture, which include breed, sex, fatness, growth promoters, slaughter procedures, and amount of connective tissue present. Other preslaughter treatments that may affect tenderness include nutritional and other conditions of stress.

Effect of Breed

In general, there is no marked difference in meat tenderness of cattle among the beef breeds (Chadwick, Cuthbertson, & Dransfield, 1979) or between the dairy and beef breeds (Brannaman, Pearson, Magee, Griswold, & Brown, 1962). Meat from *Bos indicus* is, however, usually less tender than that from *Bos taurus* (Koch, Crouse, Dikeman, Cundiff, & Gregory, 1988), which may be the result of more stability of the proteinase inhibitor calpastatin (Whipple et al., 1990). Although Dransfield, Nute, MacDougall, and Rhodes (1979) concluded that breed does not have any effect upon the tenderness of lamb, recent studies on lambs with the Callipyge gene (heavy muscled) by Koohmaraie, Shackelford, and Wheeler (1996) have shown loin chops from lambs expressing this gene to be significantly less tender than those from normal lambs. Studies with pig meat from Hampshire and Duroc pigs (Hiner, Thornton, & Alsmeyer, 1965) and the Swedish Landrace and Yorkshire breeds (Malmfors & Nilsson, 1979) suggest that breed differences should be taken into account in evaluation of texture in pork. For poultry meat, breed, sex, and diet were shown by Jones and Grey (1989) to have less influence on tenderness than stress and electrical stunning or methods of scalding, plucking, and chilling. The acceptability of fresh whitefish species seems to be dependent mainly on flavor and texture differences, with the latter being relatively small (Bennett & Hamilton, 1986). However, cross-linking of the myofibrillar proteins by the formaldehyde generated from trimethylamine oxide (TMAO) causes toughness problems in the gadoid fish group (i.e., cod, pollack, and haddock) if these fish have been stored frozen for long periods (Krueger & Fennema, 1989; Leblanc, Leblanc, & Gill, 1987).

Effects of Fatness

Even though higher grading beef tends to be more tender, there is considerable overlap between grades, and the relationship between fatness and tenderness is low, according to Campion, Crouse, and Dikeman (1975). They concluded that 2.9% of intramuscular fat in the longissimus muscle was sufficient in carcasses of young cattle to ensure acceptably tender meat. This is in essential agreement with earlier work (Blumer, 1963; Wellington & Stouffer, 1959) in which it was found that only about 10% of the variation in tenderness could be accounted for by the marbling fat. These studies suggest that only a minimal amount of marbling fat is needed to ensure tenderness in meat from young beef. Extreme fat covering was unrelated to tenderness in meat from young cattle (Campion et al., 1975).

Influence of Nutritional Stress

Hogg (1991) has summarized the effects of nutritional stress on meat quality, including tenderness. Although periods of reduced live weight gain or starvation seem to result in less tender meat (Asghar & Yeates, 1979; Kirton, Sinclair, Chrystall, Devine, & Woods, 1981; Yeates, 1964), animals that have been realimented after losing weight produce meat that is as tender as that from control unstressed animals (Morgan, 1972; Winchester & Ellis, 1956). Thus, the effects of nutritional stress followed by compensatory growth show that compensatory growth per se does not have any adverse effects on meat quality, including tenderness (Hogg, 1991).

Influence of Sex

In two extensive reviews, Purchas (1991) and Chrystall (1994) have summarized the effects of sex upon tenderness. Researchers in the United States, Ireland, and the United Kingdom have reported that even young bulls cannot be marketed with the same confidence as steers of the same ages. Dransfield, Nute, and Francombe (1984) reported that most consumers preferred meat from steers over that from young bulls in home tests, although an earlier study in European Community countries (Dransfield et al., 1982) concluded that the wide degree of variation in texture was not attributable to sex. However, comparison of twin steers and young bulls found that the bull meat was slightly tougher when roasted but that the tenderness of grilled and casserole meat was similar (Dransfield et al., 1984). Although young bull meat may be of acceptable tenderness, older bulls tend to be less tender (Liboriussen, Anderson, Buchter, Kousgaard, & Moller, 1977). As in studies comparing bulls and steers, sex differences between castrates and intact male pigs and rams were relatively small in sheep and pigs, but these comparisons have not been studied as extensively as in beef (Dransfield, 1994).

Effects of Growth Promoters

Steroid-related growth promoters have been extensively studied for their effects on beef tenderness, but the results indicate that they have little influence on texture (Ntunde, Usborne, & Ashton, 1977; Verbeke et al., 1976). Although diethylstilbestrol increased growth rate, it had little influence on meat tenderness. Other growth promoters, such as antibiotics (Hays, 1991), somatotropin (Beermann & DeVol, 1991), ionophores (Owens, Zorrilla-Rios, & Dubeski, 1991), and photo period (Petitclerc & Zinn, 1991), had little effect on texture. Hancock, Wagner, and Anderson (1991) have reviewed the influence of androgens and estrogens on meat quality. On the other hand, β agonists have been shown to markedly increase carcass leanness and to increase toughness in sheep meat (Hamby, Stouffer, & Smith, 1986), cattle (Tarrant, 1987), and, to a lesser extent, pigs (Warriss, Nute, Rolph, Brown, & Kestin, 1991). The increase in toughness observed for animals treated with β agonists appears to be due to a reduction in the levels of the proteases (cathepsins B and D and calpain I) and an increase in the concentration of calpastatin (a specific protein inhibitor), as pointed out by Kretchmar, Hathaway, Epley, and Dayton (1990). Dransfield (1994) speculated that the increase in leanness associated with administration of β agonists may be due to a reduction in myofibrillar degradation and that the increase in toughness may be the result of a decrease in the level of proteases and an increase in the amount of calpastatin. Evidence to support the calpastatin effect is found in the fact that injection of calcium salts, which activate the calpains, increases meat tenderness (Koohmaraie & Shackelford, 1991).

Effects of Other Stress Conditions

It has long been known that subjecting animals to stress prior to slaughter has an influence on meat quality, including tenderness (Briskey, 1964; Pearson & Young, 1989). Dark cutting beef and PSE pork are frequently encountered as a consequence of preslaughter stress. It is well known that exhaustive exercise prior to slaughter produces dark cutting beef and DFD pork, both conditions being associated with high muscle pH values. Generally speaking, stress depletes the muscle and liver glycogen and raises the pH, which usually results in more tender meat. However, high-pH meat has a soapy taste and a dark unattractive color, so it is undesirable (Pearson & Young, 1989). Hedrick, Boillet, Brady, and Naumann (1959) were able to produce dark cutting beef experimentally by injection of adrenaline. Although dark cutting beef and DFD pork are more tender than normal beef or pork, the soapy flavor and undesirable color preclude taking advantage of high-pH meat. Briskey (1964) reviewed the influence of PSE pork on meat quality and indicated that it had no major effects on tenderness.

Role of Connective Tissue

Ramsbottom and Strandine (1948) noted the relationship between the amount of visible connective tissue in raw meat and toughness in cooked meat. Division of beef muscles into cooking method procedures (grilling, roasting, or stewing) is based on the amount of visible collagenous tissue. Light (1987) has reviewed the relationship of collagen content and meat texture. In France, removal of collagenous tissue is often used for upgrading meat. Dransfield (1994) stated that the contribution of connective tissue changes with cooking temperature. At 40 to 50°C, the myofibrils denature, moisture is lost, and toughness increases by two- to threefold. Between 50 and 65°C, collagen shrinks, more moisture is lost, and toughness increases a further twofold. The second rise in toughness starts at 50°C in cattle 2 to 3 months in age and at about 60°C in 12-year-old cattle and has been shown to be related to the strength of the connective tissue in the muscles (Shorthose & Harris, 1990). At temperatures above 80°C, the collagen will solubilize, and the meat becomes very tender (Davey & Gilbert, 1974). As an animal becomes older, there may be but little change in total collagen content. However, the solubility of the collagen will decrease. For example, in 9-week-old calves, about 22% of the collagen was soluble in hot water, which decreased to 12% for 10-month-old steers and only 4% in old cows (Tuma, Henrickson, Stephens, & Moore, 1962). Heat-soluble collagen content, therefore, is closely related to the animal's age, but total collagen is the best predictor of tenderness among muscles. Attempts to relate toughness of lamb with the Callipyge gene to cross-linking of collagen suggests that toughness associated with this gene is not due to cross-links of collagen but is probably associated with the myofibrillar proteins (Field, McCormick, Brown, Hinds, & Snowder, 1996).

EFFECTS OF METHODS OF SLAUGHTER

Slaughter procedures, except for abnormal stress, appear to have but little effect on meat texture. Thus, slaughtering by conventional electrical stunning followed by exsanguination has no measurable effect on texture in comparison to cattle slaughtered by the kosher method (Romans, Costello, Carlson, Greaser, & Jones, 1994). Undue stress prior to slaughter can produce dark cutting beef and PSE and DFD pork (Pearson & Young, 1989) but has little effect on tenderness. Dransfield (1994) summarized the effects of the slaughtering method on fish and poultry quality. With fish, there was little consistent effect on the tenderness, except that carbon dioxide immobilization generally resulted in soft flesh. According to Azam, Mackie, and Smith (1989), the softening could be attributed to the degradation of the myofibrillar proteins at pH values of less than 6. In poultry, preslaughter electrical stunning can improve tenderness (Lee, Hargus, Webb, Rickansrud, & Hagberg, 1979), but poorer quality meat may result from using high voltages or long stunning periods. Preslaughter heat stress can also decrease tenderness, as can high temperatures or longer scalding times (Dransfield, 1994). Decreased struggling at the time of slaughter, on the other hand, has been shown to improve tenderness. All of these effects appear to be due to postmortem development of rigor mortis and its effects (Dransfield, 1994).

EFFECTS OF POSTSLAUGHTER TREATMENT

Postslaughter treatment has profound effects on the textural properties of meat. These treatments can have either beneficial or deleterious effects and vary with animal species. The major factors affecting texture are discussed below.

Influence of Conventional Chilling

Conventional chilling following slaughter is normally accomplished by moving cattle, sheep, and pig carcasses into chilling rooms held at temperatures from 0 to 5°C, where the body heat is removed to delay microbial growth. This normally occurs in about 24 to 36 hr, depending on the temperature and rate of air movement (Romans et al., 1994). Conventional chilling may cause some toughening of the outer layers of the beef round (Dransfield & Jones, 1978). Poultry carcasses are usually chilled more rapidly, either in cold water baths or by forced cold air. The more rapid chilling of poultry apparently causes no severe textural problems.

Influence of Chilling Too Rapidly

The phenomenon of "cold shortening," in which prerigor meat exposed to cold temperatures shortens and becomes tough, was first described by Locker and Hagyard (1963). Conditions that caused cold shortening were studied in greater detail by Marsh and Leet (1966). The relationship between percent shortening and shear force with the MIRINZ tenderometer is illustrated in Figure 8–4. The plot shows that prerigor muscle increased its shear value by two- to threefold on being held at 2°C until the onset of rigor. Two approaches have been shown to be effective in preventing cold shortening: less rapid chilling plus aging (Dutson & Pearson, 1985; McCrae, Seccombe, Marsh, & Carse, 1971) and electrical stimulation (Carse, 1973; Chrystall & Devine, 1985).

Electrical Stimulation

There is no question about the beneficial effects of electrical stimulation in preventing cold shortening, with an entire volume being devoted to this subject (Pearson & Dutson, 1985). There is some evidence that electrical stimulation may have an effect upon tenderness beyond prevention of cold shortening, which may be due to tissue disruption per se and also to the release of the lysosomal enzymes (Dutson & Pearson, 1985; George, Bendall, & Jones, 1980). Rapid reduction in pH of beef muscles to 5.4 in 4 hr and holding at elevated temperatures (above 30°C) can induce toughness (Marsh, Ringkob, Russell, Swartz, & Pagel, 1987), probably by exposure to conditions that induce rigor shortening. Thus, Dransfield (1994) has suggested that electric stimulation may not always improve tenderness but can cause heat shortening. Electrical stimulation on poultry and pig meat may or may not improve tenderness depending on state of rigor development and the cooling rate (Dransfield, 1994). Although usually improving tenderness, it is not always advantageous.

Effect of Carcass Suspension

The degree of muscle shortening is dependent upon the skeletal restraints placed upon individual muscles while undergoing rigor mortis, which was mentioned earlier when discussing rigor mortis. When carcasses are hung in the conventional manner by the Achilles tendon, the psoas muscle is under tension and its fibers are stretched; the increased sarcomere length plus the very low collagen content probably accounts for the tenderness of this muscle even in old cows (Dransfield, 1994). On the other hand, the longissimus muscle has the fibers in a horizontal position and is highly sensitive to shortening and the associated toughening. Pelvic suspension, hanging by the aitch bone instead of the Achilles tendon, places the longissimus muscle under tension and is the basis for the so-called "Tenderstretch" procedure (Hostetler, Link, Landmann, & Fitzhugh, 1972). The same effects on the prerigor muscle can be achieved by using the "standing posture," which places tension on the longissimus muscle (Bouton, Harris, Shorthose, & Baxter, 1973; Herring, Cassens, Suess, Brungardt, & Briskey, 1967) while it is undergoing rigor mortis. Although the Tenderstretch procedure can improve tenderness of the longissimus muscle, it has not been widely adopted because it would require modification of handling equipment.

Effects of Early-Postmortem Freezing

Freezing of postrigor meat has little or no effect on meat texture. Sharp and Marsh (1953) observed that whole muscle, frozen in the prerigor condition, lost as much as 47% of the total weight as drip fluid upon thawing.

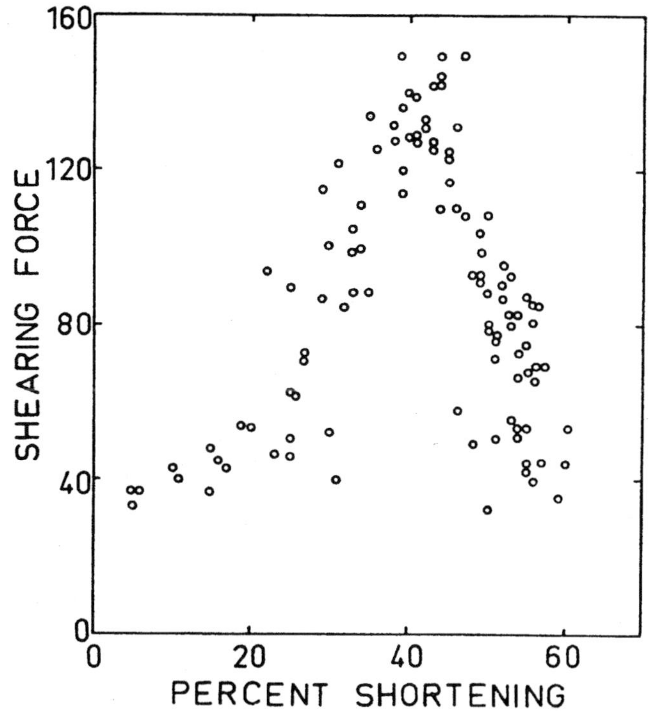

Figure 8–4 Relationship between Shear Force and Muscle Shortening. Cold shortening is expressed as percentage of initial excised length and was achieved by transferring samples from room temperature to 2°C at various time intervals. Note that maximum shortening and toughness occurred at about 40% shortening while at 20% and 60% shortening the muscles were more tender. *Source:* Reprinted with permission from B.B. Marsh, and N.G. Leet, Studies on Meat Tenderness. III, Effect of Cold Shortening on Tenderness, *Journal of Food Science*, Vol. 31, p. 453, © 1966, Institute of Food Technologists.

Marsh and Thompson (1958) noted that muscle strips frozen immediately after slaughter shortened by 72% and lost 27% drip upon thawing. However, if muscle was held at –3°C for four days, it did not shorten or lose any drip. They further observed that the amount of drip increased steeply on thawing at 5 to 10°C. According to Newbold (1966), the onset of thaw rigor occurs when the concentration of ATP is still relatively high, about 40% of rest levels. Kushmerick and Davies (1968) concluded that thaw rigor is characterized by release of excessive amounts of calcium, which activates muscle contraction. This results in extensive shortening and increased toughness, as explained by Pearson and Young (1989). It can be prevented by prerigor electrical stimulation or by allowing the meat to pass through rigor before freezing.

Influence of Postmortem Aging

Holding of postmortem meat after rigor mortis at temperatures of 0 to 5°C is known as *aging*. Dransfield (1994) concluded that the toughness of meat decreases approximately exponentially with the length of aging. Thus, tenderization occurs early, and the magnitude of change decreases over time. The rate of tenderization is lowest in beef and fastest in poultry meat and fish. It normally requires about 3 weeks' storage at chill temperatures to achieve maximum tenderness in beef (Dransfield, 1994). Although aging or ripening of meat has been used to improve meat tenderness for many years (Hoagland, McBryde, & Powick, 1917; Lehmann, 1907), the nature of the changes, which are probably due to enzymatic hydrolysis (Whitaker, 1959), is not fully understood (Pearson & Dutson, 1985). Locker (1960b) followed changes in amino-terminal groups but found little change on aging beef muscle from slaughter through 16 days postmortem. Bandack-Yuri and Rose (1961) obtained similar results on aging of chicken breast muscle. Actin and myosin remain stable postmortem, but proteolysis does occur in titin, nebulin, troponin T, and vinculin (Fritz & Greaser, 1991; Greaser & Fritz, 1995; Koohmaraie, Doumit, & Wheeler, 1996; Lusby, Ridpath, Parrish, & Robson, 1983; Ringkob, Marsh, & Greaser, 1988). Structural damage in the region of the Z lines in the myofibrils also is observed (Davey & Gilbert, 1969). Changes in the connective tissue proteins are less clear. There appears to be no increase in soluble native collagen with postmortem aging (Jeremiah & Martin, 1981). However, there must be some peptide bond cleavage, since the solubility of collagen in denaturing solvents increases markedly during aging (Stanton & Light, 1987).

EFFECTS OF PROCESSING

Processing procedures vary widely and are influenced by rigor development, which was discussed earlier in this chapter and which is influenced by postmortem temperature and pH. Hot boning, prerigor cooking, and prerigor cooling can alter meat texture and will be discussed below. Proteolytic enzymes are frequently added to improve tenderness, as are marinating and high-pressure treatments. Addition of protein and nonprotein additives can also alter tenderness by altering the water-binding capacity (Grau & Hamm, 1953). These procedures have been reviewed by Dransfield (1994) and Chrystall (1994).

Hot Boning

Boning before the body heat has been removed is known as *hot boning*. It reduces energy usage, since the bones and fat do not need to be cooled (Henrickson & Asghar, 1985). However, it must be practiced with care, and most often the meat also must be electrically stimulated before boning. In fact, the meat will still be susceptible to cold toughening unless handled carefully. Electrical stimulation will prevent shortening and should be used on all intact muscles, as explained by Cross and Seideman (1985). Meat to be ground while still hot and preblended into sausages does not need to be electrically stimulated.

Prerigor Cooking

Prerigor muscle shortening can be extreme if meat is cooked too soon after death. Paul, Bratzler, Farwell, and Knight (1952) demonstrated that small cores of beef steaks were tender when cooked in deep fat immediately after death but that on delayed cooking or more slow cooking, the cores became tougher until they reached full rigor. They then gradually increased in tenderness with further holding. Koohmaraie et al. (1996) showed that toughening associated with prerigor cooking could be prevented by restraining muscle strips from shortening during heating. The effect of prerigor cooking diminishes as the time of cooking is delayed and disappears as the meat passes through rigor (Dransfield & Rhodes, 1975). Poultry muscle shortening can also be extreme if the meat is heated soon after stunning, and this shortening can produce tough meat (Abugroun, Forrest, Marks, & Judge, 1985). However, tenderness can be acceptable if the time and rate of heating are carefully controlled. At high rates of heat, deboned muscles go into rigor very rapidly and may shorten up to 40%, but glycolysis is prevented by rapid denaturation, so that the pH remains high and the meat is still tender (Dransfield, 1994).

Addition of Proteolytic Enzymes

Proteolytic enzymes, such as papain, ficin, and bromelain, have been used to tenderize meat for a number of years. The ProTen process, which was patented by Swift and Company, used a selectively isolated solution of papain that was injected into the jugular vein of cattle 10 min prior to slaughter. The vascular system was an efficient conduit for distributing the enzyme throughout the carcasses. The fraction used became active only when the meat was heated (Dransfield, 1994). The amount of enzyme used was calculated to give a final concentration of about 2 to 5 ppm in the muscle, which was sufficient to tenderize most muscles. However, the ProTen process resulted in mushiness of the liver and created too much softening on holding of roasts on the warming table (the enzyme remained active at these temperatures and resulted in the roasts' falling apart). Proteolytic enzymes can be injected directly into cuts that need tenderization. This procedure has been favored over that of tenderizing the entire carcass, since one can adjust the process to cuts differing in toughness. Rhodes and Dransfield (1973) have been able to overcome cold shortening and its associated toughness in lamb carcasses with this approach. Another method used by Koohmaraie, Babiker, Schroeder, Merkel, and Dutson (1988) to tenderize meat was to activate the calpains by injecting calcium chloride. Although addition of calcium chloride may give an objectionable flavor, some success has been achieved in overcoming this problem by using other additives.

Marinating

Overcoming texture problems in meat, poultry, and fish by marinating these meats has been practiced by chefs and gourmet cooks for many years. Marinades are typically composed of vinegar or wine (or a combination of both) along with salt and special seasonings; the meat to be tenderized is held in these mixtures for 24 to 48 hr before cooking (Rao & Gault, 1991). Although the ingredients in the marinade alter texture, the exact mechanisms are not understood.

Pressurization

Subjecting meat to high pressures (1 hr at 1500 atm) disrupts the muscle structure and results in tenderization (MacFarlane, 1973). Although the exact mechanism by which the high-pressure treatment improves tenderness is not fully understood, it is known to speed up glycolysis in prerigor meat and to cause the pH to rise by about 0.6 units in the range of 15 to 30°C. There is also evidence that titin is broken down (Suzuki, Kim, & Ikeuchi, 1996). The high-pressure effects can be achieved in about 1 min at 1,000 atm of pressure (Kennick, Elgasim, Holmes, & Meyer, 1980). There is definitely some membrane damage; the release of calcium ions and the resultant proteolysis may be responsible for the tenderizing effects. Dransfield (1994) has suggested that the high-pressure treatment probably enhances proteolysis by release of the lysosomal enzymes into the sarcoplasm and causes rapid tenderization. The pressure treatment definitely enhances tenderization, but the difficulty of incorporating pressurization into a line operation to date has limited its usefulness to the meat industry. The effects of high pressure on meat have recently been reviewed (Cheftel & Culioli, 1997).

Hydrodyne Process

The hydrodyne process uses a small amount of explosive to generate a hydrodynamic shock wave in a liquid medium (water) to tenderize meat (Solomon, Long, & Eastridge, 1997). The shock wave passes through the meat, which is suspended and surrounded by the liquid. The meat is a close acoustical match for water, since it is approximately 75% water. The shock wave reflects off any object in the water that is not an acoustical match. Four different beef muscles were exposed to either 50, 75, or 100 grams of explosive, which was composed of nitromethane (a liquid) and ammonium nitrate (a solid). A detonating device was used to initiate the explosion. All muscles were tenderized by the Hydodyne process, with 72% increase in tenderness by the 100-g explosive. Thus, the Hydrodyne process, which is patented, was effective for tenderizing meat. Furthermore, it was effective in improving the tenderness of "cold-shortened" muscle, which tends to be resistant to some tenderizing treatments. Although the process is new, it appears to offer a promising method for meat tenderization. The mechanism and operating parameters must still be elucidated and explored further.

Curing and Meat Processing

Curing and meat processing can be manipulated by using curing salts and other additives, as explained by Pearson and Gillett (1996). Most of these procedures will improve meat texture, especially the cooking/smoking operation. The curing salts per se tend also to improve texture and palatability. Phosphates, which are commonly added to processed meats and poultry, tend to increase water binding and improve texture. However, addition of phosphates and other water binding agents may, if applied improperly, create problems in texture (Pearson & Gillett, 1996). Other additives can improve texture or create problems, depending on the additive and its application and properties. Sausage products and ground meat products require about 15% to 20% fat for maximum consumer satisfaction. The level of fat can be further reduced by use of water-binding agents, as discussed in Pearson and Dutson (1997). A number of water binders have been used and, if used properly, can compensate for the reduction in fat content.

MEASUREMENT OF TEXTURE IN FLESH FOODS

Sensory Methods

The sensory procedures used for texture assessment include the chew count (Harrington & Pearson, 1962), which simply requires panelists to count the number of chews to reduce the product to the necessary size needed for swallowing. Panel members may differ widely in the chew count, but the method attempts to place texture on a quantitative basis. Multivariate analysis has been used by Harries, Rhodes, and Chrystall (1972) to reduce eight sensory variables to a simpler expression of texture. These investigations demonstrated that juiciness contributed to texture. Jeremiah, Murray, and Gibson (1990) have used five surface properties, five first-bite impressions, 14 masticatory properties, and six after-feel properties to describe the texture of pork loins. Although these attempts to reduce texture to its component parts help us to understand texture, they are not widely used and require care in interpretation. Sensory methods of evaluating texture vary widely and use rating systems that are as varied as texture per se, but all use panel members' reactions to the tactile properties of the product in the mouth. Although such procedures can be criticized for various weaknesses, researchers must bear in mind that the acceptance of any food by consumers should be the measure of its success, including that of texture. It should be borne in mind that variability in texture, whether measured by human assessors or by instrumental methods, is common to muscle foods. Measurements within a single muscle vary by as much as 20%, according to Chrystall (1994).

Instrumental Methods

Table 8–4, which has been modified from one reproduced by Chrystall (1994), lists several objective methods for measuring texture of muscle foods. These are categorized as physical methods (which include shear devices, biting devices, compression methods, tensile methods, penetration devices, grinding methods, and fragmentation methods), structural measures, and chemical methods. References for each of the methods are included in Table 8–4. The shear devices have been the most widely used, especially the Lee-Kramer shear press and the Warner-Bratzler device. The relationship for the latter between panel tenderness and shear readings falls between 0.6 and 0.85. The configuration of the hole in the plate, the speed of shearing, and core shape all may influence the readings. Recently, Wheeler, Shackelford, and Koohmaraie (1996) have examined a number of factors that may affect shear force results. These workers found that steaks thawed at 12°C were significantly more tender than those thawed at –2°C before cooking (shear values of 5.99 vs. 7.34 kg respectively, $p > 0.05$). They also noted that the repeatability of shear force measurements was better when the cores were obtained perpendicular to the steak surface as opposed to parallel and when steaks were cooked to a constant internal temperature as opposed to a constant length of time. Electric broilers and convection oven broilers gave similar tenderness results. These results point out the need for standardization of shear force procedures for measuring tenderness of beef so that data from different laboratories can be compared.

The biting devices represent an attempt to duplicate the effects of the teeth to measuring texture. The New Zealand (MIRINZ) biting test devices have been widely used for measuring texture of meat. Perhaps the biggest problem with the biting tenderometers is that the procedures must simulate not only biting but also the grinding motion of the molars (Chrystall, 1994), which is difficult to achieve. The Instron has been used by Shorthose, Harris, Eadie, and Dickinson (1988). Several compression procedures have been used, but without much success (Chrystall, 1994). Although tensile strength has been used in fiber extensibility studies (Wang, Doty, Beard, Pierce, & Hankins, 1956), the Smith tensiometer and the L and W yield meter, only limited data are available on their relationships to texture (Bramblett, Hostetler, Vail, & Draudt, 1959; Locker & Wild, 1982). A number of penetration devices have been developed for measuring muscle foods' texture. These methods are based on the theory that there is a relationship between raw meat firmness and texture. The most widely used instrument of this type is the Armour tenderometer (Hansen, 1972), which has been used with moderate success under controlled conditions. The fragmentation index (Olson, Parrish, & Stromer, 1976)

Table 8–4 Methods for Assessment of Tenderness/Texture in Meat, Poultry, and Fish

Methods	*References*
PHYSICAL METHODS	
Shear Devices	
Shear jaw	Shockey et al. (1944)
Lee-Kramer shear press	Szczesniak and Torgeson (1965a, 1965b)
Warner-Bratzler shear	Warner (1928)
Rotating knife	Bjorksten et al. (1967)
Biting Devices	
Lehmann	Lehmann (1907)
Volodkevich	Volodkevich (1938)
MIRINZ—Mechanical	Macfarlane and Marer (1996)
MIRNIZ—Pneumatic	Frazerhurst and Macfarlane (1979)
MIT denture tenderometer	Proctor et al. (1955, 1956a, 1956b)
KT biting device	Kelly et al. (1960)
General Foods texturometer	Friedman et al. (1963)
Instron RVI	Shorthose et al. (1988)
Nip tenderometer	Purchas (1973)
Compression Methods	
Swift tenderness	Palmer (1962)
Orifice device	Howe and Bull (1927)
Sinusoidal compressor	Sale et al. (1984)
Tensile Methods	
Wang fiber extensibility	Wang et al. (1956)
Smith tensiometer	Bramblett et al. (1959)
L and W yield meter	Locker and Wild (1982)
Connective tissue–tearing device	Pool (1967)
Penetration Devices	
Christal texturemeter	Miyada and Tappel (1956)
Armour tenderometer	Hansen (1972)
Slice tenderometer	Kulwich et al. (1963)
Instron compression	Bouton and Harris (1972)
Grinding Methods	
Mincer	Miyada and Tappel (1956)

continues

Table 8–4 continued

Methods	*References*
Fragmentation Methods	
Myofibrillar fragmentation index	Olson et al. (1976)
STRUCTURAL MEASURES	
Fiber diameter	Hiner et al. (1953)
Fiber size	Herring et al. (1965)
Shortening	Locker (1960a)
CHEMICAL METHODS	
Trace elements	Vavak et al. (1976)
Collagen characteristics	Seideman (1986)
30,000-Da component	MacBride and Parrish (1977)
SENSORY RATINGS	
Chew count	Harrington and Pearson (1962)
Panel scores	Harries et al. (1972)

Source: Reprinted from B. Chrystall, in *Quality Attributes and Their Measurement in Meat, Poultry and Fish Products*, A.M. Pearson and T.R. Dutson, eds., p. 323, © 1994, Aspen Publishers, Inc.

is based on the fact that tender meat blended under standard conditions breaks down into smaller fragments than tough meat does. Measurements are followed by light scattering, with the more highly fragmented muscles decreasing the amount of light passed through the sample suspended in water. Muscle fiber diameter (Hiner, Hankins, Sloane, Anderson, & Fellers, 1953), muscle fiber size (Herring, Cassens, & Briskey, 1965), and muscle shortening (Locker, 1960a) all are related to meat tenderness, but the relationships are too weak to be useful as predictors of texture. Finally, several chemical measurements have been investigated as estimates of texture, including trace elements (Vavak, Satterlee, & Anderson, 1976), collagen characteristics (amount and solubility) (Seideman, 1986), and the 30,000-Da component (MacBride & Parrish, 1977; Uytterhaegen, Claeys, & Demeyer, 1992). Although these procedures are all related to texture, none has practical applicability for predicting tenderness.

CONCLUSION

The understanding of meat texture is complicated because of the complex nature of the structure of muscle and the large number of factors that can either harm or improve it. In addition, several species-related factors and specific postmortem treatments can drastically alter meat texture. Instrumental methods have been developed to provide quantitative information about the texture of meat, but none of the methods described to date fully simulates the biting and grinding action of the human mastication system.

REFERENCES

Abugroun, H.A., Forrest, J.C., Marks, J.S., & Judge, M.D. (1985). Effect of heating rate on shortening of pre-rigor chicken breast muscle. *Poultry Science, 64*, 1315–1321.

Amos, L., & Cross, R.A. (1997). Structure and dynamics of molecular motors. *Current Opinion in Structural Biology, 7,* 239–246.

Asghar, A., & Yeates, N.T.M. (1979). Muscle characteristics and meat quality of lambs, grown on different nutritional planes: IV. Effect on meat quality. *Agricultural and Biological Chemistry, 43*, 455–461.

Azam, K., Mackie, I.M., & Smith, K. (1989). The effect of slaughter method on quality of rainbow trout (*Salmo gairdneri*) during storage on ice. *International Journal of Food Science and Technology, 24*, 69–79.

Bailey, A.J., & Light, N.D. (1989). *Connective tissue in meat and meat products*. London: Elsevier Applied Science.

Bailey, A.J., & Sims, T.J. (1977). Meat tenderness: Distribution of molecular species of collagen in bovine muscle. *Journal of the Science of Food and Agriculture, 28*, 565–570.

Bandack-Yuri, S., & Rose, D. (1961). Proteases of chicken breast muscle. *Food Technology, 15*, 186–188.

Beermann, D.H., & DeVol, D.L. (1991). Effects of somatotropin, somatotropin releasing factor and somatostatin on growth. *Advances in Meat Research, 7*, 373–426.

Bendall, J.R. (1951). The shortening of rabbit muscle during rigor mortis: Relation to the breakdown of the adenosine triphosphate and creatine phosphate and to muscular contraction. *Journal of Physiology, 114*, 71–88.

Bendall, J.R. (1960). Post mortem changes in muscle. In G.H. Bourne (Ed.), *Structure and function of muscle* (Vol. 3, pp. 225–274). New York: Academic Press.

Bendall, J.R. (1967). The elastin content of various muscles of beef animals. *Journal of the Science of Food and Agriculture, 18*, 555–558.

Bennett, R., & Hamilton, M. (1986). Consumer acceptability of cod and whiting after chilled storage and freezing and thawing. *Journal of Food & Technology, 21*, 311–317.

Bjorksten, J., Anderson, P., Bouschart A., & Kapsalis, J. (1967). A portable rotating knife tenderometer. *Food Technology, 21*, 84–88.

Blumer, T.N. (1963). Relationship of marbling to the palatability of beef. *Journal of Animal Science, 22*, 772–778.

Boleman, S.J., Boleman, S.L., Miller, R.K., Taylor, J.F., Cross, H.R., Wheeler, T.L., Koohmaraie, M., Shackelford, S.D., Miller, M.F., West, R.L., Johnson, D.D., & Savell, J.W. (1997). Consumer evaluation of beef of known categories of tenderness. *Journal of Animal Science, 75*, 1521–1524.

Bouton, P.E., & Harris, P.V. (1972). A comparison of some objective methods used to assess meat tenderness. *Journal of Food Science, 37*, 218–221.

Bouton, P.E., Harris, P.V., Shorthose, W.R., & Baxter, R.I. (1973). A comparison of the effects of aging, conditioning and skeletal restraints on the tenderness of mutton. *Journal of Food Science, 38*, 932–937.

Bramblett, V.D., Hostetler, R.L., Vail, G.E., & Draudt, H.N. (1959). Qualities of beef as affected by cooking at very low temperatures for long periods of time. *Food Technology, 13*, 707–711.

Brannaman, G.A., Pearson, A.M., Magee, W.T., Griswold, R.M., & Brown, G.A. (1962). Comparison of the cutability and eatability of beef and dairy-type cattle. *Journal of Animal Science, 21*, 321–326.

Briskey, E.J. (1964). The etiological status and associated studies of pale, soft and exudative porcine musculature. *Advances in Food Research, 13*, 89–168.

Campion, D.R., Crouse, J.D., & Dikeman, M.E. (1975). Predictive value of USDA beef quality grade factors for cooked meat palatability. *Journal of Food Science, 40*, 1225–1228.

Carse, W.A. (1973). Meat quality and the acceleration of post-mortem glycolysis by electrical stimulation. *Journal of Food Technology, 8*, 163–166.

Chadwick, J.P., Cuthbertson, A., & Dransfield, E. (1979). Differences in meat quality and carcass composition of Friesian and beef breeds × Friesian cattle. *Proceedings of the European Meat Research Workers, Budapest, 25*, 103.

Cheftel, J.C., & Culioli, J. (1997). Effects of high pressure on meat: A review. *Meat Science, 46*, 211–236.

Chrystall, B.B. (1994). Meat texture measurement. *Advance in Meat Research, 9*, 316–336.

Chrystall, B.B., & Devine, C.E. (1985). Electrical stimulation: Its early development in New Zealand. *Advances in Meat Research, 1*, 73–119.

Craig, R. (1994). The structure of the contractile filaments. In A.G. Engel & C. Franzini-Armstrong (Eds.), *Myology* (2nd ed., Vol. 1, pp. 134–175). New York: McGraw-Hill.

Cross, H.R., & Seideman, S.C. (1985). Use of electrical stimulation for hot boning of meat. *Advances in Meat Research, 1*, 159–183.

Davey, C.L., & Gilbert, K.V. (1969). Studies on meat tenderness: 7. Changes in the fine structure of meat during aging. *Journal of Food Science, 34*, 69–74.

Davey, C.L., & Gilbert, K.V. (1974). Temperature dependent cooking toughness in beef. *Journal of the Science of Food and Agriculture, 25*, 931–938.

Dransfield, E. (1994). Tenderness of meat, poultry and fish. *Advances in Meat Research, 9*, 289–315.

Dransfield, E., & Jones, R.C.D. (1978). Effect of rate of chilling on the variability in texture of the beef round. *Journal of the Science of Food and Agriculture*, 29, 601–607.

Dransfield, E., Lockyer, D.K., & Prabhakaran, P. (1986). Changes in extensibility of raw beef muscle during storage. *Meat Science, 16*, 127–142.

Dransfield, E., Nute, G.R., & Francombe, M.A. (1984). Comparison of the eating quality of bull and steer beef. *Animal Products, 39*, 37–50.

Dransfield, E., Nute, G.R., MacDougall, D.B., & Rhodes, D.N. (1979). Effect of sire breed on eating quality of cross-bred lambs. *Journal of the Science of Food and Agriculture, 30*, 805–808.

Dransfield, E.C., & Rhodes, D.N. (1975). Texture of beef *musculus semitendinosus* heated before, during and after rigor mortis. *Journal of the Science of Food and Agriculture, 26*, 483–491.

Dransfield, E., Rhodes, D.N., Nute, G.R., Roberts, T.A., Boccard, R., Touraille, C., Buchter, L., Hood, D.E., Joseph, R.L., Schon, L., Castells, M., Cosentino, E., & Timbergen, B.J. (1982). Eating quality of European beef assessed at five research institutes. *Meat Science, 6*, 163–184.

Dutson, T.R., & Pearson, A.M. (1985). Postmortem conditioning in meat. *Advances in Meat Research, 1*, 45–72.

Farah, C.S., & F.C. Reinach. (1995). The troponin complex and regulation of muscle contraction. *FASEB Journal, 9*, 755–767.

Field, R.A., McCormick, R.J., Brown, D.R., Hinds, F.C., & Snowder, G.D. (1996). Collagen crosslinks in longissimus muscle from lambs expressing the callipyge gene. *Journal of Animal Science, 74*, 2943–2947.

Frazerhurst, L.F., & Macfarlane, P. (1979). A device for measuring the tenderness of meat. New Zealand Patent 190945.

Friedman, H.H., Whitney, J.E., & Szczesniak, A.S. (1963). The texturometer: A new instrument for objective texture measurement. *Journal of Food Science, 28*, 390–396.

Fritz, J.D., & Greaser, M.L. (1991). Changes in titin and nebulin in postmortem bovine muscle revealed by gel electrophoresis, Western blotting and immunofluorescence microscopy. *Journal of Food Science, 56*, 607–615.

George, A.R., Bendall, J.R, & Jones, R.C.D. (1980). The tenderising effect of electrical stimulation of beef carcasses. *Meat Sciences, 4*, 51–68.

Grau, R., & Hamm, R. (1953). Determination of the water binding power of muscle. *Naturwisenschaften, 40*, 29.

Greaser, M.L. (1986). Conversion of muscle to meat. In P.J. Bechtel (Ed.), *Muscle as food* (pp. 37–102). New York: Academic Press.

Greaser, M.L. (1997). Postmortem changes in muscle extracellular matrix proteins. In *50th Annual Recip. Meat Conference* (pp. 53–59). Kansas City: American Meat Science Association.

Greaser, M.L., & Fritz, J.D. (1995). Post-mortem changes in myofibrillar proteins in relation to meat texture. In A. Ouali, D.I. Demeyer, & F.J.M. Smulders (Eds.), *Expression of tissue proteinases and regulation of protein degradation as related to meat quality* (pp. 293–309). Nijmegen, The Netherlands: Audet Tijdschriften.

Hamby, P.L., Stouffer, J.R., & Smith, S.B. (1986). Muscle metabolism and real-time ultrasound measurements of muscle and subcutaneous adipose tissue growth in lambs fed diets containing a β-agonist. *Journal of Animal Science, 63*, 1410–1417.

Hancock, D.L., Wagner, J.F., & Anderson, D.B. (1991). Effects of estrogens and androgens on animal growth. *Advances in Meat Research, 7*, 255–297.

Hansen, L.J. (1972). Development of the Armour tenderometer for the tenderness evaluation of beef carcasses. *Journal of Texture Studies, 3*, 146–164.

Harries, J.M., Rhodes, D.N., & Chrystall, B.B. (1972). Meat texture: I. Subjective assessment of texture in cooked beef. *Journal of Texture Studies, 3*, 101–114.

Harrington, G., & Pearson, A.M. (1962). Chew count as a measure of tenderness of pork loins with various degrees of marbling. *Journal of Food Science, 27*, 106–110.

Hays, V.W. (1991). Effects of antibiotics. *Advances in Meat Research, 7*, 299–320.

Hedrick, H.B., Boillet, J.B., Brady, D.E., & Naumann, H.D. (1959). Etiology of dark-cutting beef. *Missouri Agricultural Experiment Station Research Bulletin Columbia, 717*.

Helander, E. (1957). On quantitative muscle protein determination. *Acta Physiologica Scandinavica, 41*, (Suppl.), 141, 1–99.

Henrickson, R.L., & Ashgar, A. (1985). Cold storage energy aspects of electrically stimulated hot boned meat. *Advances in Meat Research,* 1, 237–276.

Herring, H.K., Cassens, R.G., & Briskey, E.J. (1965). Sarcomere length of free and restrained bovine muscles at low temperatures as related to tenderness. *Journal of the Science of Food and Agriculture, 16*, 379–384.

Herring, H.K., Cassens, R.G., Suess, G.G., Brungardt, V.H., & Briskey, E.J. (1967). Tenderness and associated characteristics of stretched and contracted bovine muscle. *Journal of Food Science, 32*, 317–323.

Hiner, R.L., Hankins, O.G., Sloane, H.S., Anderson, E.E., & Fellers, C.R. (1953). Fiber diameter in relation to tenderness of beef muscle. *Food Research, 18*, 364–376.

Hiner, R.L., Thornton, J.W., & Alsmeyer, R.H. (1965). Palatability and quantity of pork as influenced by breed and fatness. *Journal of Food Science, 30*, 550–555.

Hoagland, R., McBryde, C.N., & Powick, W.C. (1917). *Changes in fresh beef during cold storage above freezing* (Bulletin No. 433). Washington, DC: U.S. Department of Agriculture.

Hogg, B.W. (1991). Compensatory growth in ruminants. *Advances in Meat Research, 7*, 103–134.

Hostetler, R.L., Link, B.A., Landmann, W.A., & Fitzhugh, H.A. (1972). Effect of carcass suspension on sarcomere length and shear force of some major bovine muscles. *Journal of Food Science, 37*, 132–135.

Howe, P.E., & Bull, S. (1927). *A study of the factors which influence the quality and palatability of meat.* Washington, DC: U.S. Department of Agriculture Cooperative Project of the Bureau of Animal Industry.

Jeremiah, L.E., & Martin, A.H. (1981). Intramuscular collagen content and solubility: Their relationship to tenderness and alteration by postmortem ageing. *Canadian Journal of Animal Science, 61*, 53–61.

Jeremiah, L.E., Murray, A.C., & Gibson, L.L. (1990). The effects of differences in inherent muscle quality and frozen storage on the flavor and texture profiles of pork loin roasts. *Meat Science, 27*, 305–327.

Jones, J.M., & Grey, T.C. (1989). Influence of processing on product quality and yield. In G.C. Meade (Ed.), *Processing of poultry* (pp. 127–149). London: Elsevier.

Kabsch, W., Mannherz, H.G., Suck, D., Pai, E.F., & Holmes, K.C. (1990). Atomic structure of the actin:DNAse I complex. *Nature, 347*, 37–44.

Kalder, K. (1994). Extracellular matrix: 1. Fibril-forming collagens. *Protein Profiles, 1*, 519–638.

Kellermayer, M., Smith, S., Granzier, H., & Bustamante, C. (1997). Folding-unfolding transitions in single titin molecules characterized by laser tweezers. *Science, 276*, 1112–1116.

Kelly, R.F., Taylor, J.C., & Graham, P.P. (1960). Preliminary comparisons of a new tenderness measuring device with objective and subjective evaluations of beef [Abstract]. *Journal of Animal Science, 19*, 645.

Kennick, W.H., Elgasim, E.A., Holmes, Z.A., & Meyer, P.F. (1980). The effect of pressurization of pre-rigor muscle on post rigor meat characteristics. *Meat Science, 4*, 33–40.

Kirton, A.H., Sinclair, P.D., Chrystall, B.B., Devine, C.E., & Woods, E.G. (1981). Effect of plane of nutrition on carcass composition and the palatability of pasture fed lamb. *Journal of Animal Science, 52*, 285–291.

Koch, R.M., Crouse, M.E., Dikeman, M.E., Cundiff, L.V., & Gregory, K.E. (1988). Effects of marbling on sensory panel tenderness in *Bos taurus* and *Bos indicus* crosses [Abstract]. *Journal of Animal Science, 66*, 505.

Koohmaraie, M., Babiker, A.S., Schroeder, R.A., Merkel, R.A., & Dutson, T.R. (1988). Acceleration of post-mortem tenderization in ovine carcasses through activation of Ca^{2+} dependent proteases. *Journal of Food Science, 53*, 1638–1641.

Koohmaraie, M., Doumit, M.E., & Wheeler, T.L. (1996). Meat toughening does not occur when rigor shortening is prevented. *Journal of Animal Science, 74*, 2935–2942.

Koohmaraie, M., & Shackelford, S.D. (1991). Effect of calcium infusion on the tenderness of lambs fed β-adrenergic agonist. *Journal of Animal Science, 69*, 2465–2471.

Koohmaraie, M., Shackelford, S.D., & Wheeler, T.L. (1996). Effects of β-adrenergic agonist (L-644,969) and male sex condition on muscle growth and meat quality of *Callipyge* lambs. *Journal of Animal Science, 74*, 70–79.

Kretchmar, D.H., Hathaway, M.R., Epley, R.J., & Dayton, W.R. (1990). Alteration of postmortem degradation of myofibrillar proteins in muscle of lambs fed beta-adrenergic agonist. *Journal of Animal Science, 68*, 1760–1772.

Krueger, D.J., & Fennema, O.R. (1989). Effect of chemical additives on toughening of fillets of frozen Alaska pollack (*Theragra chalcogramma*). *Journal of Food Science, 54*, 1101–1106.

Kulwich, R., Decker, R.W., & Alsmeyer, R.H. (1963). Use of a slice-tenderness evaluation device with pork. *Food Technology, 17*(2), 83–85.

Kushmerick, M.J., & Davies, R.E. (1968). The role of phosphate compounds in thaw contraction and the mechanism of thaw rigor. *Biochimica Biophysica Acta, 153*, 279–287.

Labeit, S., & Kolmerer, B. (1995). Titins: Giant proteins in charge of muscle ultrastructure and elasticity. *Science, 270*, 293–296.

Leblanc, E.L., Leblanc, R.J., & Gill, T.A. (1987). Effects of pressure processing on frozen stored muscle proteins of Atlantic cod (*Gadus morhua*) fillets. *Journal of Food Processing and Preservation, 11*, 209–235.

Lee, Y.B., Hargus, G.L., Webb, J.E., Rickansrud, D.A., & Hagberg, E.C. (1979). Effects of electrical stunning on postmortem biochemical changes and tenderness in broiler breast muscle. *Journal of Food Science, 44*, 1121–1128.

Lehmann, K.B. (1907). Studien uber die Zahigkeit des Fleishes und ihre Ursuchen. *Archives Hygiene, 63*, 134.

Liboriussen, T., Anderson, B.B., Buchter, L., Kousgaard, K., & Moller, A.J. (1977). Crossbreeding experiments with beef and dual-purpose sire breeds on Danish/dairy cows: IV. Physical, chemical and palatability characteristics of longissimus dorsi and semitendinosus muscles from cross-bred young bulls. *Livestock Production Science, 4*, 31–43.

Light, N.D. (1987). The role of collagen in determining the texture of meat. *Advances in Meat Research, 4*, 87–107.

Locker, R.H. (1960a). Degree of muscular contraction as a factor in tenderness of beef. *Food Research, 25*, 304–307.

Locker, R.H., (1960b). Proteolysis in the storage of beef. *Journal of the Science of Food and Agriculture, 11*, 520–526.

Locker, R.H., & Hagyard, C.J. (1963). A cold-shortening effect in beef muscles. *Journal of the Science of Food and Agriculture, 14*, 787–793.

Locker, R.H., & Wild, D.J.C. (1982). A machine for measuring yield point in raw meat. *Journal of Texture Studies, 13*, 71–82.

Lusby, M.L., Ridpath, J.F., Parrish, F.C., Jr., & Robson, R.M. (1983). Effect of postmortem storage on the degradation of the myofibrillar protein titin in bovine longissimus muscle. *Journal of Food Science, 48*, 1787–1795.

MacBride, M.A., & Parrish, F.C., Jr. (1977). The 30,000-dalton component of tender bovine longissimus muscle. *Journal of Food Science, 42*, 1627–1629.

MacFarlane, J.J. (1973). Pre-rigor pressurization of muscle: Effects on pH, shear value and taste panel assessment. *Journal of Food Science, 38*, 294–298.

Macfarlane, P.G., & Marer, J.M. (1996). An apparatus for determining the tenderness of meat. *Food Technology, 20*, 838–839.

Malmfors, M., & Nilsson, R. (1979). Meat quality traits in Swedish Landrace and Yorkshire pigs with emphasis on genetics. *Acta Agriculturae Scandinavica*, Suppl. 1, 81.

Marsh, B.B., & Leet, N.G. (1966). Studies in meat tenderness: III. The effects of cold shortening on tenderness. *Journal of Food Science, 31*, 450–459.

Marsh, B.B., Ringkob, T.P., Russell, R.L., Swartz, D.R., & Pagel, L.A. (1987). Effects of early-postmortem glycolytic rate on beef tenderness. *Meat Science, 21*, 241–248.

Marsh, B.B., & Thompson, J.F. (1958). Rigor mortis and thaw rigor in lamb. *Journal of the Science of Food and Agriculture, 9*, 417–424.

McCormick, R.J. (1994). The flexibility of the collagen compartment of muscle. *Meat Science, 36*, 79–91.

McCrae, S.E., Seccombe, C.G., Marsh, B.B., & Carse, W.A. (1971). Studies in meat tenderness: 9. The tenderness of various lamb muscles in relation to their skeletal restraint and delay before freezing. *Journal of Food Science, 36*, 566–570.

Miyada, D.S., & Tappel, A.L. (1956). Meat tenderization: Two mechanical devices for measuring texture. *Food Technology, 10*, 142–145.

Morgan, J.H.L. (1972). Effect of plane of nutrition in early life on subsequent live-weight gain, carcass and muscle characteristics and eating quality of meat in cattle. *Journal of Agricultural Science (Cambridge) 78*, 417–423.

Newbold, R.P. (1966). Changes associated with rigor mortis. In E.J. Briskey, R.G. Cassens, & J.C. Trautman (Eds.), *The physiology and biochemistry of muscle as a food* (pp. 213–224). Madison: University of Wisconsin Press.

Ntunde, B.N., Usborne, W.R., & Ashton, G.C. (1977). Responses in meat characteristics of Holstein-Friesian males to castration and diets. *Canadian Journal of Animal Science, 57*, 449–458.

Olson, D.G,. Parrish, F.C., Jr., & Stromer, M.H. (1976). Myofibril fragmentation and shear resistance of three bovine muscles during postmortem storage. *Journal of Food Science, 41*, 1036–1041.

Owens, F.N., Zorrilla-Rios, J., & Dubeski, P. (1991). Effect of ionophores on metabolism, growth and meat quality. *Advances in Meat Research, 7*, 321–342.

Palmer, W.E. (1962). Tenderness testing device. Canadian Patent 639 364.

Paul, P.C., Bratzler, L.D., Farwell, E.D., & Knight, K. (1952). Studies on beef: I. Rate of heat penetration. *Food Research, 17*, 504–510.

Pearson, A.M., & Dutson, T.R. (Eds.). (1985). Electrical stimulation. *Advances in Meat Research, 1*.

Pearson, A.M., & Dutson, T.R. (Eds.). (1997). *Production and processing of healthy meat, poultry, and fish products*. London: Blackie Academic & Professional.

Pearson, A.M., & Gillett, T.A. (1996). *Processed meats* (3rd ed.). New York: Chapman & Hall.

Pearson, A.M., & Young, R.B. (1989). *Muscle biochemistry*. San Diego, CA: Academic Press.

Perry, S.V. (1950). Studies on the rigor resulting from the thawing of frozen frog sartorius muscle. *Journal of General Physiology, 33*, 563–577.

Petitclerc, D., & Zinn, S.A. (1991). Effects of photoperiod on growth and development of ruminants. *Advances in Meat Research, 7*, 427–453.

Pool, M.F. (1967). Objective measurements of connective tissue tenacity of poultry meat. *Journal of Food Science, 32*, 550–553.

Prockop, D.J., & Kivirikko, K.I. (1995). Collagens: Molecular biology, diseases, and potential for therapy. *Annual Review of Biochemistry, 64*, 403–434.

Proctor, B.E., Davison, S., & Brody, A.L. (1956a). A recording strain gauge denture tenderometer for foods: II. Studies on the masticatory force and motion, and force penetration relationship. *Food Technology, 10*, 327–331.

Proctor, B.E., Davison, S., & Brody, A.L. (1956b). A recording strain gauge denture tenderometer for foods: III. Correlation with subjective tests and the Canco Tenderometer. *Food Technology, 10*, 344–346.

Proctor, B.E., Davison, S., Malecki, G.J., & Welch, M. (1955). A recording strain gauge denture tenderometer for foods: I. Instrument evaluation and initial tests. *Food Technology, 9*, 471–492.

Purchas, R.W. (1973). Some aspects of raw meat tenderness. *Journal of Food Science, 38*, 556–559.

Purchas, R.W. (1991). Effect of sex and castration on growth and composition. *Advance in Meat Research, 7*, 203–254.

Ramsbottom, J.M., & Strandine, E.J. (1948). Comparative tenderness and identification of muscles in wholesale beef cuts. *Food Research, 13*, 315–330.

Rao, M.V., & Gault, N.F.S. (1991). Acetic acid marinading: The rheological characteristics of some raw and cooked beef muscles which contribute to changes in meat tenderness. *Journal of Texture Studies, 21*, 455–477.

Rayment, I., Rypniewski, W.R., Schmidt-Base, K., Smith, R., Tomchick, D.R., Benning, M.M., Winkelmann, D.A., Wesenberg, G., & Holden, H.M. (1993). Three-dimensional structure of myosin subfragment-1: A molecular motor. *Science, 261*, 50–58.

Rhodes, D.N., & Dransfield, E. (1973). Effect of preslaughter injections of papain on toughness in lamb muscles induced by rapid chilling. *Journal of the Science of Food and Agriculture, 24*, 1583–1588.

Ringkob, T.P., Marsh, B.B., & Greaser, M.L. (1988). Change in titin position in postmortem bovine muscle. *Journal of Food Science, 53*, 276–277.

Romans, J.R., Costello, W.J., Carlson, C.W., Greaser, M.L., & Jones, K.W. (1994). *The meat we eat.* Danville, IL: Interstate.

Rowe, R.W.D. (1989). Electron microscopy of bovine muscle. I. The native state of post rigor sarcolemma and endomysium. *Meat Science, 26*, 271–279.

Sale, P., Noel, Y., Lasteyras, A. & Oleon, C. (1984). A sinusoidal compression system to study rheological properties of foods in the transient state. *Journal of Texture Studies, 15*, 103–114.

Satyshur, K.A., Pyzalska, D., Greaser, M., Rao, S.T., & Sundaralingam, M. (1993). Structure of chicken skeletal muscle troponin C at 1.78Å resolution. *Acta Crystallographica, D50*, 40–49.

Scopes, R.K. (1970). Characterization and study of sarcoplasmic proteins. In E.J. Briskey, R.G. Cassens, & B.B. Marsh (Eds.), *Physiology and biochemistry of muscle as a food* (pp. 471–492). Madison: University of Wisconsin Press.

Seideman, S.C. (1986). Methods of expressing collagen characteristics and their relationship to meat tenderness and muscle fiber types. *Journal of Food Science, 51*, 273–276.

Sharp, J.G., & Marsh, B.B. (1953). *Whalemeat: Production and preservation.* (Special Rep. No. 58 of the Food Investigation Board). London.

Sheterline, P., Clayton, J., & Sparrow, J. (1996). *Actins: A complete data resource* (3rd ed.). New York: Academic Press.

Shockey, C.F., McKee, L.G., & Hamm, W.S. (1944). Instrument for measuring changes in texture of dehydrated fish. *Industrial Engineering and Chemistry (analytical ed.), 16*, 638.

Shorthose, W.R., & Harris P.V. (1990). Effects of animal age on the tenderness of selected beef muscles. *Journal of Food Science, 55*, 1–14.

Shorthose, W.R., Harris, P.V., Eadie, L.M., & Dickinson, R.F. (1988). Prediction of the tenderness of cooked meat from measurements made on the raw meat. *Proceedings of the International Congress of Meat Science and Technology, 34 A*, 194–196.

Sjostrand, F. (1962). The connections between A- and I-band filaments in striated frog muscle. *Journal of Ultrastructural Research, 7*, 225–246.

Solomon, M.B., Long, J.B., & Eastridge, J.S. (1997). The Hydrodyne: A new process to improve beef tenderness. *Journal of Animal Science, 75*, 1534–1537.

Squire, J.M. (1997). Architecture and function of the muscle sarcomere. *Current Opinion in Structural Biology, 7*, 247–257.

Stanton, C., & Light, N. (1987). The effects of conditioning on meat collagen: Part 1. Evidence for gross *in situ* proteolysis. *Meat Science, 21*, 249–265.

Suzuki, A., Kim, K., & Ikeuchi, Y. (1996). Proteolytic cleavage of connectin/titin. *Advances in Biophysics, 33*, 53–64.

Swatland, H.J. (1997). Observations on the rheological, electrical, and optical changes during rigor development in pork and beef. *Journal of Animal Science, 75*, 975–985.

Szczesniak, A.S., & Torgeson, K.W. (1965a). Methods of meat texture measurements. In A. Kramer & A.S. Szczesniak (Eds.), *Texture measurements of foods* (pp. 71–104). Dordrecht, the Netherlands: D. Reidel.

Szczesniak, A.S., & Torgeson, K.W. (1965b). Methods of meat texture measurement viewed from the background of factors affecting tenderness. *Advances in Food Research, 14*, 33–165.

Tarrant, P.V. (1981). The occurrence, causes and consequences of dark-cutting beef. In D.E. Hood & P.V. Tarrant (Eds.), *The problem of dark cutting in beef* (pp. 3–36). The Hague: Nijhoff.

Tarrant, P.V. (1987). Muscle biology and biochemistry. *Proceedings of the 33rd International Conference on Meat Science and Technology, Finland*, pp. 1–5.

Tskhovrebova, L., Trinick, J., Sleep, J., & Simmons, R. (1997). Elasticity and unfolding of single molecules of the giant muscle protein titin. *Nature, 387*, 308–312.

Tuma, H.L., Henrickson, R.L., Stephens, D.F., & Moore, R. (1962), Influence of marbling and animal age on factors associated with beef quality. *Journal of Animal Science, 21,* 848–851.

Uytterhaegen, L., Claeys, E., & Demeyer, D. (1992). The effect of electrical stimulation on beef tenderness, protease activity, and myofibrillar fragmentation index. *Biochimie, 74*, 275–281.

Vavak, L.D., Satterlee, L.D., & Anderson, P.C. (1976). The relationship of cardiac shear and trace element content to beef muscle tenderness. *Journal of Food Science, 41*, 729–731.

Verbeke, R., Debackere, M., Hicquit, R., Lauwers, H., Pottie, G., Stevenes, J., van Moer, D., van Hoof, J., & Veermershch, G. (1976). Quality of the meat after the application of anabolic agents in young calves. In F.C. Rendell & J. Rendell (Eds.), *Anabolic agents in animal production* (pp. 123–143). Stuttgart: George Theime.

Volodkevich, N.N. (1938). Apparatus for measurement of chewing resistance or tenderness of food stuff. *Food Research, 3*, 221–225.

Wang, H., Doty, D.M., Beard, F.J., Pierce, J.C., & Hankins, O.G. (1956). Extensibility of single beef muscle fibers. *Journal of Animal Science, 15*, 97–108.

Warner, K.F. (1928). Progress report of the mechanical tenderness of meat. *Proceedings of the American Society for Animal Products, 21*, 114.

Warriss, P.D., Nute, G.R., Rolph, T.P., Brown, S.N., & Kestin, S.C. (1991). Eating quality of meat from pigs given the beta-adrenergic agonist salbutamol. *Meat Science, 30*, 75–80.

Wellington, G.W., & Stouffer, J.R. (1959). *Beef marbling: Its estimation and influence on tenderness and juiciness* (Bulletin No. 941). Ithaca, NY: Cornell Agricultural Experiment Station.

Wheeler, T.L., Shackelford, S.D., & Koohmaraie, M. (1996). Sampling, cooking, and coring effects on Warner-Bratzler shear force values in beef. *Journal of Animal Science, 74*, 1553–1562.

Whipple, G., Koohmaraie, M., Dikeman, M.E., Crouse, J.D., Hunt, M.C., & Klemm, R.D. (1990). Evaluation of attributes that affect longissimus muscle tenderness in *Bos taurus* and *Bos indicus* cattle. *Journal of Animal Science, 68*, 2716–2728.

Whitaker, J.R. (1959). Chemical changes with aging of beef with special emphasis on the proteins. *Advances in Food Research, 9*, 1–60.

Winchester, C.F., & Ellis, N.R. (1956). *Delayed growth of beef cattle.* (USDA Tech. Bull. No. 1159). Washington, DC: U.S. Department of Agriculture.

Yates, L.D., & Greaser, M.L. (1983). Quantitative determination of myosin and actin in rabbit skeletal muscle. *Journal of Molecular Biology, 168*, 123–141.

Yeates, N.T.M. (1964). Starvation changes and subsequent recovery of adult beef muscles. *Journal of Agricultural Science (Cambridge), 62*, 267–271.

Chapter 9

Vegetables and Fruit

Michael Edwards

Fruits and vegetables are derived from living organisms, which are composed of cells. The structure of these cells is one of the major contributors to the characteristic texture of a food. The breakdown pattern of the food as it is chewed and the size, shape, and composition of the particles produced by chewing are responsible for the food texture that is perceived in the mouth (Christiansen, 1984). These factors are determined by the initial structure of the food material and the changes produced by processing and cooking in modifying the planes of weakness in the tissues. It is part of the aim of food processing to maintain the characteristics of that structure through the treatments that are applied, such as freezing and thermal processes. The texture for which the food processor is aiming will also depend on the expectations of the consumer. For example, the desired texture of a frozen raspberry will be as close as possible to that of the fresh fruit, whereas in processing beans or potatoes, which are normally eaten cooked, much more disruption of the structure may be acceptable or even desirable.

Many of the changes discussed in earlier chapters of this book have an influence on the behavior of cells during food processing, but it is primarily their structure that determines how this influence is manifested. This chapter will examine changes in the cell structure caused by food processing of plant tissues and discuss the effects of these processes on the mechanisms of structural breakdown. The measurement of fruit and vegetable texture by instrumental and sensory methods will then be considered.

PLANT AND CELL STRUCTURE

Plant Structure

Fruits and vegetables are the edible parts of plants. Fruits are usually regarded as the reproductive organs of plants, containing the seeds. These often have a high sugar and acid content and also aromatic flavors. Vegetables are generally classified as the nonreproductive parts of plants, such as roots, leaves, or stems. However, the distinction between a fruit and a vegetable is often unclear, with common usage of the terms being at variance with strict botanical classification. For example, rhubarb is often regarded as a fruit because its sweetness results in its being used in puddings and sweets, although botanically it is the leaf stalk or petiole, and hence a vegetative part of the plant. In contrast, celery, which is also a petiole, is treated as a vegetable. Marrows and courgettes, on the other hand, are treated as vegetables in the kitchen, although they are in fact fruits. Figure 9–1 shows a schematic diagram of some of the various edible parts of plants.

Each of these plant parts is a plant organ, which is composed of a number of different tissues. Each tissue in turn is composed of large numbers of cells of similar structure to each other. The presence of a range of different tissues in a plant organ is in contrast to meat and fish, which consist primarily of a single tissue, usually muscle. This is one of the main factors making the processing and preparation of plant foods more difficult than that of meat: the method used must take account of all the different tissues present. For example, the rib tissue in a cabbage leaf may require more cooking than the main blade of the leaf to make it tender.

The tissues present in fruit and vegetables vary considerably depending on the organ concerned. Winton and Winton (1935) carried out a very comprehensive survey of the structure and composition of foods that is still of great value to today's food scientists. Most plant organs have a skin or epidermis. This consists of a single layer of cells in the case of leaves, stems, and fruits and usually has an external layer of cutin and wax to reduce the rate of water loss. The epidermis may develop into a more substantial corky layer, as

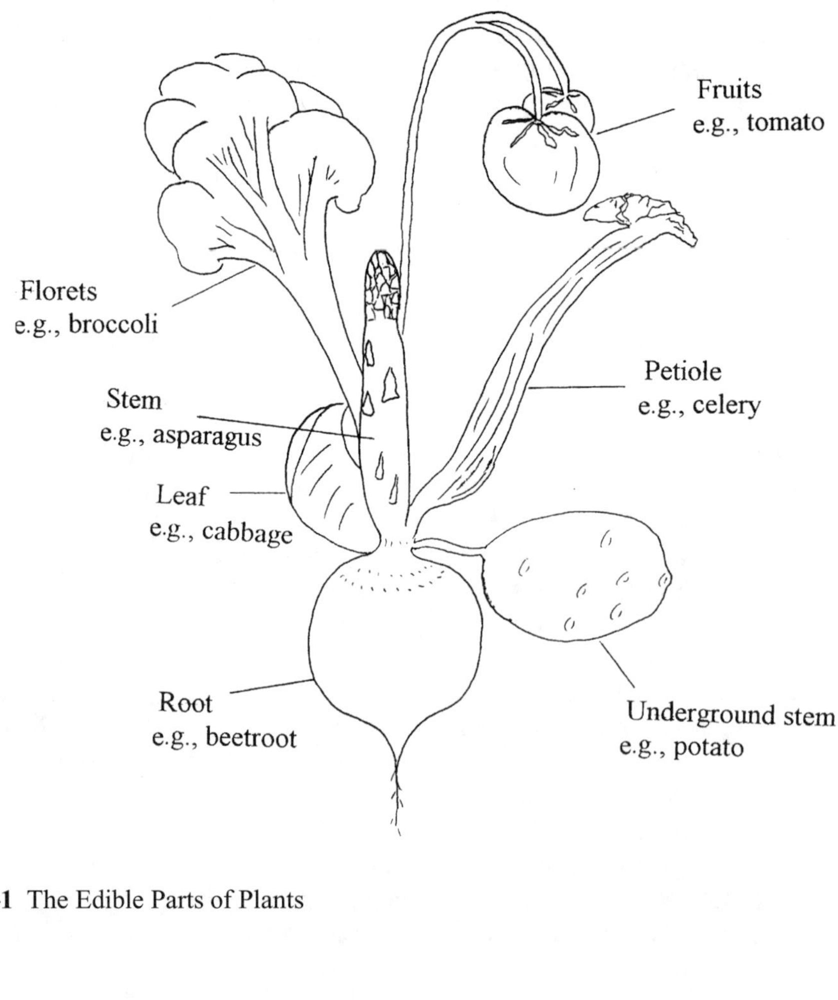

Figure 9–1 The Edible Parts of Plants

in the case of potatoes. Leaves, such as cabbage or lettuce, consist principally of a thin layer of tissue whose function is photosynthesis. The products of photosynthesis are transported away from the leaf, so the sugar and starch content is quite low. In contrast, most bulky fruits and vegetables, such as apples, tomatoes, onions, potatoes, and carrots, are designed to store plant foods and have a higher carbohydrate content. In the case of fruits, the stored sugar or starch is provided to attract animals to spread the seeds of the plant. In consequence, these organs are often designed to have a relatively short life, and this influences the structure and storage characteristics of the organs. However, if the food item is the seed itself, such as beans or peas, it may have a much longer shelf life, although possibly in a dried form. In this case, the storage material is usually starch stored in the endosperm or seed leaves within the seed. During germination, these resources are converted back to sugars and mobilized. The development of the seed is a dynamic process. This is illustrated by the importance of harvesting peas at the correct point in their maturation for processing as frozen, canned, or dried peas. Each of these products requires the peas to be at a different stage of maturity, and much effort has been expended in developing new and improved methods of measuring maturity in peas.

Vegetative storage organs, such as the swollen underground stem or tuber that is the potato, are intended in nature to act as overwintering plant food storage organs. Photosynthetic products are transported as sugars from the leaves to the tuber, where they are converted to starch for storage. If left in the ground, the potato will convert the starch back to sugar in the spring and transport it to the growing shoots. When potato crisps are made, the presence of high concentrations of sugars may lead to unacceptable levels of browning. Hence, the stage of development reached by potatoes has a profound effect on their suitability for particular food uses at different times of year unless measures are taken to control their behavior.

Many plant organs may be modified to become storage organs. The potato tuber is a swollen stem, the onion bulb is composed of swollen leaf bases, and carrots and beetroot are swollen roots. Each of these organs therefore contains all of the tissues normally composing those organs, and some of the most important of these, from a food-processing point of view, are the various types of fibers. Roots and stems contain vascular bundles consisting of xylem and phloem cells and surrounding fibers. These function to transport food and water around the plant, but they also have an important structural function. Many of these cells may have thickened walls that also become lignified as the plant matures. This makes them much tougher and more resistant to cooking, presenting problems for the food processor. Other important types of fiber include sclerenchyma, which makes up much of the fibrous tissue in celery, for example.

Cell Structure

All living organisms are composed of cells (Figure 9–2), and each cell is composed of a living protoplast bounded by a fragile semipermeable membrane, the plasmalemma. The protoplast contains the cytoplasm of the cell, which in turn contains the cell nucleus, mitochondria and other organelles, and the cell vacuole. Each of the various organelles has a particular function in life and to this end is surrounded by a semipermeable membrane of its own, which controls the movement of molecules into and out of the or-

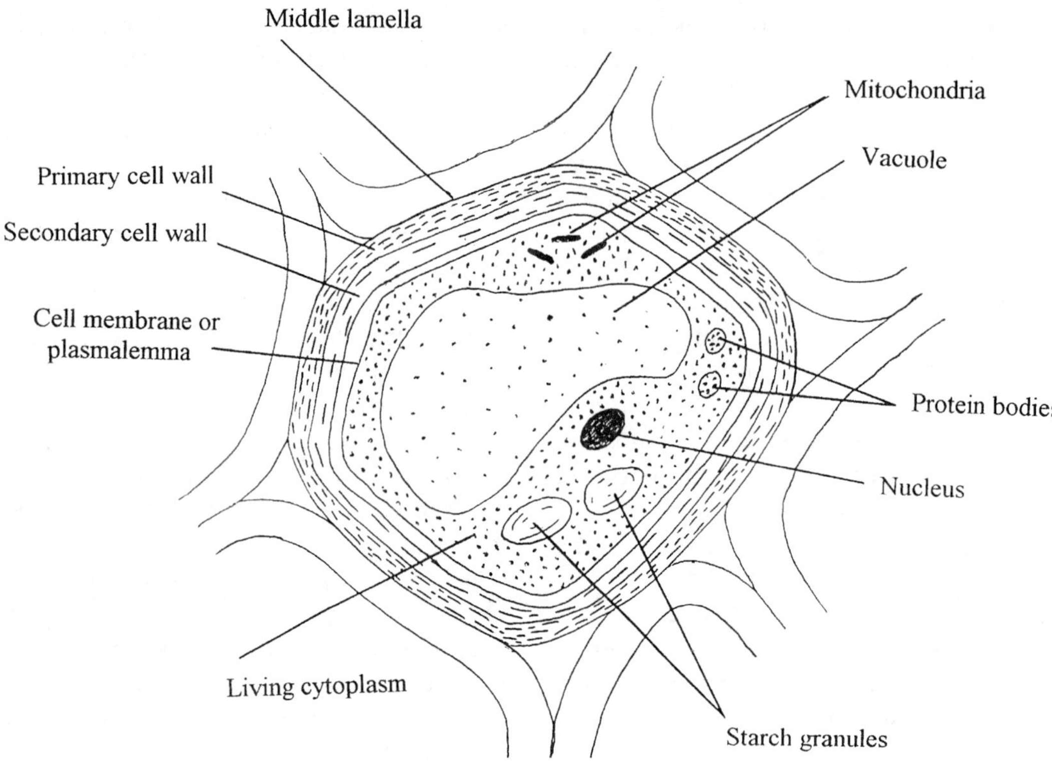

Figure 9–2 Diagram Showing the Main Structural Features of a Plant Cell

ganelle. One of the defining characteristics of the plant kingdom is that the plasmalemma is further surrounded by a cell wall. The wall is composed of cellulose and hemicellulose fibers in a matrix of water and pectins. Pectins (polymers of galacturonic acid) also cement neighboring cell walls to each other along the junction between them (the middle lamella). Cell walls cemented together impart a degree of rigidity, but the main means of maintaining shape in plant tissue is the turgor pressure within individual cells. Turgor pressure is the hydrostatic pressure produced by osmotically active constituents such as sugars and salts. When turgor pressure is lost, the structure collapses to a greater or lesser degree, as illustrated in a plant that wilts for lack of water.

Different plants and plant organs have different degrees of susceptibility to such loss of structure, depending on the relative importance of a range of other factors. These may include the inherent rigidity of the cell wall, the strength of the middle-lamella bond, and the degree of support from within the cell due to structures such as starch granules and from other tissues such as fibers. Hence, a thin-walled, largely two-dimensional structure such as a lettuce leaf loses its characteristic texture much more readily than a potato, which has a greater three-dimensional structure of cell walls giving each other mutual support. In addition, each potato cell is to some extent supported internally by the starch granules within it, and the tissue is much less susceptible to water loss as a result of a much smaller surface-to-volume ratio.

MECHANICS OF STRUCTURAL BREAKDOWN

Principles

The main structural material in fruit and vegetables is cellulose, a polymer of glucose monomers, and the fracture behavior of this material is a vital part of the fracture behavior pattern of plant material. However, the behavior of cellulose is dramatically affected by the presence of water. Most plant tissues used for food contain 75 to 95% water, and the behavior of this component is a vital factor in the effects of food processing on structure and texture.

One of the most important structural functions of water in plant materials is its action as a plasticizer. Many changes in plant structure are mediated by changes in water content, such as the dehydration of seed pods from a rubbery to a brittle glass condition in order to scatter the seeds (Blanshard, 1995). Glassy materials are brittle and have no ordered crystalline structure. In hydrophilic polymers such as cellulose or starch, the transition to the glassy state is achieved by the removal of water. Thus, in plant material, water acts as a plasticizer, reducing the glass transition temperature from about 200° to about −10°C, without which the material would be glassy. Using composite mechanics, Vincent (1982) showed that the longitudinal sclerenchyma fibers in the leaves govern the tensile and fracture properties of grass. The stiffness of grass leaves changes with water content, increasing rapidly below a water content of 20% of dry weight, owing to the disappearance of "free" (i.e., unbound liquid) water, which acts as a plasticizer. A fibrous plant material such as grass is "notch-insensitive": that is, its strength is directly proportional to the load-bearing cross-sectional area. Therefore, a small notch cut in the edge does not reduce its strength markedly. In contrast, "notch-sensitive" materials such as fresh apples, carrots, or potatoes lose a significant part of their strength if even a small notch is cut in the side. Notch insensitivity results from the matrix material binding the continuous parallel elements together so that shear is not transmitted effectively. Hence, if a single fiber breaks, stress is redistributed evenly across the other fibers rather than being concentrated at any one point. At high moisture contents, fractures pass straight across laminae between notches. Below about 50% water content, the fracture takes a more complex course and is more readily deflected by fiber bundles. This is readily demonstrated by an attempt to snap a carrot in half. A fresh, turgid carrot will break straight across, whereas an old partially dehydrated carrot will feel rubbery and will bend and then split unevenly.

Effects of Food Processing on Cell Structure

Food-processing methods that have the effect of killing the cells destroy the integrity of the plasmalemma and the ability of the cell to maintain turgor. Food processing by heat will also result in changes to the cell wall, particularly the middle lamella, and gelatinization of the starch. The final texture will therefore depend on the relative importance of each factor contributing to texture and the degree to which that factor has been changed by the processing method used.

Effects of Moist Heat

As the temperature rises from 20° to 60°C, cell walls thicken, with disruption of the plasmalemma. Loss of cellular integrity becomes evident at around 60°C in carrots (Grote & Fromme, 1984). The rigidity of the tissue is closely related to the shear modulus, which declines rapidly at about 55° to 60°C, in contrast to the slow, steady decline in the modulus of elasticity (Ramana, Wright, & Taylor, 1992). On prolonged heating, hemicellulose and cellulose both undergo some breakdown (Reeve, 1954a).

In mature cellulose walls, two principal layers may be distinguished by microscopy: primary and secondary. The more mature primary cell wall is less susceptible to breakdown than the younger secondary wall. Here, the typically reticular microscopic structure of the raw material is lost on heating and is replaced by an amorphous structure embedded within a fibrillar network. The central zone of the secondary cell wall, seen as densely packed tubules in raw tissue, is transformed into a mass of dense amorphous matter after cooking (Shomer, Paster, Lindner, & Vasiliver, 1990). The cytoplasm of the cells forms dense clumps of granules, which are usually arranged along the cell walls (Grote & Fromme, 1984). Mealy potatoes have also been shown to have more cell wall material per unit cell surface than nonmealy potatoes (Marle et al., 1997), apparently because the cell walls are denser.

Recent research has suggested that the retention of a crisp texture in Chinese water chestnuts may be due to cross-linking of hemicelluloses in the cell wall by ferulic acid (Parker & Waldron, 1995). It is quite possible that this mechanism may operate in other plant cell material, particularly where phenolics are known to be present in the cells.

Cell Separation. As well as breakdown of the cell walls themselves, breakdown of the pectin in the middle lamellae of some cells begins, leading to cell separation (Burton, 1989). This results in a change in the fracture properties of the tissue (Figure 9–3). In raw or lightly cooked plant tissue, the weakest plane, through which the tissue fractures, is across cell walls, so a fractured surface of the tissue reveals details of the internal structure of the cells (Figure 9–4). As middle-lamella breakdown proceeds, the weakest plane becomes the line of the middle lamella, so a fracture surface consists of the outer surface of the cells (Figure 9–5) (Edwards, 1994; Marle, Clerkx, & Boekestein, 1992; Rose, Edwards, Rendell, & Roberts, 1989). Middle-lamella breakdown occurs earlier in potato varieties giving a mealy texture than in those giving a firmer texture (Reeve, 1954a, 1954b, 1963, 1967a, 1967b; Sterling, 1966), and more pectin is released (Marle, Dijk, Voragen, & Biekman, 1994). More release of pectins may be the cause or the effect of more cell sloughing. Cells in the fracture planes of firm potatoes have large intercellular contact areas; most of the cell walls are flat, and cell surfaces show folds and cracks. In mealy potatoes, cells in fracture planes have much smaller intercellular contact areas, and the cells are rounded and turgid, with smooth surfaces. This suggests that the final texture in potatoes is determined by a combination of the structure of the cell wall and middle lamella, combined with distension of the cell caused by starch swelling (see previous section "Cell Structure") (Marle et al., 1992). Release of amylose from ruptured cells results in a sticky texture (Reeve, 1954a).

Middle-lamella breakdown is believed to be due to the breaking of hydrogen bonds in pectins and other cell wall polysaccharides (Sterling, 1966). High-methoxyl pectins, which have a high degree of substitution of hydroxyl groups with methoxyl groups, tend to be less water soluble and hence more resistant to this breakdown than low-methoxyl pectins. Therefore, the ratio of high- and low-methoxyl pectins in the tissue influences the rate at which cell separation occurs. Changes in these ratios are often responsible for increasing tenderness during ripening of fruits,

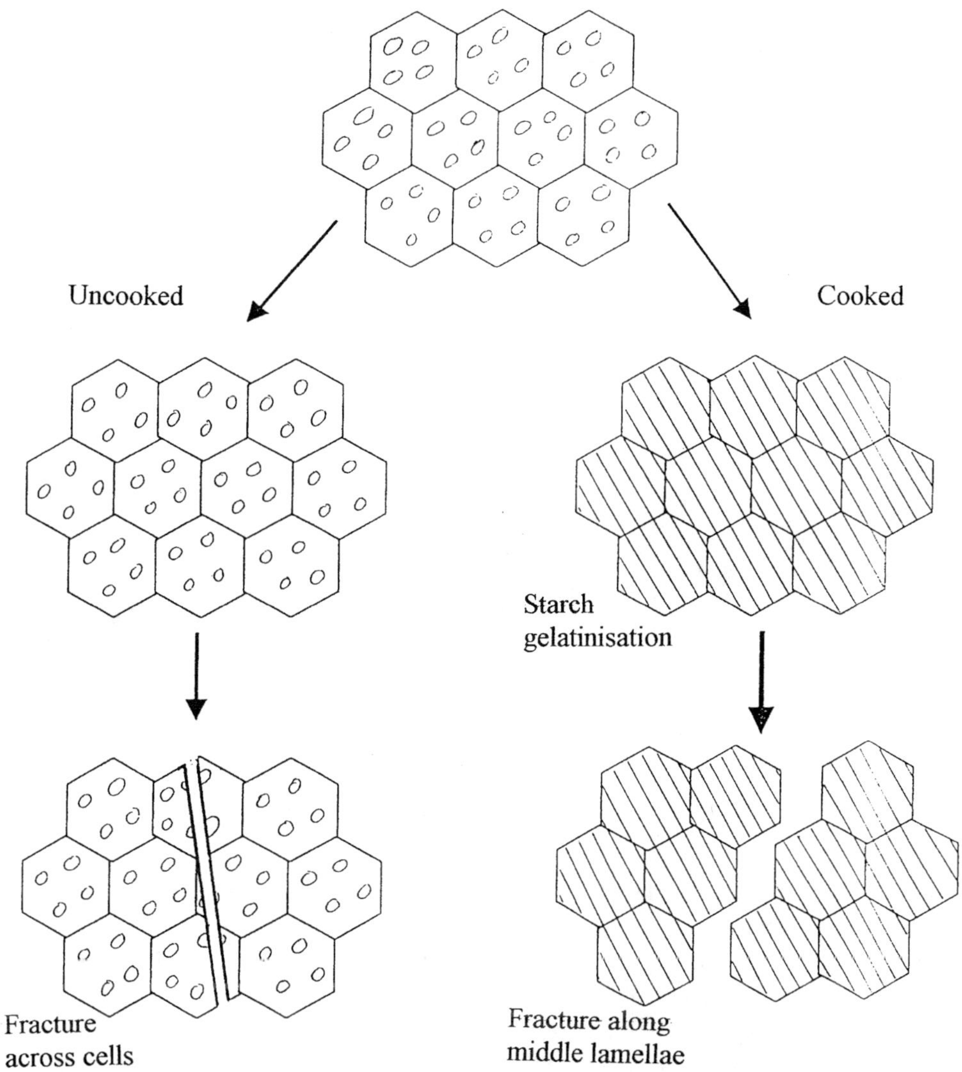

Figure 9–3 Diagram Showing Changes in Fracture Properties of Plant Tissues as a Result of Heat Treatment

with cells becoming increasingly separated as the tissue matures. In cherries, an increase in pectin solubility leads to cell sliding and increasing elasticity of tissues (Batisse, Buret, Coulomb, & Coulomb, 1996). The biochemical cell wall process differs between soft and crisp fruits of the same cultivar grown under different conditions. In crisp fruits, the middle lamella is solubilized and the primary cell wall restructured during ripening. In contrast, the middle lamella of soft fruits is better preserved, and the primary cell walls are thin and degrade as the fruit matures. These processes are controlled by the native enzyme polygalacturonase; genetic engineering has been used to regulate the activity of this enzyme in order to produce a tomato that ripens very slowly and hence has an extended supermarket shelf life (Schuch, 1991; Smith et al., 1988). Pectinase enzyme activity can also be

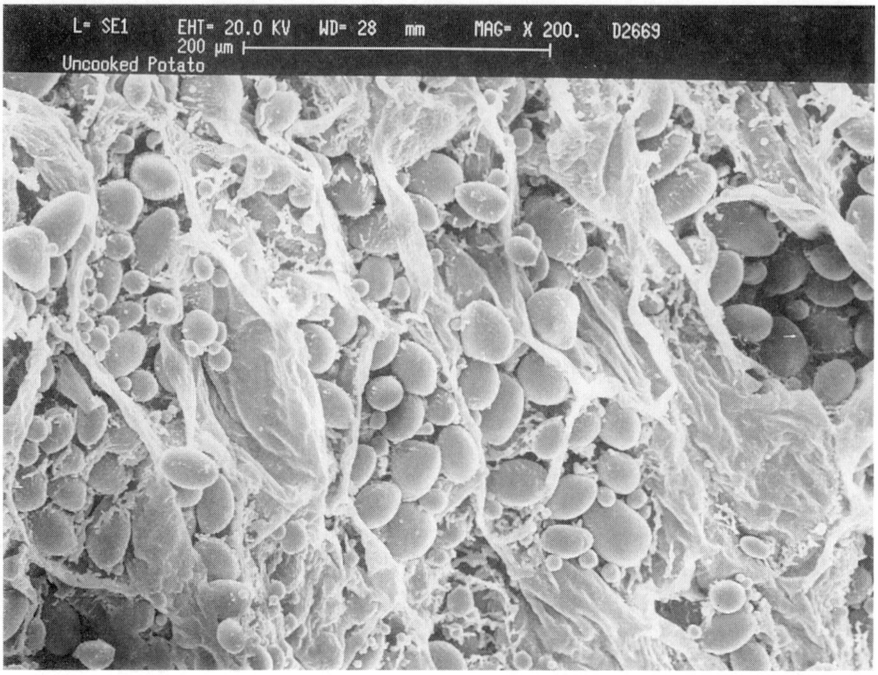

Figure 9–4 Scanning Electron Microscope (SEM) View of the Fracture Surface of Uncooked Potato Tissue. Note that the tissue has fractured across the cells, revealing cell contents such as starch granules. Freeze-dried specimen.

manipulated to improve the texture of frozen vegetables such as green beans, which are normally given a hot-water blanch to deactivate potentially damaging enzymes before freezing. Instead of a single blanch at about 93°C, the beans are given a preliminary 70°C blanch, during which pectinase enzymes partially demethylate the pectins. This leaves hydroxyl sites free on the pectin chain to cross-link with other pectin molecules via a calcium ion bridge, resulting in a firmer texture (Adams & Robertson, 1987; Moledina, Haydar, Ooraikul, & Hadziyev, 1981). The beans are then given a second, high-temperature blanch to destroy enzyme activity. This must be controlled very carefully; pectinase activity that occurs in products kept warm for too long during processing may destroy sufficient pectin to reduce the final product to a soft paste. Extraction of pectins into the liquor in canned apricots and peaches can result in a softer final fruit texture (Dobias, Curda, Vana, & Zakova, 1993), although this phenomenon has also been related to fungal pectinases that are heat-resistant. In some canned vegetables, calcium and magnesium may be added to soft water used in canning to give a firmer texture.

Hard-to-Cook Legumes. In navy beans (*Phaseolus vulgaris*), the so-called "hard-to-cook" (HTC) phenomenon has been attributed to a failure of the middle lamella to break down during cooking, preventing cell separation (Shomer et al., 1990). A possible mechanism by which the middle lamella becomes stabilized was proposed by Jones and Boulter (1983), who showed that during storage under conditions known to lead to HTC,

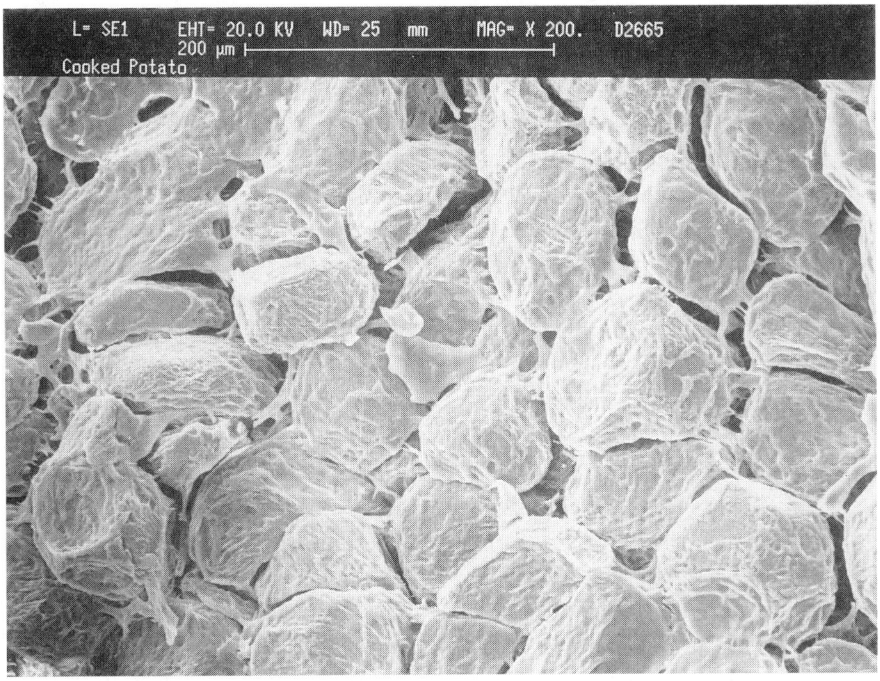

Figure 9–5 SEM View of the Fracture Surface of Cooked Potato Tissue. Note that the tissue has fractured across the middle lamellae, revealing the outside of each cell wall. Freeze-dried specimen.

phytate is partially hydrolyzed and pectin is deesterified. This results in the formation of calcium magnesium pectate, rendering the middle-lamella pectins insoluble. Although Hincks and Stanley (1987) provided some tentative evidence for the lignification of cell walls in HTC beans, Srisuma et al. (1989) could detect no significant differences in lignin content between normal and HTC beans.

Liu, Hung, and Phillips (1993) found evidence of decreases in solubility and the thermal stability of intracellular proteins in HTC cowpeas during storage and suggested that this prevented water reaching the starch granules. However, they also showed a lack of cell separation at the middle lamella, and their work failed to demonstrate clearly that the protein changes were the cause rather than a consequence of the phenomenon. Garcia and Lajolo (1994) linked changes in the gelatinization temperature of starch to starch alterations (resulting in an increased resistance of starch granules to enzyme attack) and the hard-to-cook phenomenon in beans.

In a similar phenomenon, some samples of dry peas fail to take up water when soaked prior to heat processing, leading to hard, dry patches in the final product in affected peas. These are known as "nonsoakers." Work at the Campden & Chorleywood Food Research Association and elsewhere (Hohlberg & Stanley, 1987) has shown that this is due to a failure of the middle lamella to soften and separate during soaking, thereby preventing cells from swelling fully as water is taken up. This has been linked to differences in the relative contents of low- and high-methoxyl pectins. At the opposite end of the scale, some pea samples produce a gelled brine in the can after processing; this is related to high relative proportions of water-soluble pectins in the peas, leached out during processing.

Effects of Processing on Starch. The two most important factors other than turgor are the cell walls and the effect of starch, and the effect of heat processing on these is seen in starchy foods such as the pea (Edwards, 1994) and the potato (Marle et al., 1992; Rose et al., 1989). The first starch granules gelatinize as the temperature reaches about 60°C. However, some granules may remain ungelatinized until the temperature reaches 70°C (Reeve, 1967a, 1967b).

Kaczynska, Autio, and Fornal (1993) studied heat-induced structural changes and starch swelling in fava bean starch paste. Amylose is released from the granules as they are heated, and the amylopectin inside the granules is highly swollen.

In potatoes and other starch-based tissues, some workers have concluded that swelling of the gelatinized starch causes distension of the cell wall, resulting in splitting (Jarvis, Mackenzie, & Duncan, 1992; Reeve, 1954a, 1954b). Reeve (1954a, 1954b, 1954c) has related the degree of starch swelling and cell wall rupture to stickiness in mashed and processed potato products. However, other workers have suggested that although gelatinized starch may swell to fill the space available, it does not cause cell separation and splitting (Bretzloff, 1970; Warren & Woodman, 1974).

The gelatinization of starch is heavily dependent on water availability. Gelatinization of potato granules in a solution of amylose or amylopectin results in less swelling of the granules than in water (Svegmark & Hermansson, 1993). Much higher enthalpy levels than in high-moisture gelatinization are required to produce proper gelatinization of starch at low water content (Vainionpaeae, Forssell, & Virtanen, 1993). In baked products, with relatively low water activity, swelling of granules is limited, and many granules do not gelatinize, even at quite high temperatures.

Effects of Dry Heat

The effect of dry heat on plant tissue has been most intensively studied in relation to the manufacture of bakery products and hence has concentrated on the behavior of gluten. However, some parallels can be drawn with dry cooking of vegetables and with oil frying (see below). Milling destroys much of the structure of grains used in bakery products, although fragments of cell structure can still be clearly identified after milling (Gallant, Monredon, Bouchet, Tacon, & Delort-Laval, 1993). The addition of water to the flour to produce dough results in significant structural changes. The protein gluten is responsible for much of the structural development in the baking of bread; of the proteins composing gluten, only glutenin swells in contact with water, and gliadin dissolves completely (Eckert, Amend, & Berlitz, 1993). Individual gluten networks already existing in flour particles amalgamate to form a coherent whole during dough making, and kneading results in the formation of protein films. These structures form the gas cell walls in the final product. After heating of gluten at different moisture contents, the first major change in moisture contents below 20% is a decrease in extractability of large glutenin aggregates and hydrophobicity of some proteins. At higher moisture contents, disulphide bonding and irreversible conformational changes can also be observed (Weegels, Verhoek, Groot, & Hamer, 1994). After heat treatment, proteins are stretched in elongated sheets, and starch granules are gelatinized (Gallant et al., 1993). Endosperm cell walls become highly fragmented during baking and, along with the starch, help to form a continuous matrix (Parkkonen, Haerkoenen, & Autio, 1994; Virtanen & Autio, 1993).

Effects of Oil Frying

Cooking in oil is widely used in the production of snack foods of various kinds and results in drastic changes to the structure of the food. Because of the high temperatures and dehydration involved, comparisons can be drawn with the effects of baking on cell structure. Structural changes to the cytoplasmic network and protein bodies and cell separation occur in oil-cooked peanuts more rapidly than in oven-cooked peanuts (Young & Schadel, 1993), and this is probably typical of most such products.

Penetration of the oil is particularly important for structure development in products such as potato crisps (Gamble & Rice, 1987), tortilla chips (McDonough, Gomez, Lee, Waniska, & Rooney, 1993), and potato chips (Selman & Hopkins, 1989). In products that become totally hard and crisp, the oil first coats the surface and begins to move into the chip. As the heat builds up inside, moisture turns to steam and exits the chip, leaving a spongelike network of tunnels that fill with oil. This process occurs within the first 20 seconds of frying. Starch granules gelatinize inside the chip but not at the surface. The interior becomes smooth and plastic as protein, starch, and lipids interact to form a continuous phase that hardens upon dehydration. In products with a soft interior, such as potato chips (French fries), the dehydrated oil–infused layer is confined to the peripheral four or five layers of cells (Selman & Hopkins, 1989). The hard, crisp surface pellicle develops at the same time as the interior of the chip is changed to a structure comparable to that produced by moist heat.

Effects of Microwaving

Although many acceptable food products are prepared using microwave energy, results are less than satisfactory with some food products, particularly starch-based foods. Relatively few studies have been carried out on the effects of microwave heating on plant cell structures and food texture. The poor quality of some starch-based microwave-treated foods may be related to fast heating rates, difference in heat and mass transfer mechanisms, or specific interactions of the components of the food with microwave radiation (Goebel, Grider, Davis, & Gordon, 1984). Microwave energy can produce comparable softening of potato tissue in about one-third of the time required by boiling water (Collins & McCarty, 1969). A more mealy texture results in microwave-treated potatoes. This may suggest that more complete cell separation occurs or that there is less leakage of amylose from damaged cells, reducing the apparent stickiness.

Effect of High Pressure

Although the potential benefits of high-pressure treatments have been known for a long time, the commercial use of high-pressure processing is a very recent and, so far, limited development. There appear to be advantages in retention of a crisp texture, and it has been found that gelatinized starch resulting from high-pressure processing retrogrades more slowly than heat-gelatinized starch (Jones & Beckett, 1995).

Effects of Freezing on Plant Cell Structure

The principal damage caused to plant cell structure by freezing is the result of ice crystal formation; hence, the position and size of ice crystals is critical. Ice crystal formation is a phase change in which water molecules stop moving and form an ordered crystalline structure. In so doing, they give up energy as heat to their surroundings. This slows down the freezing of other water molecules around, so that during freezing the temperature of the water or the food drops only slowly, but once freezing is complete there is a sudden fall in temperature (Figure 9–6).

Plant material is generally recognized as being more difficult to freeze satisfactorily than animal tissue. This is due, in the main, to the wide range of tissues in a fruit or vegetable, as discussed above in the section "Cell Separation." One of the consequences of this is that they also have a variety of enzyme systems and substrates, so a wide range of enzyme reaction sequences is possible.

There are two processes during freezing that cause damage to cellular structure and lead directly to a loss of firmness in plant tissue. First, ice crystals puncture the cell membrane, leading to loss of turgor pressure. Second, there is damage to the cell wall structure caused by growing ice crystals. Since fruits tend to have thin-walled cells, they rely heavily on turgor pressure for firmness of texture and so are easily damaged. Vegetable tissues generally have somewhat thicker cell walls and are more resistant. Ice crystals tend to form first outside the cells, in the intercellular spaces, where solute concentrations are lowest. When freezing is slow, these crystals grow very large, and water is withdrawn from within the cells to add to the crystals. Fast freezing, however, prevents the water from moving before it is frozen and therefore gives smaller crystals both inside and outside the cells. This results in less damage to the tissue.

The formation of ice crystals also punctures internal cell membranes, allowing enzymes and substrates to mix that are normally separated. This results in a wide range of chemical reactions leading to breakdown of the cells and the development of off-flavors and colors. There is a range of potentially damaging enzyme reactions, listed in Table

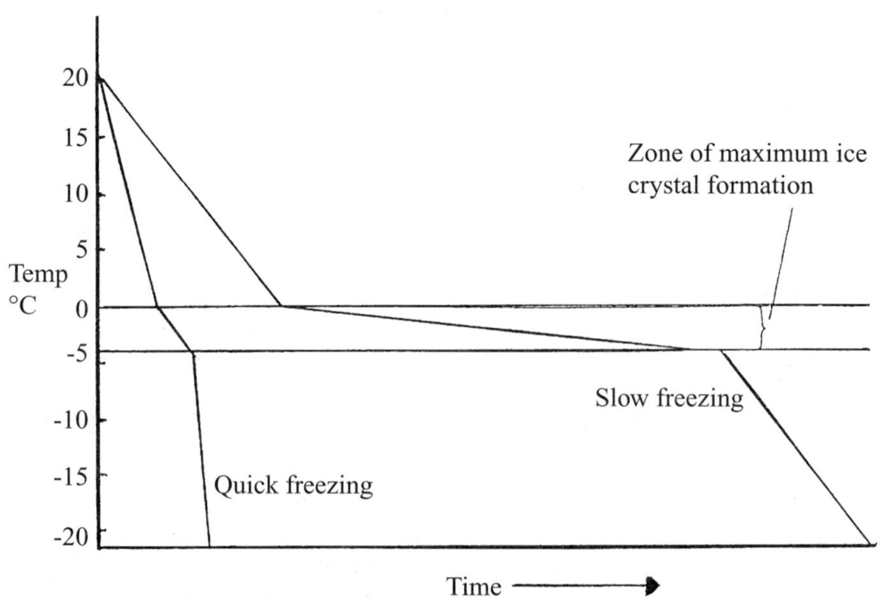

Figure 9–6 Diagram Showing the Effect of the Water-Ice Phase Change on the Change in Temperature of a Food Sample during Freezing

Table 9–1 Potentially Damaging Enzyme Reactions in Frozen Vegetables and Fruit

Enzyme	*Substrate*	*Product*
Catalase	Hydroperoxides	Oxygen
Lipase	Triglycerides	Monoglycerides, glycerol, fatty acids
Lipoxygenase	Unsaturated fatty acids	Fatty acid hydroperoxides, carbonyl compounds
Pectic enzymes	Pectins	Pectic acids, uronides, glacturonic acid
Peroxidase	Peroxides and other electron donors	Oxidized donors
Polyphenol oxidase	Phenols	Melanins

9–1. Where these enzyme reactions are important—particularly, for example, in green vegetables—a brief heat treatment (blanching) can be used to inactivate the enzyme prior to freezing (see previous section "Cell Separation").

Freezing of plant tissues slows down chemical reactions. However, because withdrawal of water to form ice crystals concentrates the remaining solutions, reaction rates during frozen storage may be increased in some cases. Proteins may be denatured by alteration of their tertiary structure or by breakage into smaller units that do not reassemble on thawing. Vitamin C (ascorbic acid) may be converted to dehydroascorbic acid, which is active as vitamin C, and to 2,3-diketogulonic acid, which is not active. Chlorophyll may be converted to pheophytin, and other pigments may also be degraded.

The main physical damage occurring during frozen storage is dehydration as a result of fluctuating storage temperatures. As the temperature rises, water evaporates from the food. As the temperature falls again, water vapor lost from the food condenses on the packaging because this cools before the food. This process results in noticeably dry and discolored areas on the food, known as "freezer burn."

Starch granules in vegetables such as potatoes may be partly gelatinized during blanching and may therefore absorb some water. Starch gelatinization may be reversed to some degree during frozen storage due to progressive loss of water as a result of fluctuating temperature or to competition for water between the different components of the food. This retrogradation of starch results in poor texture and a loss of water-holding capacity of the starch gel. Starch pastes can be used as protective systems of solid elements in precooked frozen foods, minimizing dehydration and chemical changes in storage. Rapid freezing of starch pastes results in smaller ice crystals and the absence of retrogradation. Retrogradation and ice crystallization during storage of frozen pastes contribute to degenerative changes such as spongy structure, decreased apparent viscosity, and increased syneresis. Addition of xanthan gum minimizes this deterioration but affords no protective effect on ice crystal size and amylopectin retrogradation (Ferrero, Martino, & Zaritzky, 1993).

It is not always appreciated that the most damaging stage in processing frozen food is often thawing. During thawing of plant tissue, the temperature rise moves inwards and can cause dehydration in the same way as fluctuations in storage temperature. As ice crystals melt, they undergo the same phase change as in freezing but in reverse, so they take in heat energy. This cools their surroundings, and there is partial refreezing, resulting in the growth of very large ice crystals during slow thawing, and causing further damage to cell walls and membranes. Due to puncturing of cell walls and membranes, much water is lost as "drip," and any enzyme reactions that were retarded by freezing will speed up again. This may include any of the enzyme reactions mentioned earlier, and since more membranes will have been ruptured during freezing and thawing, more mixing of enzymes and substrates is possible.

RHEOLOGICAL PROPERTIES OF VEGETABLES AND FRUIT

Rheology, the study of the deformation and flow of matter, has been extensively applied to fruit and vegetables in an effort to understand the relationships between structure, texture, and the changes induced by processing. The mechanical properties of vegetables and fruit have been widely studied using testing methods developed for nonbiological materials. This approach can yield a very useful understanding of the mechanical behavior of foods in certain circumstances, but, as will be demonstrated later, there are certain disadvantages when these methods are used to measure the texture of foods. These relate principally to the fact that human perceptions of food texture are based on our own mechanical testing apparatus, the mouth, which does not always work in ways that are exactly paralleled by classical materials science testing methods.

Texture is measured in fruit and vegetables for two main purposes: (1) to determine the maturity of crops in order to predict the optimum harvest date, and, in the case of peas, to determine payment of growers; and (2) to determine the quality of the processed product. Many techniques used are destructive, but there is increasing interest in the use of nondestructive techniques. Most of the tests are designed to measure texture related to the overall or macrostructure of the crop, although these may be determined by microscopic structural features, and in some cases measurements have been used to determine the textural influences of specific structures.

Destructive Methods of Testing

Penetrometers

Penetrometers have been widely used to measure the texture of fruits such as apples, peaches, and cherries. Multiprobe devices such as the Pea Maturometer and Christel Texture Meters, and more recently the multiple pea attachment for the SMS Texture analyzer, have enabled multiple samples of individual peas to be assessed simultaneously. At the other end of the scale are many simple hand-operated instruments where the force required to push the probe into the fruit is indicated by a spring scale. Plant breeders originally used the test to evaluate fruit and vegetable maturity. They have subsequently used it for a wide range of purposes, including assessment of the penetration force for potato tubers; skin development in tomatoes; resistance of cherries, blackberries, and raspberries to mechanical injury; and penetration of sweet corn pericarp. The method is commonly used to assess the strength of the skin. Holt (1970) used a puncture test device with a tapered tip so that the shaft did not exert a frictional effect during penetration of the fruit. The technique was used to identify the effect of the exocarp, vascular strand, mesocarp, endocarp, and locular juice of tomato on the force/distance curve and hence the contribution of each tissue to the overall fruit texture. The variation in skin strength was also determined over the whole surface. Mohsenin and Goelich (1962) devised an apparatus for punching holes in an excised piece of apple skin to assess the contribution of apple skin to the resistance of apples to shearing and puncturing forces. Results showed that in a tough-skinned apple the skin contributed more to the results of a compression puncture test than in an apple with a less tough skin.

Shear Tests

Another widely used approach in the measurement of fruit and vegetable texture is the shear test. The Pea Tenderometer was introduced in 1937. It consists of a grid of blades rotated at constant speed through a second grid of blades; as the peas are cut by the blades, the maximum force is indicated by a pointer. The instrument is still widely used to assess the maturity of peas at harvest. The structural changes being measured by the instrument have not been widely studied; the factor that appears to relate most closely to tenderometer values is alcohol-insoluble solids (AIS).

The Kramer Shear Press was also developed to test peas but has since been much more widely used in texture assessment of many fruits and vegetables. The original concept was applied to the complete machine, consisting of a ram press forcing a probe through the sample. However, the name is now applied in particular to refer to a multiblade shear compression system designed to produce a similar action to the Pea Tenderometer without the same mechanical complications, and a number of similar instruments have been developed by other manufacturers. Although the test is referred to as a shear test, it in fact involves both extrusion and compression as well as shear (Szczesniak, Humbaugh, & Block, 1970). Single-blade shearing has been shown to be useful for determining the texture of potatoes and carrots.

Shear stresses act parallel to the shearing plane, in contrast to compressive and tensile stresses, which act normal to a plane. It is important to distinguish between true shear tests and cutting tests. Shear tests have been widely used to measure ripening of fruit and cooking of vegetables. In an unripe fruit, cells are held firmly together by the pectins in the middle lamellae, so that the plane of weakness runs through the cell walls rather than between them. As the fruit ripens, the pectin content changes slowly from high methoxyl to low methoxyl, resulting in a weaker cell-to-cell bond, such that the cells slide across each other instead of being fractured. As this happens, shear strength decreases. Similarly, shearing in raw potato or carrot tissue also results in breaking of the cell walls during the test. When the tissue is cooked, the middle lamellae pectins are gradually dissolved, again resulting in softening of the tissue (Speiss, Gierschner, & Grunewald, 1987).

The shear strength of the surface layers contributes significantly to the perceived texture of vegetables and fruit. As a potato chip (French fry) is cooked, the internal tissue softens as outlined above, but the surface crust develops and can be measured by shear tests using a multibladed Kramer shear press (Rose et al., 1989).

Extrusion Tests

The Ottawa Texture Measuring System was also developed as a replacement for the Pea Tenderometer. The sample is extruded by being pushed through a grid of heavy wires at the bottom of the test cell, although other test cells have been developed for the instrument that rely on other principles such as shear or compression. Another example of a method based on extrusion, but this time combined with compression, is the back extrusion cell, which consists of a piston working in a blind cylinder with a clear gap between piston and cylinder, through which the compressed sample is forced back.

Another application of extrusion is the succulometer, which is used to measure the volume of juice that can be pressed from fresh sweet corn and is used as an index of maturity and quality (Kramer & Smith, 1946; Meyer, 1929; Sayre & Morris, 1931, 1932).

Texture Profile Analysis

The Texture Profile was first developed by the General Foods Corporation and combines measurements of a number of parameters. The sample is compressed twice, and analysis of the force-time curve produced gives figures for hardness, cohesiveness, springiness, adhesiveness, fracturability (brittleness), chewiness, and gumminess (see Chapter 1). Frozen strawberries showed decreased fragility, hardness, and cohesiveness values compared to fresh berries (Szczesniak & Smith, 1969). Results for stored and irradiated berries indicated that they would be more susceptible to mechanical damage during handling than the stored fruits (Ahmed, Dennison, & Fluck, 1972).

The technique has also been used with mangoes and peaches, including irradiated products (Ahmed & Dennison, 1971). The mechanical characteristics of irradiated fruits decreased during storage, with the exception of cohesiveness of mangoes, which increased during storage.

Bourne (1968) found that most of the textural parameters of pears decreased during ripening, apart from adhesiveness, which was zero.

Other Destructive Tests

Compression Tests. Compression tests have been more widely reported for vegetables and fruit material than tensile tests, possibly because of the greater ease of carrying out such tests. The presence of air cells within many fruits and vegetables has a significant effect upon the results of such tests. Sharma and Mohsenin (1970) carried out compression tests on apples, and Finney and Hall (1967) worked on potatoes. More recently, Tu, Baerdemaeker, Deltour, and Barsy (1996) used compression tests on apples as a control method of texture assessment in their investigations into acoustic impulse techniques.

Tension Tests. Tension tests have not been widely used with fruit and vegetable materials. The reasons for this are twofold. First, it is quite difficult to arrange means of clamping the samples for tests, although it should be noted that Tu et al. (1996) solved this problem elegantly by the use of a ring of tissue cut from apple. Second, the process of mastication involves compression, not tension, of the food between the teeth. Nevertheless, the point of failure of the sample in many shear tests is actually in tension. However, tensile tests have been used in plant materials in attempts to explain factors such as skin splitting in cherries and cracking of potatoes during handling. In apples, good correlation was found between sensory evaluation, ultimate tensile strength, and fracture resistance, and tensile testing was found to be a good means of evaluating shelf life (Holt & Schoorl, 1984; Tu et al., 1996).

Bending Tests. Bending tests have not been extensively used to measure food texture, possibly because specimens of constant size and shape are required. They have been used on stems or stalks, including the midrib of leaves such as cabbages and the petiole of celery, and in carrots (Kapsalis, Segars, & Krizik, 1970). Bending has also been used to measure flaccidity in strips cut from tissue such as apple or potato, where the resistance to bending decreases as water is lost from the tissue and its rigidity is lessened. The sagging or drooping of a group of chips (French fries) supported on cantilevers has been measured using a similar approach (Anon, 1966; Ross & Porter, 1968; Voisey, Larmond, & Seark, 1974).

Poisson's Ratio. Poisson's ratio is defined as the absolute value of the ratio of transverse strain to the corresponding axial strain resulting from uniformly distributed axial stress below the proportional limit. A technique has been developed for measuring this parameter in biological materials (Hammerle & McClure, 1971). However, these dimensional changes are difficult to measure, so the Poisson effect is not often considered in fruits and vegetables. In vegetable and fruit material, the tissue will rarely be isotropic, so Poisson's ratio will have more than one value depending on the orientation. The value of Poisson's ratio is influenced by the tissue structure. Chappell and Hamann (1968) demonstrated this in apples. The ratio for a plug of apple, for example, was calculated to be less than that for a plug of potato tissue because of the higher air content of apple tissue.

Viscoelastic Behavior. Morrow and Mohsenin (1966) studied fruits and vegetables as viscoelastic materials, suggesting that samples with high water content behave very similarly to the mechanical models used to represent them. The cells behave in a nearly elastic manner, and the cellular fluids represent the viscous element in the model. Relaxation and creep data obtained from apples showed that they could be considered as viscoelastic materials.

Stress Relaxation Behavior. Baerdemaeker, Wouters, and Tijskens (1987) compressed and relaxed cylinders of apple flesh and used the data to characterize the viscoelastic behavior of apple flesh. The relaxation data could also be used as an alternative test to estimate failure strength and possibly bruise susceptibility of apples.

Nondestructive Methods

Nondestructive methods include the compression of the sample below the yield stress (see "Stress Relaxation Behavior," above), elasticity (bouncing of peas and apples), and sonic and ultrasonic techniques, as well as near infrared spectroscopy (NIRS). Finney (1971) used a sonic resonance technique to produce a nondestructive firmness index in apples. This was found to be related to changes in firmness of the apple fruit during ripening on the tree and could be used to determine the optimum time for harvesting (Finney, 1976). Baerdemaeker and Wouters (1987) used a similar approach on excised cylinders of apple tissue as well as on whole apples, comparing the results with data from uniaxial compression tests on cylinders of tissue. They found that a stiffness factor derived from the nondestructive resonance data showed a sharp decrease around the time apples reach maximum maturity. They suggested that combining the stiffness factors with failure stress measurements gives good information about changes in fruit firmness and degree of maturity and thus about the sensory acceptability to the consumer.

Work is being carried out at CCFRA to develop an alternative to the tenderometer employing NIRS, results from which appear to relate most closely to AIS and to correlate well with many sensory attributes (Bedford & Scotter, 1996).

Monitoring the impact force when a fruit or vegetable is dropped from a known height onto a flat plate can provide useful information. Analyses of the frequencies produced showed that more of the high frequencies were present when the fruits were hard (Baerdemaeker, Lemaitre, & Meire, 1982).

A penetrometer can be modified to measure the deformability of fruit and vegetables, provided that they are neither too firm nor too small (Bourne, 1973), by replacing the needle or cone with a flat disc. Voisey and Crête (1973) devised a method to assess the squeezing of fruit and vegetables by hand during consumers' assessments of texture.

SENSORY PROPERTIES

The sensory assessment of food texture is a complex process achieved by a range of different sensing mechanisms and dependent on a number of different structural features and breakdown factors. Although this is a discussion of texture, it is important to realize that the sensory appraisal of a product is often influenced by a sequence of visual cues such as color, surface appearance, and softness to the touch. Voisey and Crête (1973) recognized that the consumer often carries out the initial assessment of texture in fruit and vegetables by squeezing the product, and they attempted to measure the force and the rate at which squeezing is applied to the product. They found that the rates and maximum forces used by males and females were different and depended on the firmness of the product. Their data also suggested that the rate of force application in sensory testing of fruit is significantly greater than that customarily used in sensory tests.

Once the fruit or vegetable is in the mouth, different regions of the mouth, teeth, tongue, and palate operate over different scales of physical force to assess the food. The stages of consumption are also important: for example, a pear of satisfactory softness on initial bite may have an unacceptably gritty or woolly texture on further chewing. The forces produced between the teeth and the rates of movement involved are very important in the assessment of texture and are discussed by Bourne (1982). The wearing of dentures has a profound effect on the way in which food is chewed (Heath, 1988, see also Chapter 2).

The perceived texture of fruits and vegetables is influenced by all of the structural properties discussed earlier in this chapter. Unprocessed fruits and vegetables that retain their turgor remain firm and are often described as crisp: they are able to resist an applied mechanical stress from the teeth at initial bite until a yield point is reached, at which there is a sudden loss of resistance as the structure breaks. This may be accompanied by an audible sound as the breakage occurs, and perception of this is often an important factor in the overall perception of crispness (Vickers & Bourne, 1976a, 1976b).

Another important characteristic of unprocessed fruits and vegetables is often juiciness, which is dependent on the amount of cell sap released during fracture. As discussed earlier, uncooked cell walls tend to fracture across to reveal the internal structure of the cell, thus releasing the sap. In contrast, cooked cells tend to separate along the middle lamella, so cell sap is not released in the same way.

Mealiness in apples is described as the breakdown of flesh into small pieces, which tend to be dry, in the mouth (Harker & Hallett, 1992). It is related to increases in water-soluble pectins and decreases in insoluble pectins during maturation. Thus, when the apple is eaten, cells separate easily, without the release of cell sap, and the mouth perceives the outside surfaces of clumps of cells rather than cleaved cells leaking sap (Reeve, 1953). In nectarines, the presence or absence of juice on the fracture surface after a tensile test was the main difference between nonmealy and mealy fruit (Harker & Sutherland, 1993), emphasizing the importance of lubricant liquids provided by tissue breakdown on the sensory assessment of food texture. Sponginess in apples is related to the presence of large intercellular air spaces making up 20% to 30% of the tissue volume (Reeve, 1953). Thus, the flesh is able to absorb some of the strain applied by the teeth by compression of the air spaces before any tissue fracture occurs.

Loss of turgor pressure as a result of heating or freezing reduces the initial resistance to the bite. The damage to cell structure caused by food processing results in further loss of cell contents as pressure is applied by chewing, and it increases as the pressure is increased. The resulting texture may be described as soft or mushy.

Some support tissues such as fibers or the thick-walled cells in green bean pods can provide a degree of support for surrounding tissues, resulting in a material that is perceived as firm or even crisp, even in the absence of turgor. The support provided by starch was discussed at length earlier in this chapter. Swollen, gelatinized starch can fill the volume of cells and provide resistance to deformation but does not impart crispness, which is characterized by the buildup of stress culminating in sudden fracture.

The key to evaluating sensory properties in fruit and vegetables is obtaining an accurate description of their sensory properties. In the profile procedure, terms are generated and agreed between assessors. Some terms used in texture profiling may be based on an association with other well-known foods. Beyond this, the texture may be assessed in different states of the product, such as the analysis of apple with or without peel to assess the contribution of the peel to the overall texture. With the peel intact, terms such as *crispness*, *flouriness*, and *hardness* might be used, whereas in the absence of peel, *hardness* might be replaced by *toughness* to reflect the removal of the influence of the peel in imparting a hard surface texture. This is particularly marked in dates, where softening of the fruit flesh on ripening results in strong contrasts between the skin texture and the soft flesh. These differences may become more extreme under certain conditions of post-harvest softening and processing (Coggins, Knapp, & Riker, 1968).

For more detail on the practical aspects of sensory analysis, see Chapter 3 of this volume or Lyon, Francombe, Hasdell, and Lawson, (1992).

CONCLUSION

The structural changes to vegetable and fruit cells caused by food processing are the result of complex interactions between the response of their chemical components to heat; cold; treatment with chemicals such as salt, sugar, or acid; and the overall structure of the cells. The most important factors in texture and structure development are cell separation and starch gelatinization. Further improvements in the control of food quality during commercial processing are likely to derive from a better understanding of these interactions. There is a great need for further information on the structural effects of newer methods of processing such as high pressure.

Disappointingly few studies have succeeded in using microscopic studies to explain the differences in the texture of fruits and vegetables. In a detailed study, Szczesniak and Smith (1969) failed to explain the behavior of strawberry texture on the basis of the structure. However, C.B. Holt's (1970) study did use texture measurements to identify the structures responsible for texture in tomato fruit.

Great strides have been made in the development of microscopic techniques for studying plant structures, in the evolution of new instrumental methods of measuring texture, and in methods of sensory analysis. However, the challenge still remains to develop concepts to link all three areas in a much more fundamental way than has been achieved hitherto.

REFERENCES

Adams, J.B., & Robertson, A. (1987). *Instrumental methods of quality assessment: texture* (Tech. Mem. No. 449). Chipping Campden, UK: Campden Food Preservation Research Association.

Ahmed, E.M., & Dennison, R.A. (1971). Texture profile of irradiated mangoes and peaches. *Journal of Texture Studies, 2*, 489–496.

Ahmed, E.M., Dennison, R.A., & Fluck, R.C. (1972). Textural properties of stored and irradiated Tioga strawberries. *Journal of Texture Studies, 3*, 80–88.

Anon. (1966). A new instrument for measuring the texture of potato French fries. *American Potato Journal, 43*, 175–176.

Baerdemaeker, J. de, Lemaitre, L., & Meire, R. (1982). Quality detection by frequency spectrum analysis of the fruit impact force. *Transactions of the American Society of Agricultural Engineers, 25*, 175–178.

Baerdemaeker, J. de, & Wouters, A. (1987). Mechanical properties of apples: 2. Dynamic measurement methods and their use in fruit quality evaluation. In R. Jowitt, F. Escher, M. Kent, B. McKenna, & M. Roques (Eds.), *Physical properties of foods 2* (pp. 417–427). London: Elsevier.

Baerdemaeker, J. de, Wouters, A., & Tijskens, L. (1987). Mechanical properties of apples: 1. Results of the collaborative study. In R. Jowitt, F. Escher, M. Kent, B. McKenna, & M. Roques (Eds.), *Physical properties of foods 2* (pp. 403–416). London: Elsevier.

Batisse, C., Buret, M., Coulomb, P.J., & Coulomb, C. (1996). Ultrastructure of different textures of Bigarreau Burlat cherries during maturation. *Canadian Journal of Botany, 74*, 1974–1981.

Bedford, L., & Scotter, C. (1996). *Use of NIR to measure maturity and quality of vining peas* (Pisum sativum *L.) Phase II.* (Research Summary Sheet No. 52). Chipping Campden, UK: Campden and Chorleywood Food Research Association.

Blanshard, J.M.V. (1995). The glass transition, its nature and significance in food processing. In S.T. Beckett (Ed.), *Physico-chemical aspects of food processing* (pp. 17–48). London: Blackie Academic & Professional.

Bourne, M.C. (1968). Texture profile of ripening pears. *Journal of Food Science, 33*, 223–226.

Bourne, M.C. (1973). Use of the penetrometer for deformation testing of foods. *Journal of Food Science, 38*, 720–721.

Bourne, M.C. (1982). *Food texture and viscosity: Concept and measurement*. New York: Academic Press.

Bretzloff, C.W. (1970). Some aspects of cooked potato texture and appearance. II. Potato cell size stability during cooking and freezing. *American Potato Journal, 47*, 176–182.

Burton, W.G. (1989). *The potato* (3rd ed.). New York: Longman.

Chappell, T.W., & Hamann, D.D. (1968). Poisson's ratio and Young's modulus for apple flesh under compressive loading. *Transactions of the American Society of Agricultural Engineers, 11*, 608–610.

Christiansen, C.M. (1984). Food texture perception. *Advances in Food Research, 29*, 159–199.

Coggins, C.J., Jr., Knapp, J.C.F., & Riker, A.L. (1968). Post-harvest softening studies of Deglet Noor dates: Physical, chemical and histological changes. *Date Growers Institute Report, 45*, 3.

Collins, J.L., & McCarty, I.E. (1969). Comparison of microwave energy with boiling water for blanching white potatoes. *Food Technology, 23* (3), 63–66.

Dobias, J., Curda, D., Vana, V., & Zakova, P. (1993). Influence of pectic substance extraction into the brine on the texture of canned fruit. *Potravinarske Vedy, 11*, 233–242.

Eckert, B., Amend, T., & Berlitz, H.D. (1993). The course of the SDS and Zeleny sedimentation tests for gluten quality and related phenomena studied using the light microscope. *Zeitschrift für Lebensmittel-Untersuchung und Forschung, 196*, 122–125.

Edwards, M.C. (1994). The measurement of pea texture and its interpretation. In V.D. Arthey (Ed.), *Proceedings of the symposium "The Measurement of Maturity in Vining Peas."* Chipping Campden, UK: Campden Food and Drink Research Association.

Ferrero, C., Martino, M.N., & Zaritzky, N.E. (1993). Stability of frozen starch pastes: Effect of freezing, storage and xanthan gum addition. *Journal of Food Processing and Preservation, 17*, 191–211.

Finney, E.E., Jr. (1971). Dynamic elastic properties and sensory quality of apple fruit. *Journal of Texture Studies, 2*, 62–74.

Finney, E.E., Jr. (1976). Mechanical resonance within Red Delicious apples and its relation to fruit texture. In J.J. Gaffney (Ed.), *Quality detection in foods* (pp. 146–149). St. Joseph, MI: American Society of Agricultural Engineers.

Finney, E.E. Jr., & Hall, C.W. (1967). Elastic properties of potatoes. *Transactions of the American Society of Agricultural Engineers, 10*, 4–8.

Gallant, J.D.J., Monredon, F. de, Bouchet, B., Tacon, P., & Delort-Laval, J. (1993). Cytochemical study of intact and processed barley grain. *Fermentation, 6*, 111–114.

Gamble, M.H., & Rice, P. (1987). Effect of pre-fry drying on oil uptake and distribution in potato crisp manufacture. *International Journal of Food Science and Technology, 22*, 535–548.

Garcia, E., & Lajolo, F.M. (1994). Starch alterations in hard-to-cook beans (*Phaseolus vulgaris*). *Journal of Agricultural and Food Chemistry, 42*, 612–615.

Goebel, N.K., Grider, J., Davis, E.A., & Gordon, J. (1984). The effects of microwave energy and convection heating on wheat starch granule transformations. *Food Microstructure, 3*, 73–82.

Grote, M., & Fromme, H.G. (1984). Electron microscopic investigations of the cell structure in fresh and processed vegetables (carrots and green bean pods). *Food Microstructure, 3*, 55–64.

Hammerle, J.R., & McClure, W.F. (1971). The determination of Poisson's ratio by compression tests of cylindrical specimens. *Journal of Texture Studies, 2*, 31–39.

Harker, F.R., & Hallett, I.C. (1992). Physiological changes associated with development of mealiness of apple fruit during cool storage. *HortScience, 27*, 1291–1294.

Harker, F.R., & Sutherland, P.W. (1993). Physiological changes associated with fruit ripening and the development of mealy texture during storage of nectarines. *Postharvest Biology and Technology, 2*, 269–277.

Heath, M.R. (1988). The relevance of food structure: A dental clinical perspective. In J.M.V. Blanshard and J.R. Mitchell (Eds.), *Food structure: Its creation and evaluation* (pp. 1–5). London: Butterworths.

Hincks, M.J., & Stanley, D.W. (1987). Lignification: Evidence for a role in hard-to-cook beans. *Journal of Food Biochemistry, 11*, 41–58.

Hohlberg, A.I., & Stanley, D.W. (1987). Hard-to-cook defect in black beans: Protein and starch considerations. *Journal of Agricultural and Food Chemistry, 35*, 571–576.

Holt, C.B. (1970). Measurement of tomato firmness with a Universal Testing machine. *Journal of Texture Studies, 1*, 491.

Holt, J.E., & Schoorl, D. (1984). Mechanical properties and texture of stored apples. *Journal of Texture Studies, 15*, 377–394.

Jarvis, M.C., Mackenzie, E., & Duncan, H.J. (1992). The textural analysis of cooked potato. 2. Swelling pressure of starch during gelatinisation. *Potato Research, 35*, 93–102.

Jones, H.F., & Beckett, S.T. (1995). Fruits and vegetables. In S.T. Beckett (Ed.), *Physicochemical aspects of food processing* (pp 292–314). London: Blackie Academic & Professional.

Jones, P.M.B., & Boulter, D. (1983). The cause of reduced cooking rate in *Phaseolus vulgaris* following adverse storage conditions. *Journal of Food Science, 48*, 623–626, 649.

Kaczynska, B., Autio, K., & Fornal, J. (1993). Heat-induced structural changes in fava bean starch paste: The effect of steaming fava bean seeds. *Food Structure, 12*, 217–224.

Kapsalis, J.G., Segars, R.A., & Krizik, J.G. (1970). An instrument for measuring rheological properties by bending: Application to food materials. *Journal of Texture Studies, 3*, 31–50.

Kramer, A., & Smith, H.R. (1946). The succulometer: An instrument for measuring the maturity of raw and canned whole kernel corn. *Canning Trade, 68*, 5–6.

Liu, K., Hung, Y-C., & Phillips, R.D. (1993). Mechanism of hard-to-cook defect in cowpeas: Verification via microstructure examination. *Food Structure, 12*, 51–58.

Lyon, D.H., Francombe, M.A., Hasdell, T.A., & Lawson, K. (Eds.). (1992). *Guidelines for sensory analysis in food product development and quality control*. London: Chapman & Hall.

Marle, J.T. van, Clerkx, A.C.M., & Boekestein, A. (1992). Cryo-scanning electron microscopy investigation of the texture of cooked potatoes. *Food Structure, 11*, 209–216.

Marle, J.T. van, Dijk, C. van, Voragen, A.G.J., & Biekman, E.S.A. (1994). Comparison of the cooking behavior of the potato cultivars Nicola and Irene with respect to pectin breakdown and the transfer of ions. *Potato Research, 37*, 183–195.

Marle, J.T. van, Stolle-Smits, T., Donkers, J., Dijk, C. van, Voragen, A.G.J., & Recourt, K. (1997). Chemical and microscopic characterisation of potato (*Solanum tuberosum* L.) cell walls during cooking. *Journal of Agricultural and Food Chemistry, 45*, 50–58.

McDonough, C., Gomez, M.H., Lee, J.K., Waniska, R.D., & Rooney, L.W. (1993). Environmental scanning electron microscopy evaluation of tortilla chip manufacture during deep-fat frying. *Journal of Food Science, 58*, 199–203.

Meyer, G.S. (1929). Some critical comments on the methods employed in the expression of leaf saps. *Plant Physiology, 4*, 103–111.

Mohsenin, N.N., & Goelich, H. (1962). Techniques for determination of mechanical properties of fruits and vegetables as related to design and development of harvesting and processing machinery. *Journal of Agricultural Engineering Research, 7*, 300–315.

Moledina, K.H., Haydar, M., Ooraikul, B., & Hadziyev, D. (1981). Pectin changes in the pre-cooking step of dehydrated mashed potato production. *Journal of the Science of Food and Agriculture, 32*, 1091–1102.

Morrow, C.T., & Mohsenin, N.N. (1966). Consideration of selected agricultural products as viscoelastic materials. *Journal of Food Science, 31*, 686–698.

Parker, M.L., & Waldron, K.W. (1995). Texture of Chinese water chestnut: Involvement of cell wall phenolics. *Journal of the Science of Food and Agriculture, 68*, 337–346.

Parkkonen, T., Haerkoenen, H., & Autio, K. (1994). Effect of baking on the microstructure of rye cell walls and protein. *Cereal Chemistry, 71*, 58–63.

Ramana, S.V., Wright, C.J., & Taylor, A.J. (1992). Measurement of firmness in carrot tissue during cooking using dynamic, static and sensory tests. *Journal of the Science of Food and Agriculture, 60*, 369–375.

Reeve, R.M. (1953). Histological investigations of texture in apples. II: Structure and intercellular spaces. *Food Research, 18*, 604.

Reeve, R.M. (1954a). Histological survey of conditions influencing texture in potatoes. I. Effects of heat treatments on structure. *Food Research, 19*, 323–332.

Reeve, R.M. (1954b). Histological survey of conditions influencing texture in potatoes. II. Observations on starch in treated cells. *Food Research, 19*, 333–339.

Reeve, R.M. (1954c). Histological survey of conditions influencing texture in potatoes. III. Structure and texture in dehydrated potatoes. *Food Research, 19*, 340–349.

Reeve, R.M. (1963). Estimation of extracellular starch of dehydrated potatoes. *Journal of Food Science, 28*, 198–206.

Reeve, R.M. (1967a). A review of cellular structure and texture qualities of processed potatoes. *Economic Botany, 21*, 294–308.

Reeve, R.M. (1967b). Starch and texture of potato products. In *Proceedings of the 16th Annual Potato Utilization Conference* (ARS 74-40). Fort Collins, CO: U.S. Dept. of Agriculture.

Rose, D.J., Edwards, M.C., Rendell, E., & Roberts, C. (1989). *The effect of process variables on the structure and texture of vegetables using the potato as a model* (Tech. Mem. No. 550). Chipping Campden, UK: Campden Food and Drink Research Association.

Ross, L.R., & Porter, W.L. (1968). Interpretation of multiple-peak shear force curves obtained with French fried potatoes. *American Potato Journal, 45*, 461–471.

Sayre, J.D., & Morris, V.H. (1931). Use of expressed sap in physiologic studies of corn. *Plant Physiology, 6*, 139–148.

Sayre, J.D., & Morris, V.H. (1932). Use of expressed sap in determining the composition of corn tissue. *Plant Physiology, 7*, 261–271.

Schuch, W. (1991). Improving fruit quality through biotechnology. *AgBiotech News and Information, 3*, 249–252.

Selman, J.D., & Hopkins, M. (1989). *Factors affecting oil uptake during the production of dried potato products* (Tech. Mem. No. 475). Chipping Campden, UK: Campden Food and Drink Research Association.

Sharma, M.C., & Mohsenin, N.N. (1970). Mechanics of deformation of a fruit subjected to hydrostatic pressure. *Journal of Agricultural Engineering Research, 15* (1), 65–74.

Shomer, I., Paster, N., Lindner, P., & Vasiliver, R. (1990). The role of cell wall structure in the hard-to-cook phenomenon in beans (*Phaseolus vulgaris* L.). *Food Structure, 9*, 139–149.

Smith, C.J.S., Watson, C.F., Ray, J., Bird, C.R., Morris, P.C., Schuch, W., & Grierson, D. (1988). Antisense RNA inhibition of polygalacturonase gene expression in transgenic tomatoes. *Nature, 334*, 724–726.

Speiss, W.E.L., Gierschner, K., & Grunewald, T. (1987). Studies of the rheological behavior of carrots and potatoes during cooking. In R. Jowitt, F. Escher, M. Kent, B. McKenna, & M. Roques (Eds.), *Physical properties of foods 2* (pp. 471–478). London: Elsevier.

Srisuma, N., Hammerschmidt, R., Uebersax, M.A., Ruengsakulrach, S., Bennink, M.R., & Hosfield, G.L. (1989). Storage induced changes of phenolic acids and the development of hard-to-cook in dry beans (*Phaseolus vulgaris* var. Seafarer). *Journal of Food Science, 54*, 311–314, 318.

Sterling, C. (1966). Anatomy and histology of the tuber with respect to processed quality. In *Proceedings: Plant science symposium* (pp. 11–25). Campden, NJ: Campbell Institute of Agricultural Research.

Svegmark, K., & Hermansson, A.M. (1993). Microstructure and rheological properties of composites of potato starch granules and amylose: A comparison of observed and predicted structures. *Food Structure, 12*, 181–193.

Szczesniak, A.S., Humbaugh, P.R., & Block, H.W. (1970). Behavior of different foods in the standard shear compression cell of the shear press and the effect of sample weight on peak area and maximum force. *Journal of Texture Studies, 1*, 356–378.

Szczesniak, A.S., & Smith, B.J. (1969). Observations on strawberry texture. *Journal of Texture Studies, 1*, 65–89.

Tu, K., Baerdemaeker, J. de, Deltour, R., & Barsy, T. de (1996). Monitoring post-harvest-quality of Granny Smith apple under simulated shelf-life conditions: Destructive, non-destructive and analytical measurements. *International Journal of Food Science and Technology, 31*, 267–276.

Vainionpaeae, J., Forssell, P., & Virtanen, T. (1993). High-pressure gelatinisation of barley starch at low moisture levels and elevated temperature. *Starch/Staerke, 45*, 19–24.

Vickers, Z.M., & Bourne, M.C. (1976a). Crispness in foods: A review. *Journal of Food Science, 41*, 1153–1157.

Vickers, Z.M., & Bourne, M.C. (1976b). A psychoacoustical theory of crispness. *Journal of Food Science, 41*, 1158–1164.

Vincent, J.F.V. (1982). *Structural biomaterials.* Basingstoke, UK: Macmillan.

Virtanen, T., & Autio, K. (1993). The microscopic structure of rye kernel and dough. *Carbohydrate Polymers, 21*, 97–98.

Voisey, P.W., & Crête, R. (1973). A technique for establishing experimental conditions for measuring food firmness to simulate consumer evaluations. *Journal of Texture Studies, 4*, 371–377.

Voisey, P.W., Larmond, E., & Seark, R. (1974). Sensory and instrumental measurement of French fry limpness. *American Potato Journal, 51*, 78–89.

Warren, D.S., & Woodman, J.S. (1974). The texture of cooked potatoes: A review. *Journal of the Science of Food and Agriculture, 25*, 129–138.

Weegels, P.L., Verhoek, J.A., Groot, A.M.G. de, & Hamer, R.J. (1994). Effects on gluten of heating at different moisture contents. II. Changes in physicochemical properties and secondary structure. *Journal of Cereal Science, 19*, 39–47.

Winton, A.L., & Winton, K.B. (1935). *The structure and composition of foods: Vol. 2. Vegetables, legumes, fruits.* New York: John Wiley.

Young, C.T., & Schadel, W.E. (1993). A comparison of the effects of oven roasting and oil cooking on the microstructure of peanut (*Arachis hypogaea* L. cv. Florigiant) cotyledon. *Food Structure, 12*, 59–66.

CHAPTER 10

Hydrocolloids

Jo Smewing

Thickeners, stabilizers, and gelling agents are major food ingredients that significantly control the textural properties of foods. Polysaccharides and proteins, the two most important biopolymers that are used to create and modify the texture in many processed or formulated food products, are frequently incorporated in one form or another for their gelling properties. Collectively, these food-structuring ingredients can be categorized as food hydrocolloids. Each gelling hydrocolloid produces a gel with unique textural attributes, although the manner in which it imparts its specific texturizing role to any particular food product may vary according to the ingredient in question and will depend, for example, on type, source, and concentration as well as on the physical properties of the particular system (Glicksman, 1969).

Hydrocolloids commonly used in foods are starches, gelatin, guar gum, locust bean gum, pectin, algin, carrageenan, xanthan, gum arabic, cellulose derivatives, and, to a much lesser extent, agar and gums ghatti, karaya, and tragacanth. More recently, the bacterial polysaccharide gellan gum has been approved for use in some food products in the European Union. Hydrocolloids seldom constitute an entire finished product but are used most often as additives to improve or control the properties of a food product. All applications depend on the properties provided by very large molecules in various states of hydration but mostly depend on the properties that these impart to solutions and gels (Whistler, 1993). The principal properties of hydrocolloids are their ability to produce aqueous solutions possessing high viscosities at low gel concentrations to supply required consistency and to produce gels of different strengths and stability.

This chapter is concerned with the sensory and rheological measurement of the thickening and gelling properties of hydrocolloids specifically designed as ingredients that control, enhance, or modify products for the food industry. As such, it will not deal with the detailed structure or functional properties of each of the hydrocolloids that foodstuffs may contain inherently; rather it will mainly discuss those texture assessment techniques that are considered more practical and have a major role in this rapidly advancing area.

GENERAL STRUCTURE OF HYDROCOLLOIDS

Several basic types of food gels can be distinguished. There are extreme types of gels of

I would like to thank Professor John Mitchell for his help with this chapter.

cross-linked flexible macromolecules (e.g., gelatin gels) and gels of hard particles (e.g., margarine). In between, there are, for instance, gels made of cross-linked relatively stiff macromolecules (most polysaccharide gels) and gels of deformable particles (e.g., milk gels) (van Vliet, 1995). Each hydrocolloid has special properties that are a result of its individual unique molecular structure. Although texturizing effects may vary according to the hydrocolloid chosen, a common structural feature among food hydrocolloids is that they are all high–molecular-weight polymeric compounds, and it is the molecular associations and entanglements of these polymer molecules that give rise to the thickening and gelling properties desired (Clegg, 1995).

Structurally, hydrocolloid gels are formed by a continuous colloidal (tridimensional) network, which holds the product together, with a more or less viscous medium embedded in it. Interaction between this colloidal network (arising from the cross-linking of neighboring macromolecules) and the viscous medium determines the mechanical behavior and stability of the system and therefore gives rise to its characteristic properties (Duran, Costell, & Fiszman, 1987). The formation of a colloidal network from solutions of long-chain polymeric substances (pectin, agar, gelatin, starch, etc.) can result from linked-together particles' forming strands, enveloping pores and/or droplets or inclusions. The dimensions (size and shape) of the particles, strands, and pores vary, thus creating different product properties (Langton, Astrom, Stading, & Hermansson, 1997). The spacing of cross-links or the frequency of their occurrence along the polymeric chains may also play an important role in relation to the gel's mechanical characteristics (Harvey, 1960).

In categorizing types of colloidal structure, a useful distinction can be made between reversible and irreversible aggregation processes. The former are primarily determined by thermodynamic considerations and the latter by kinetic considerations. Both are influenced by the nature of the interparticle interactions. Net repulsive interactions imply colloidal stability and hence inhibition of aggregation (Dickinson, 1997).

DYNAMICS AND MECHANICS OF STRUCTURAL FORMATION AND BREAKDOWN

Structural Formation

When segments of linear molecules in solution collide, they form an association over several chain units (Whistler, 1993). This entanglement of a number of monomer units between different polymer chains gives rise to "junction zones." The stability (or strength) of these helical aggregates is directly related to the "rupture strength" of the macroscopic gel and is dependent mainly on the length—that is, the number of intermolecular bonds that develop rather than the number of junction zones (Ainsworth & Blanshard, 1979; Rees, 1972). In addition, interjunction connections exert an influence on gel strength (Ainsworth & Blanshard, 1979).

The bonds are of various kinds, but hydrogen bonds contribute the major associative forces (Whistler, 1993). When two chains form a junction zone at one place along their chain lengths, other, perhaps distant reaches of each chain can collide and form a junction zone with other molecules (Whistler, 1993), and the chain segments from different polymer chains become packed in an ordered array of non–covalently linked chain segments (Clegg, 1995). The remaining disordered sections of polymer chains then link the ordered junction zones together, as depicted schematically in Figure 10–1.

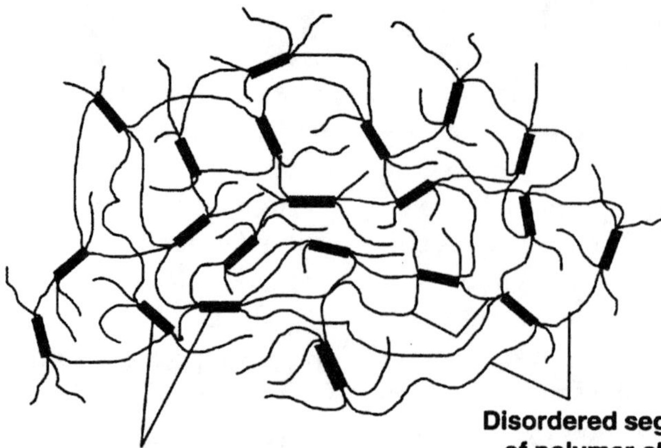

Figure 10–1 Schematic Illustration of a Cross-Linked Hydrocolloid Gel, with Non–Covalently Bonded Ordered Junction Zones Linking the Disordered Segments of Polymer Chains into a Three-Dimensional Network Structure. *Source:* Reprinted with permission from S.M. Clegg, *Physco-Chemical Aspects of Food Processing*, S.T. Beckett, ed., p.131, © 1995, Aspen Publishers, Inc.

This process of random junction zone formation between different molecules can continue until many, if not most, of the polymer molecules become involved in a great three-dimensional network that may fill the entire vessel (Whistler, 1993). The nature of the ordered junction zones varies between different gelling systems, but they are usually composed of chain segments with conformationally ordered structures, which are the same as those found in the solid state of the hydrocolloid (Morris, 1990; Rees, Morris, Thom, & Madden, 1982).

Structural Breakdown

As already stated, the gel will possess a strength dependent on the strength of intermolecular bonding at the junction zones. If the bonding is weak, it may be broken and the total gel structure disrupted by mild stirring. In this case, the weak gel is said to be *thixotropic*. The forces of stirring supply sufficient energy to break many, or even most, of the junction zones, and the viscosity of the dispersion decreases. When stirring is stopped, the molecules may again collide in normal translation, or Brownian-like motion, and junction zones may reestablish (Whistler, 1993).

When the junction zones are of greater strength, a more identifiable gel forms that may not be easily broken by stirring and may be strong enough that, under force, the gel will split, as seen with gelatin gels. Such a gel may be reliquified if the junction zone energies are low enough so that simple heating energizes the molecules sufficiently to cause them to tear apart from each other.

Once formed, a gel may undergo lengthening of its junction zones by the molecules' sliding over each other or simply moving together at the junction zone ends so as to increase the length of the zone. This will cause a tightening of the overall network structure and a decrease in the water-filled spaces between the molecules (Whistler, 1993).

During mastication, the material breaks down, and the structure is perceived as texture. Knowledge about the relationship between microstructure and texture can consequently be used to optimize food production processes, as well as to develop new products with desired sensory properties (Langton et al., 1997).

RHEOLOGICAL BEHAVIOR AND MEASUREMENT

Over the last decade, many investigators have contributed to the knowledge base describing the fundamental properties of polymer gels. In particular, the rheological behavior of these structures has been studied extensively in both time and frequency domains, using small strain and stress regimes (Lelievre, Mirza, & Tang, 1992a)—that is, small deformation. However, during the processing, manufacture, and consumption of foods, these systems are subject to large deformations. The large strains may cause the food to fail or fracture.

Rheological measurements can therefore be divided into those that induce small or large deformations, and essentially different instrumentation has been developed for these. These two types of tests give complementary information that does not necessarily correlate. By adjusting the concentration, for example, many gels can be made that all have the same small deformation properties; however, their large-deformation properties will usually vary greatly, illustrating that small- and large-deformation properties need not be related (van Vliet, 1995) and that different types of information can be obtained.

Small-deformation testing generally investigates viscoelastic parameters and frequently uses specialized dynamic mechanical analysis equipment. Parameters often used to describe the rheological behavior of gels are the shear or Young's moduli determined in the so-called *linear region* (i.e., the strain region where the stress required to deform a material is proportional to the strain) (van Vliet, 1995). These moduli are independent of the geometry of both the measuring system and the gel sample; consequently, the parameters are invaluable for comparing the elasticity of gelling (or "curing") systems.

For large-deformation experiments, devices such as compression and tension testers (e.g., texture analyzers) are employed. These more often measure stress, strain, and failure properties of fully gelled systems that have greater relevance to the ultimate (failure) properties of the material (Ross-Murphy, 1995).

Rheological test methods for the assessment of gel characteristics can generally be grouped into three types: fundamental, imitative, and empirical.

Fundamental Methods: Small Deformation

Oscillatory rheometry has been frequently used to measure elastic and viscoelastic properties of gels and foodstuffs. Test methods used in oscillatory rheometry fall into two categories (Marin, 1988). The first type involves steady rotation of boundary surfaces at the same angular velocity; this includes balance rheometers and orthogonal rheometers. The second type uses conventional controlled-strain or controlled-stress rheometers.

There are a number of oscillatory controlled-stress rheometers on the market, such as the Rheotech Viscoelastic Analyser, the Carrimed, Bohlin, and the Deer Rheometer (Ross-Murphy, 1988). The controlled variable is the torque to the driving motor. This rheometer is controlled stress rather than controlled strain. There are several measuring geometries available for use on rotational rheometers, the most widely used being cone and plate, parallel plates, and cup and bob (see Chapter 4).

For viscoelastic gels, the storage modulus G' (a measure of the energy that is stored elastically in the structure during a periodic application of stress) and the loss modulus G'' (a measure of the energy dissipated or the viscous response), as determined by dynamic mechanical spectroscopy, give a good characterization of the small-deformation properties (van Vliet, 1995). Their ratio $\tan \delta = G''/G'$ is calculated, and the phase (or loss) angle, δ, is a measure of how much the stress and strain are out of phase with each other. Under sinusoidal oscillation, a perfectly elastic material would have its stress and strain exactly in phase. A perfectly viscous liquid, under the same conditions, would have exactly out-of-phase stress and strain—that is, 90° out of phase (Ferry, 1970). The time scale of the experiment is the time during which a stress of a certain size and direction is applied to the gel. In a dynamic experiment, it is equivalent to the reciprocal frequency (ω^{-1}) of the sinusoidally oscillating stress or strain (van Vliet, 1995).

These parameters are commonly derived by dynamic oscillatory testing in the linear viscoelastic region of the test material. At low strains, many gels can be thought of as elastic materials, obeying Hooke's law, which states that the stress applied to a material is proportional to the resultant strain in that material. Mitchell (1976) put limits of about 0.13 on the amount of strain that may be used in such a test.

Dynamic measurements allow gelation to be monitored easily, since the induced deformations are usually so small that their effect on structure is negligible (Ross-Murphy, 1984).

Oscillatory rheological methods have been widely applied. For example, they have been used to measure the gel formation potential of water-soluble pentosans from wheat (Izydorczyk, Biliaderis, & Bushuk, 1990) and to characterize perceived texture of thickened systems (Richardson, Morris, Ross-Murphy, Taylor, & Dea, 1989). They have recently been used to study the rheological properties of sodium caseinate–carrageenan mixtures (Keogh, Laine, & O'Connor, 1995), soy isolate (Chronakis, Kasapis, Richardson, & Doxastakis, 1995), surimi gels (Chan, Gill, Thompson, & Singer, 1995), ovalbumin (Savoie & Arntfield, 1996), and galactomannan/κ-carrageenan gels (Fernandes, Goncalves, & Doublier, 1994).

Other rheological methods that can be carried out using a rotational rheometer to evaluate elastic and viscoelastic properties of hydrocolloids are creep and stress relaxation. In a creep experiment, a constant stress is applied, and the strain is followed as a function of time. Frequently, strain recovery is followed after removal of the stress. Recently, creep experiments have been used to measure viscoelastic properties of the following food gels: gellan (Nussinovitch, Ak, Normand, & Peleg, 1990), κ-carrageenan gel mixed with various gums (Murayama, Ichikawa, & Kawabata, 1995), gelatin (Kamata & Kinsella, 1989), soy isolate (Chronakis et al., 1995), heat-induced soy bean gel (Moss, Mohri, Artik, & Matsumura, 1989), and transglutaminase-induced globulin gels (Chanyongvorakul, Matsumura, Nonaka, Motoki, & Mori, 1995). These experiments relate to creep measurements in compression (i.e., using a texture analyzer), but the equivalent experiments can be carried out in rotation using a controlled stress instrument.

In stress relaxation, a constant strain is applied, and the decay in stress is followed with time. This type of experiment has been performed recently on gelatin (Maekaji & Kawamura, 1985), agar-gelatin mixed gels (Yano & Shiinoki, 1986; Yano, Shiinoki,

Miyawaki, & Sakiyama, 1987), milk gels (Zoon, van Vliet, & Walstra, 1989a, 1989b), agar (Nussinovitch, Peleg, & Normand, 1989; Purkayastha & Peleg, 1987), gellan gels (Nussinovitch et al., 1990), alginate (Nussinovitch et al., 1989), κ-carrageenan (Maekaji & Kawamura, 1985), and BSA and Maillard protein gels (Armstrong, Hill, & Mitchell, 1994; Armstrong, Hill, Schrooyen, & Mitchell, 1994). In rotation, a controlled strain instrument (e.g., a Bohlin VOR or a Rheometrics) would be required for stress relaxation measurements.

Fundamental Methods: Large Deformation

Observation of the microstructural and fracture properties of gels reveals clear distinctions that cannot be determined by small-strain rheological analysis (Foegeding, Bowland, & Hardin, 1995). Despite the industrial and social significance of foods, there is relatively little information concerning the fundamental behavior of food gels during large-deformation or failure tests (Lelievre, Mirza, & Tang, 1992a).

Characterization of the fracture behavior of food is more complex than that of the small-deformation rheological properties (Luyten, van Vliet, & Walstra, 1992), but in many practical situations this behavior is of key importance. Large-deformation properties are important for the behavior of food materials during production, further handling, and consumption; permanent changes in structure occur, often accompanied by yielding or fracturing of the material, which represent important features in consumer perception and acceptance (Keetels & van Vliet, 1993; Szczesniak, 1983).

When a gel-type food is consumed, the perception of texture is, in part, an evaluation of deformation and fracture properties of a gel matrix that is viscoelastic (Foegeding et al., 1995). Fracture starts at defects in the material that cause local stress and strain concentrations (Atkins & Mai, 1985). Large defects in the material have a far stronger effect on fracture initiation than on the behavior at small deformations, and this fracture behavior cannot therefore be predicted from small-deformation behavior (Luyten & van Vliet, 1995).

Fundamental rheological properties determined at large ("fracture-causing") deformations are called fracture, failure, or rupture stress and strains and sometimes also the "work of fracture" (as illustrated in Figure 10–2). These parameters are usually determined by constant-speed experiments such as uniaxial compression, uniaxial tension, and three-point bending (Rohm, Jaros, & Benedikt, 1995) performed on equipment such as the Instron, Stevens, Lloyd, or Stable Micro Systems TA.XT2 Texture Analyser.

In general, stress reflects the firmness (hardness) of the gel, and strain is an indication of cohesive properties (Foegeding et al., 1995). Small-strain moduli (shear stress/shear strain) show little correlation with sensory texture (Montejano, Hamann, & Lanier, 1985) and only in certain cases is there a direct relation between the rheological properties determined at small deformations and the parameters describing fracture.

For an ideally elastic material exhibiting brittle fracture, a direct relation may be expected between the modulus and the fracture stress as long as the fracture strain does not change (van Vliet, 1995). In the case of viscoelastic materials, a direct relation can sometimes be observed between the loss tangent and the dependence of the strain at fracture on the deformation rate (van Vliet, Luyten, & Walstra, 1993). However, there is often no simple relation between small deformation, large deformation, and the fracture

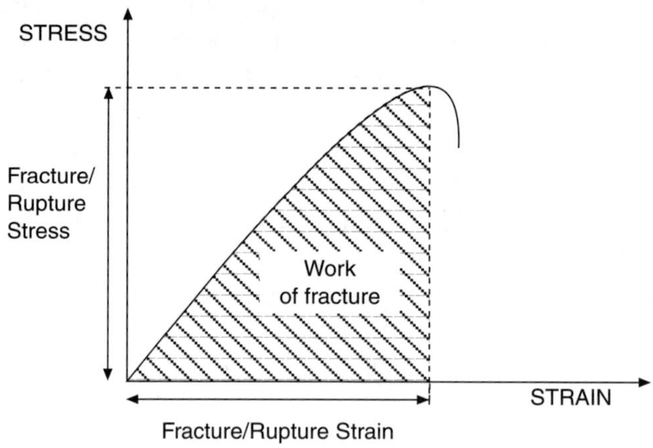

Figure 10–2 Stress/Strain Curve of Gel Compression up to Rupture.

properties. The prime reason for this is the inhomogeneity of most materials (van Vliet, 1995).

Food technologists involved in product development or quality assurance are interested in the mechanical properties of the gel in relation to shaping, stand-up, handling, cutting/slicing, or eating characteristics. Most of these properties relate to large-deformation behavior, including fracture characteristics at various deformation rates, and only to a lesser extent to those determined at small deformations. Table 10–1 lists some of the relevant mechanical properties that are of practical importance (as described by van Vliet, 1995).

For most foods, the ratio between the stress and strain is not constant at large deformations—that is, the stress-strain relation becomes nonlinear. The deviation from linearity is a product-dependent characteristic. In this nonlinear region, products can still be characterized by a modulus, but its magnitude will depend on the strain applied and often also on the strain history. It is this difference in nonlinear characteristics that makes margarine and gelatin gels so different (van Vliet, 1995).

Although compression tests are more common, tensile tests give a clearer picture of the stresses in a sample (Luyten, 1988). Tensile properties of foods are potentially useful, since the complex deformation modes during food mastication will include development of tensile stresses. The total stress during compression is the sum of both tensile and shear stresses, whereas the shear stresses in tensile testing are negligible. Other advantages of tensile testing are that the energy incurred is used only for deformation and not for friction, the start of fracture can be determined more precisely, and it is possible to study the notch sensitivity of the material (Luyten, 1988; Stading & Hermansson, 1991). Notch sensitivity, crack propagation rate, and the difference between the critical stresses for initiation and propagation of cracks are important but lesser known characteristics (Luyten et al., 1992; van Vliet, Luyten, & Walstra, 1991, 1993).

Although there are well-established methods of studying the failure of gels in compressive (Bagley, Wolf, & Christianson, 1985) and torsional modes (Diehl, Hamann, & Whitfield, 1979), this is not the case for

Table 10–1 The Main Practical and Relevant Mechanical Properties of Gels

Property Desired	*Relevant Parameter*	*Relevant Conditions*
Thickness	Apparent viscosity after yielding	Stress/strain rate
Firmness	Modulus, fracture stress, yield stress	Time,* strain, strain rate
Stand-up	Yield stress (apparent viscosity)	Time, stress
Shaping	Yield stress and restoration time on setting	Strain rate, stress
Immobilization of solvent	Permeability, tan δ, modulus, yield stress	Stress, time
Handling, slicing	Fracture stress and strain, work of fracture	Strain rate
Eating characteristics	Yield or fracture properties, modulus at large deformation	Strain rate

*Time = the time period that a stress is exerted on a product.

Source: Reprinted from T. van Vliet, *Food Macromolecules and Colloids*, E. Dickinson, ed., p. 448, © 1995.

tension (Hamann, 1983). One problem is that the gel samples may not be self-supporting. Another major problem when performing tensile testing of gels (particularly if they are weak) concerns mounting the samples in the testing apparatus (usually by pneumatic grips or screw clamps). In tension, cracks and inhomogeneities in the sample can result in stress concentrations that lead to failure at relatively low apparent stresses or strains (Lelievre, Mirza, & Tang, 1992a). Considerable care is necessary to attach the gel sample in the test rig without prestressing the gel and inducing preferential failure. Langley, Millard, and Evans (1986) described a technique designed to avoid these problems. The method uses matching semicylindrical molds to prepare the gels with inserts at either end to mount the sample in the test rig (Lelievre, Mirza, & Tang, 1992b). Extending rings over dowel pins so that tension measurements can be made without the need to clamp samples has been performed (Tschoegl, Rinde, & Smith, 1970), but this again is not without handling and mounting problems.

Empirical Methods: Small Deformation

The most common objective measure of gel quality is gel strength (Pomeranz & Meloan, 1994) for which numerous methods are available. The choice of method is dependent upon the purpose of the test, tradition, and the kind of material to be evaluated. Some empirical methods—namely, nondestructive methods (small deformation), as described below—measure the gel strength within the elastic limits of the gel; other methods measure breaking strength. Unfortunately, results obtained with different methods are not easily compared (Beach, Davis, Ikkala, & Lundbye, 1986). The choice of instrument for gel measurement will depend upon whether a single-point or multiparameter analysis is required and will also be controlled by such deciding factors as high-speed data acquisition, precision and accuracy of results, and the subtleties of sophisticated software interfacing (programmability) for ease of use and analysis.

Compression (or penetration) tests constitute the basis of many small-deformation empirical tests used to measure gel strength. They use a variety of instruments, such as the Boucher Electronic Jelly Tester (BEJT), the Bloom Gelometer, Instron, Stevens LFRA, and TA.XT2 Texture Analysers. The operating principle is that a standard gel sample is compressed by a standard cylindrical plunger (or "probe") descending to a preset distance, which for gelatin is a 4-mm deformation (i.e., small strain), with a constant load increase over the time used. The final load (or load at a 4-mm penetration) is read as gel strength (or Bloom strength in the case of standard gelatin testing), indicating rigidity, and will be proportional to the elastic modulus. The greater the force (or load) required to penetrate, the higher the strength of the gel (Gelatin Manufacturers Institute of America, 1993). This represents one of the simplest and most commonly accepted methods for objective measurement of gels and has long been used for measuring the firmness of gels and the quality of gelatin, agar, and pectin.

Empirical Methods: Large Deformation

While the texture analyzers already mentioned are often used for measuring gel firmness at small strain levels, they are also used to determine rupture strength—that is, the force required to break the gel. Break-strength measurements are made by compressing or penetrating (puncturing) a standard gel, again using a standard probe, which descends at constant speed until the gel ruptures. Although failure characteristics of gels may be measured in compressive, tensile, and torsional modes, uniaxial compression is most appropriate, since in many practical situations food systems are subject to compression (or penetration). Further, sample preparation is easier (Lelievre, Mirza, & Tang, 1992) and is well suited to the jelly industry because gels may be tested in the containers in which they are to be sold (given that the same probe and sample geometries are used for comparison purposes).

Apart from texture analyzer–type equipment that provides rupture or fracture terms (as shown previously in Figure 10–2), usually on a basic force-versus-time/distance plot, there are several other methods for determining rupture strength of gels (see below). These are simple methods still in use, and although they have been extensively used in the past, they are being gradually superseded by more sophisticated and accurate texture analyzers.

Marine Colloids Gel Tester

The Marine Colloids Gel Tester, which evolved from the Cherry-Burrell Curd Firmness Tester developed by Marine Colloids, Inc., has been used to measure the strength of gels made from carrageenan and other refined seaweed extracts. A probe of 0.845 or 0.431 inches in diameter is driven downward by a synchronous motor at 18 cm min^{-1}, and the maximum force is measured on the Chatillon scale and taken as the measure of firmness of carrageenan and various other gels (Bourne, 1982).

Stress-Strain Curve (SSC) Method

Screw-topped, 65-cm^3 ointment jars are filled to the 50-cm^3 level with a 1.6% solids solution at 45°C, sealed and held for one hour in a stirred water bath maintained at 19.5 to 20.5°C. A 1-cm^2 cylindrical plunger is pressed into the gel without lateral movement at 2-mm/s. Load and depth of gel depression are simultaneously recorded to rupture. Rupture stress is a measure of strength; depression depth at failure varies with resilience, and curve slope is a function of elasticity (Selby & Whistler, 1993). Variation of this method with texture analyzer measurements is now in general use for gel strength determination.

FIRA Gel Tester

The FIRA instrument operates with a form of shear rather than compressive deformation, in that a paddle immersed in the gel is subject to a constantly increasing torque (Therkelsen, 1993).

Ridgelimeter (or Pectin SAG Method)

The easy-to-use pectin SAG method was developed by Cox and Higby (Rolin, 1993) for judging the grade of fruit pectins and the stiffness of pectin jellies and is used in the method of pectin standardization approved by the Institute of Food Technology (IFT) (1959). The instrument is essentially a height-measuring gauge. The gel to be tested is prepared in a tapered glass tumbler of standardized dimensions. After curing (20–24 hr at 25 ± 3°C), the gel is carefully removed from the glass and allowed to stand without support. The height of the gel deformed by its own weight is measured with a precision instrument (ridgelimeter) after exactly 2 min. Results are reported as percent SAG (i.e., loss of height as percent of original height). This gives a good measure of the shape retention of the gel and is also used for testing of the soft ι-type carrageenan gels.

Nikkankyo—Rupture at Constant-Stress (CS) Method

In Japan, the Nikkankyo method has been used in official grading of agar. Ten plungers (cylindrical probes) loaded in an ascending series of mass are simultaneously lowered on 10 replicate samples of a 1.5% gel (air-dry basis) that have been aged for 15 hr at 20°C. The maximum stress in g/cm^3 for 20 s without rupture is reported as the solidity of the sample (Selby & Whistler, 1993).

Although the Nikkankyo and ridgelimeter tests are measured in the region before rupture, generally they could be accepted as large-deformation techniques.

Lüer/Lochmüller Method

Modifications of the Lüer/Lochmüller method are used to some extent, especially in Germany (Steinhauser, Otterbach, & Gierschner, 1979; Uhlenbrock, 1983). A body is embedded in the gel that is to be tested. The force required to pull the body out is taken as the breaking strength. A precision instrument, the Herbstreith Pectinometer, has been developed for this purpose (Rolin, 1993).

In addition to the determination of gel strength, it may be desirable to measure adhesiveness (stickiness). This can be again measured by texture analyzers and is usually determined by the force required to separate the probe (usually cylindrical) from the sample upon probe withdrawal.

Imitative Methods

Traditionally, single-point measurements such as "gel strength" have been used by suppliers and users to characterize gelled systems. However, these single-point measurements, often based on rupture tests, are not representative of the overall mechanical behavior of gels (Pomeranz & Meloan, 1994). In recent decades, much work has been done on the development of mechanical tests to imitate the sensory evaluation of food texture (Peleg, 1983). A much more comprehensive understanding of gel texture is obtained by analysis of the force/deformation curve generated by compressing a gel sample using a texture analyzer. This instrumental technique, known as Texture Profile Analysis (TPA), was originally developed by General Foods in the 1960s during the course of their studies on correlation between sensory evaluation and objective measurements of food texture (Friedman, Whitney, & Szczesniak, 1963; Szczesniak, 1963; Szczesniak, Brandt, & Friedman, 1963). TPA has been applied as a useful method for a wide range of foods but is particularly useful for analysis of gelled systems. However, it is often erroneously applied without correct definition of its parameters or selection of suitable experimental conditions (Pomeranz & Meloan, 1994).

The procedure entails compression of a gel sample twice in succession using a texture analyzer. A general texture profile for a gel and the parameters calculated from the trace are shown in Figure 10–3.

The *modulus* is the initial slope of the force-deformation curve. This is a measure of how the sample behaves when compressed a small amount. The modulus usually correlates very closely with the sensory perception of the sample's firmness and is expressed in units of force per unit area (usually in Newtons per square centimeter). This "initial" modulus is not a parameter from the original TPA work but instead has developed from looking more closely at the portion of the curve before fracture.

Hardness is defined as the maximum force that occurs at any time during the first compression cycle. It may occur when the gel initially breaks or later in the test (as shown in the general plot) as the sample is flattened and deformed. In most cases, the hardness is correlated to the rupture strength of the material. It is similar to gel strength measured on a Marine Colloids Gel Tester and is expressed in units of force (usually grams or Newtons).

Fracturability

The first significant drop in the force-deformation curve during the first compression cycle is defined as the fracturability (sometimes termed the *brittleness*). It is the point of first fracture or cracking of the sample. A gel that fractures very early in the compression cycle is considered to be more brittle or fragile than one that breaks later. Depending upon the extent of deformation, this "yield" may not exist in gelled systems, and many papers do not mention this parameter in TPA tests on gel systems (Konstance, 1993; Tang, McCarthy, & Munro, 1995; Trzeciecki, Swiderski, & Fedorowicz, 1992). It is important to emphasize that in most cases, parallel to TPA without rupture of the sample (where "recovery" is possible), a test of single compression up to rupture is often done to obtain parameters such as rupture force, rupture deformation, and rupture work of energy (Pons & Fiszman, 1996).

Adhesiveness

Following the first compression cycle, the force is removed from the sample as the probe moves back to its original position. If the material is at all sticky or adhesive, the force becomes negative. The area of this negative peak is taken as a measure of the adhesiveness of the sample. There are no real units for this parameter, which is expressed in the internal integrator units of the computer (e.g., gram seconds), based on a force-versus-time graph.

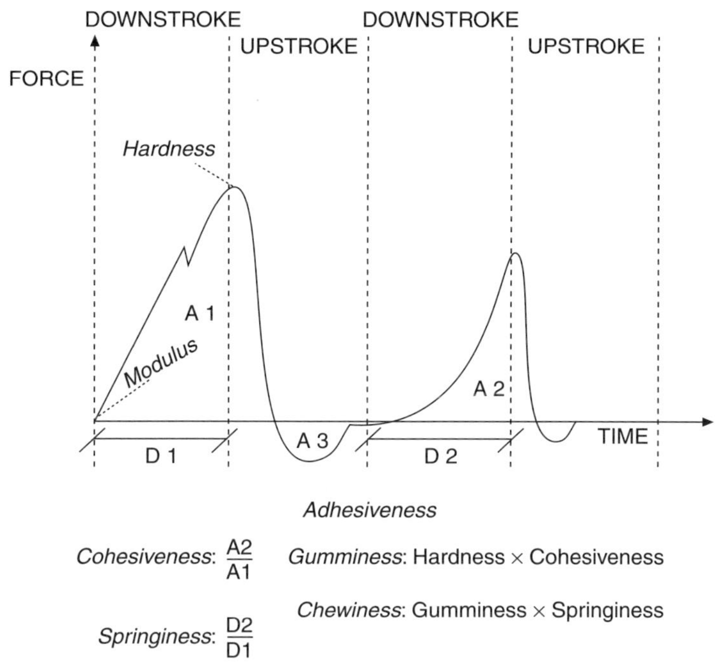

Figure 10–3 Typical Texture Profile for a Gel

Springiness

As the second compression cycle begins, elasticity of the sample is determined. By noting where the force begins to increase during this second compression cycle, a measure of the sample height may be obtained. If the sample returned to its original height, the elasticity will be 100%. The elasticity is a measure of how much the original structure of the sample was broken down by the initial compression. In sensory terms, it can be thought of as how "rubbery" the sample will feel in the mouth. The units are dimensionless (a length divided by another length) and are usually expressed as a percentage.

Cohesiveness

Cohesiveness is measured by taking the total work done on the sample during the second cycle and dividing it by the work done during the first cycle. Work is measured as the area under the respective curves. This ratio is usually expressed as a percentage. Samples that are very cohesive will have high values and will be perceived as tough and difficult to break up in the mouth. As with elasticity, there are no units, since the number is dimensionless.

Figure 10–4 illustrates the usefulness of certain TPA results in highlighting different textural parameters obtained from various gellan gums gels.

SENSORY PERCEPTION

In production situations, the quality of gels is often assessed simply by shaking, squeezing, or prodding with the finger. These simple subjective techniques usually provide the required information, especially when used by experienced personnel (Sanderson, Bell, Burgum, Clark, & Ortega, 1988).

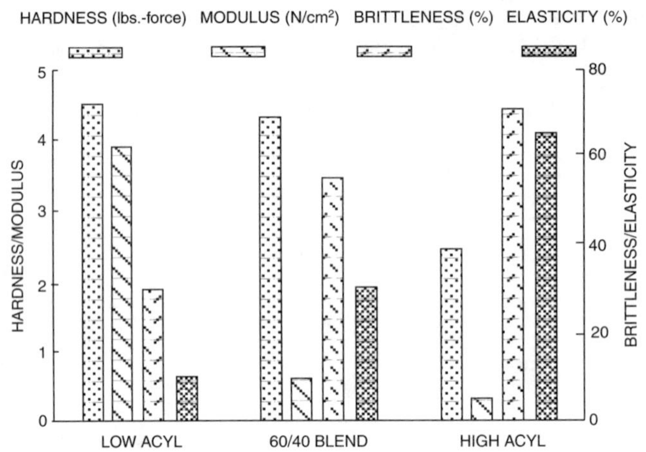

Figure 10–4 Comparison of the Textures of Low-Acyl, High-Acyl, and Low-Acyl/High-Acyl Gellan Gum Gels. *Source:* G.R. Sanderson et al., *Gums & Stabilizers for the Food Industry 4*, p. 228, © 1988, by permission of Oxford University Press.

In sensory assessment, a combination of rheological parameters is always perceived (van Vliet, 1995). Low sensory-mechanical correlations can occur when one measurement is much more sensitive than the other and/or when one responds to multiple (e.g., sensory) product attributes while the other records a single (e.g., mechanical) dimension (Munoz, Pangborn, & Noble, 1986). This makes it difficult to predict consumers' sensory assessments from results of measurements yielding fundamental rheological parameters. Therefore, an empirical (or imitative) method, by which mostly the effect of a combination of parameters is determined (as described previously in the section "Rheological Behavior and Measurement"), may work better (van Vliet, 1995), and the extensive use of TPA measurements (particularly in the gelled-systems area) reflects the importance of this interest.

Testing of sensory perceptions of texture has a problem similar to that of mechanical testing in that it is sensitive to the test principle used, so that different attributes and descriptors are influenced by different microstructural parameters. When the texture of particulate whey protein gels was evaluated by Langton et al. (1997), two main sets of attributes were found. Principal-component analysis of the textural sensory data identified the two sets as (1) grainy appearance, gritty, creamy, and "falling apart," and (2) soft, springy, and sticky (as described in Table 10–2). In the second set, the attributes "soft" and "springy" were negatively correlated with each other and were evaluated by the application of a light pressure with the forefinger (i.e., small deformation). The perceptions of "soft" and "springy" were found to be related to the strand characteristics of particulate gels. The perception of stickiness (how much the sample adheres to the teeth after gently chewing) effectively forms a group of its own. The attributes in the first set were all perceived by using a larger force (large deformation) during the analysis—namely, cutting and chewing. This group of descriptors was found to be sensitive to the variation in the overall particulate network dimensions, such as pore size and particle size (Langton et al., 1997).

Wood (1979) reported that oral hardness of gels correlated with Instron measures of

Table 10–2 Descriptors Used for the Sensory Profiling of Whey Protein Gels

Attribute	*Descriptor*	*Explanation*
Manual texture (touching)—measured by a light pressure with forefinger	Soft	How much the sample resists light pressure with forefinger
	Springy	How rapidly the sample regains its initial shape after light pressure with the forefinger
Visual texture (visualization)—observations of a newly cut surface	Surface moisture	How much water is released from the newly cut surface
	Grainy	How grainy the newly cut surface appears
Oral texture (mouth-feel)—perception during the manipulation of the sample in the mouth	Gritty	How gritty the sample is perceived to be during chewing
	Creamy	How creamy and smooth the sample is perceived to be during chewing
	Sticky	How much the sample adheres to the teeth after chewing
	Falling apart	How easily the sample falls apart or disintegrates during chewing

Source: Reprinted with permission from M. Langton et al., *Food Colloids*, E. Dickinson and B. Bergenstall, eds., p. 21, © 1997, The Royal Society of Chemistry.

maximum yield force, while hardness to touch correlated highly with Bloom Gelometer values.

Much of the discussion to this point has been with regard to semisolid or solid-state gels. Consistency and viscosity are also of great importance to the perception of gels, although mainly to the perception of dilute gels or starch pastes. In liquid systems, consistency depends mainly on the viscosity. The quantity perceived in the mouth is the shearing stress or shear rate. It has been suggested that for thick foods, it is the stress that is perceived, whereas for lower viscosity foods, it is the shear rate (Shama & Sherman, 1973). Consistency is noted when a foodstuff is taken on the upper surface of the tongue and the tongue is raised to the roof of the mouth with a shearing motion. Work by Richardson et al. (1989) has shown that for weak gel structures, the complex viscosity measured in oscillation at a frequency of around 50 rad/s correlates better with perceived thickness than the steady shear viscosity. This may be a reflection of the stress overshoot experienced when these structures are sheared. This overshoot is not measured in continuous shear experiments but does contribute to small-amplitude oscillatory measurements.

It has been suggested that the shape of the flow curve correlates with perceived sliminess in the mouth (Figure 10–5). Newtonian fluids are perceived as more slimy than non-Newtonian fluids. It seems more likely that it is the viscosity under the shear stress and shear rate conditions perceived in the mouth that correlates with sliminess. The original work (Szczesniak & Farkas, 1962) compared solutions that all had the same viscosity at a low shear rate. If the viscosities of these same so-

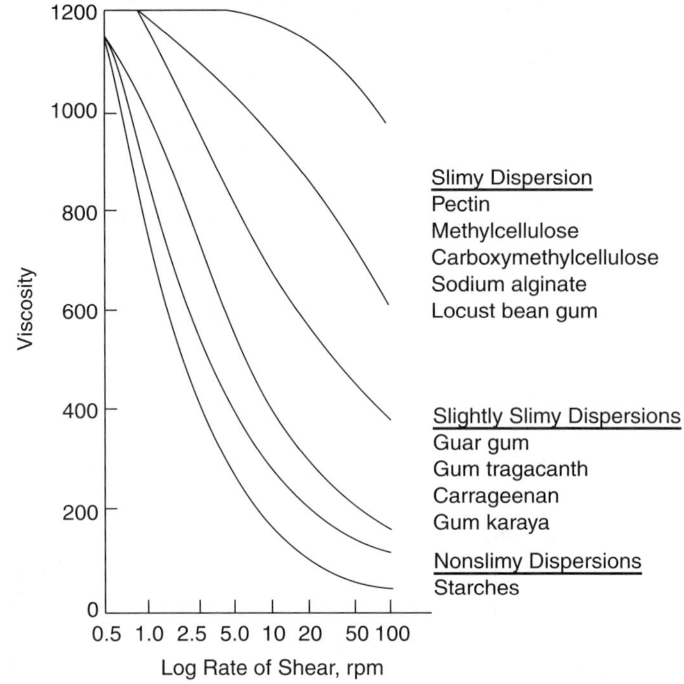

Figure 10–5 Relation of Rheological Behavior of Gum Solutions to Organoleptic Properties. *Source:* Reprinted with permission from A. Imeson, ed., *Thickening and Gelling Agents for Food*, p. 34, © 1992, Chapman & Hall.

lutions are compared at the shear rate relevant to perception in the mouth (10 to 50 s^{-1}), the correlation between viscosity and sliminess becomes apparent (Richardson et al., 1989).

GENERAL TEXTURAL PROPERTIES AND HOW THEY ARE AFFECTED

The stability and rheological properties of food colloids are dependent on the ways in which the constituent particles and macromolecules are assembled to form colloidal structures. Different kinds of interparticle interactions lead to different aggregation mechanisms, and different aggregation processes lead to different kinds of colloidal structures and hence variation in rheological behavior (Langton et al., 1997).

Gelling biopolymers are widely used in food products to confer structure and texture attributes that appeal to the consumer. Each of the currently used gelling agents, such as gelatin, agar-agar, alginate, carrageenan, pectin, starch, xanthan/locust bean gum, and gellan, has a unique set of gel characteristics (Marrs, 1997). Factors that influence the performance of a food hydrocolloid in any particular application include temperature, shear, pH, and ionic strength, and the effects that these factors can have on the functionality of a food hydrocolloid are specific to the hydrocolloid in question. Consequently, care must be taken by food processors when using such ingredients to ensure that the maximum functionality of thickening and gelling systems is achieved, while bearing in mind the likely effect that the process itself may have on the system (Clegg, 1995).

In the final analysis, the choice of gelling agent for a particular product will be driven by processing considerations, organoleptic attributes, and economic factors (Marrs,

1997). Though factors affecting the properties of food hydrocolloids are specific to each particular hydrocolloid, these systems have many properties in common (Clegg, 1995).

The following is not intended to be an extensive list of physical characteristics and effects of all conditions on each hydrocolloid. Rather, it gives a brief description of textural properties and certain factors that affect the performance and sensory perception of some of the more popular hydrocolloids.

Gelatin

Gelatin gels are soft, elastic and vibrant, clear, and mostly color-free and melt in the mouth to provide excellent mouth-feel and flavor perception (Marrs, 1997). Gelatin, which must be chilled for gelation, has a low susceptibility to acid breakdown and produces gels with a low melting point, which facilitates its preparation and its dissolution in the mouth (Jones, 1977).

Agar

Agar-agar forms very rigid, short-textured, crumbly, brittle gels that do not melt in the mouth (Marrs, 1997). Agar and agaran are among the most potent gel-forming agents known, for gelation is perceptible at concentrations as low as 0.04%. Threshold gels are valuable for their protective action, diffusion prevention, and texture enhancement effects. Stronger gels are of value because of their strength, resilience, elasticity, relative transparency, relative permanence, and reversibility. Agaran gels are firmer but less elastic than are gels from the parent agar (Whistler, 1993). Pure agar-water gels are rather stable but age slowly. At 25°C, the gel strength increases at about 1% per hour for the first 8 hr or so, then the rate of increase lessens, and the gel strength becomes stable in 15 hr (Selby & Whistler, 1993).

Alginate

Two different blocks of monomers present in the alginate will give different gel strength depending on the amount of mannuronic acid (M) and guluronic acid (G) present in the alginate solution. High G-block alginates are more calcium ion–reactive and therefore form higher gel strength than the M-block type for the same concentration of alginate (Clare, 1993). High-G alginates produce strong, brittle gels that possess good heat stability, whereas the high-M alginates produce gels that are weaker, more elastic, and less prone to syneresis.

Carrageenan

Gelation of κ and ι carrageenan occurs in the presence of cations such as K+, Ca++, or NH4+, each of which produces gels with different textures (Guiseley, Stanley, & Whitehouse, 1980). The strength of carrageenan gels is greatly dependent on the carrageenan concentration and on the type and concentration of monovalent cation. Cation concentration beyond a certain limit (~0.2 M) will weaken the gel, but this limit is rarely reached in any application. Whereas ι-carrageenan produces transparent, soft, and elastic gels that resemble gelatin gels, κ-carrageenan tends to form stiff, brittle gels (it is the strongest of all carrageenans, as shown in Figure 10–6, although textures vary with the chosen cation; see Christensen & Trudsoe, 1980).

Pectin

Pectin forms gels at high sugar solids and low pH, and these gels are soft, inelastic, and short textured (Marrs, 1997).

Starch

Starch forms highly milk-white gels having a short texture and relatively heavy-bodied

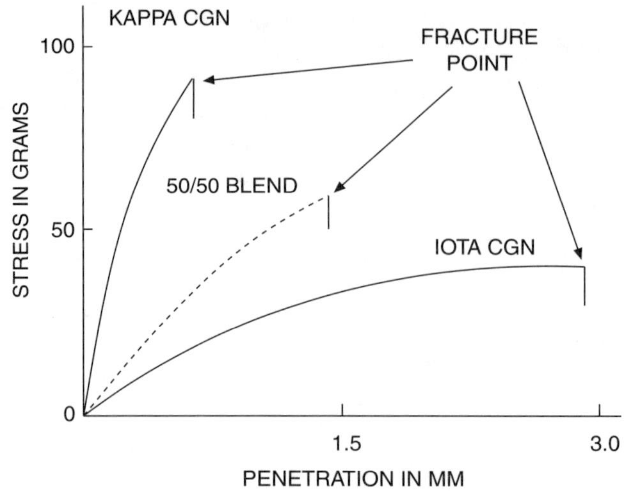

Figure 10–6 Gel Texture of Carrageenan. *Source:* Reprinted from W.R. Thomas, *Thickening and Gelling Agents for Food*, A. Imeson, ed., p. 34, © 1992, Aspen Publishers, Inc.

mouth-feel (Marrs, 1997). Starch-based foods are dealt with in more detail in Chapter 6.

Whey Protein Gels

Whey proteins can form either thermo-reversible or thermo-irreversible gels when heated above 65 to 70°C, depending on the initial composition and pH of the protein solution (Kinsella & Whitehead, 1989; Rector, Kella, & Kinsella, 1989). Thermally induced WPI gels can be formed that are transparent and brittle and exhibit low syneresis or that are opaque and rubbery and exhibit high syneresis (McClements & Keogh, 1995).

Gellan

Gellan gels are similar in texture to agar and carrageenan gels. All three gums form firm, brittle gels. Gellan gels exhibit a marked melting/setting temperature hysteresis, as do agar gels. Textures of gellan gels can be modified by the concentration and type of cation and by the addition of other polysaccharides.

Deacylated polysaccharide produces a firm, brittle gel similar to an agar gel. However, with a varying degree of acylation, gellan can provide a range of gel textures. For this reason, gels can be used to mimic textures achieved with other gellants or to create new textures (Kang & Pettitt, 1993).

Mixed Gels

So far, discussion has mentioned the textural attributes provided by any one biopolymer, a particulate gel system consisting of compact particles and liquid-rich pores. Over the years, a good deal of research has been performed on pure polymer gels in which a suitable polymer can be selected to provide the desired texture but the range of attributes from this single polymer may be relatively limited. However, the increased demand for new products with special properties—for example low-fat products—has led to research of mixed gels, since a wider range of textural characteristics and opportunities can be obtained. For mixed-gel systems, a second polymer may be located within the pores of the first polymer, forming a bicontinuous system and a transition in rheological properties (measured by small- and large-deformation tests). In mixtures, the density of the particles as well as the density of the pores can vary more gradually, and this will have an impact on the textural characteristics. The perception of a gritty texture is likely to depend on the differences in the density as well as in the particle size.

Combined effects of gums are often used to obtain desired gel characteristics. For instance, pure κ-type carrageenan gives rigid, brittle, and syneresing gels. Some modification of texture is needed for greater resilience, cohesiveness, softness, and improved water retention. These properties can be imparted by blending in ι- and λ-types. In such mixtures, texture modification is achieved at the expense of gel strength.

Another example is xanthan/locust bean gum systems, which form relatively soft, elastic gels that melt at 57 to 60°C. They tend to have a chewy, unyielding texture because of their high elasticity (Marrs, 1997).

Gellan/gelatin blends have perhaps the greatest potential amongst commercial applications, since gelatin provides body and "melt-in-the-mouth" characteristics for products in all major food areas. Under suitable conditions of total and relative polymer concentrations, pH, and ionic strength, useful gels can be produced without refrigeration that exhibit thermal stability and a "synergistic" increase in gel strength (Shim, 1985).

Gellan has an attractive feature in that its combination with widely used food hydrocolloids produces a range of gels that satisfy traditional and novel textural requirements (Sanderson et al., 1988). Thus, optimal stability and texture are achieved when small amounts of the gum replace carrageenan or agar in pet foods and fabricated jellies. Starch-based fillings and pastes also show better performance after cooking and shearing, in terms of viscosity and water-holding capacity, when gellan is added to the formulation. The interactions between alginates and gellan have been rationalized to a certain extent in the production of water dessert gels prepared by mixing and cooling hot preparations of the two gums (Ortega, Rakitsky, & Sanderson, 1994; Sanderson & Ortega, 1993). Depending on polymer composition and the interplay of salt (type and ionic strength), either an alginate continuous phase with gellan inclusions is formed, imparting weak and elastic properties to the system, or a bicontinuous network is created, resulting in firm composite gels with shorter mouth-feel (Papageorgiou, Kasapis, & Richardson, 1994).

Another route to texture control adopted by the starch industry has been to extend gelling functionality through molecular weight control. It has long been recognized that substantial improvements to the physical and functional properties of starches can be achieved by a controlled reduction in the molecular weight of amylose and amylopectin. Although manufacturers of other hydrocolloids have traditionally offered a limited range of products having different molecular weights, only in recent times has the importance of molecular weight in relation to gel properties, texture, and other attributes been properly recognized and explored in depth (Marrs, 1997).

Thus, the effects of different process variables and ingredients can be evaluated by combining quantified microstructural data and textural data into models that enable prediction of textural behavior. The knowledge about the relationship between microstructure and texture should lead to optimized food production processes as well as to the development of new products with desirable properties (Langton et al., 1997).

REFERENCES

Ainsworth, P.A., & Blanshard, J.M.V. (1979). *Journal of Food Technology, 14*, 141.

Armstrong, H.J., Hill, S.E., & Mitchell, J.R. (1994). Enhancement of the gelation of food macromolecules using the Maillard reaction and elevated temperatures. *Maillard Reactions in Chemistry, Food and Health*, 159–163.

Armstrong H.J., Hill, S.E., Schrooyen, P., & Mitchell, J.R. (1994). A comparison of the viscoelastic properties of conventional and Maillard protein gels. *Journal of Texture Studies, 25*, 285–298.

Atkins, A.G., & Mai, Y.W. (1985). *Elastic and plastic fracture*. Chichester, UK: Ellis-Horwood.

Bagley, E.B., Wolf, W.J., & Christianson, D.D. (1985). Effect of sample dimensions, lubrication and deformation rate on uniaxial compression of gelatin gels. *Rheological Acta, 24*, 265–271.

Beach, P., Davis, E., Ikkala, P., & Lundbye, M. (1986). *ACS Symposium Series, 310*, 103.

Bourne, M.C. (1982). *Food texture and viscosity*. New York: Academic Press.

Chan, J.K., Gill, T.A., Thompson, J.W, & Singer, D.S. (1995). Herring surimi during low temperature setting: Physicochemical and textural properties. *Journal of Food Science, 60*, 1248–1253.

Chanyongvorakul, Y., Matsumura, Y., Nonaka, M., Motoki, M., & Mori, T. (1995). Physical properties of soy bean and broad bean 11S globulin gels formed by transglutaminase reaction. *Journal of Food Science, 60*, 483–488.

Christensen, O., & Trudsoe, J. (1980). Effect of other hydrocolloids on the texture of kappa-carrageenan gels. *Journal of Texture Studies, 11*, 137–142.

Chronakis, I.S., Kasapis, S., Richardson, R.K., & Doxastakis, G. (1995). Characterisation of a commercial soy isolate by physical techniques. *Journal of Texture Studies, 26*, 371–389.

Clare, K. (1993). Algin. In R.L. Whistler & J.N. BeMiller (Eds.), *Industrial gums* (3rd ed., pp. 105–143). New York: Academic Press.

Clegg, S.M. (1995). Thickeners, gels and gelling. In S.T. Beckett (Ed.), *Physico-chemical aspects of food processing* (pp. 117–141). Glasgow, Scotland: Blackie Academic & Professional.

Dickinson, E. (1997). Aggregation processes, particle interactions, and colloidal structure. In E. Dickinson & B. Bergenstahl (Eds.), *Food colloids* (pp. 107–126). Cambridge, UK: Royal Society of Chemistry.

Diehl, K.C., Hamann, D.D., & Whitfield, J.K. (1979). Structural failure in selected raw fruits and vegetables. *Journal of Texture Studies, 10*, 371–400.

Duran, L., Costell, E., & Fiszman, S.M. (1987). Collaborative compression tests on gels. In R. Jowitt, F. Escher, M. Kent, B. McKenna, & M. Roques (Eds.), *Physical properties of foods 2* (pp. 429–443). London: Elsevier Applied Science.

Fernandes, P.B., Goncalves, M.P., & Doublier, J.L. (1994). Rheological behaviour of kappa-carrageenan/galactomannan mixtures at a very low level of kappa-carrageenan. *Journal of Texture Studies, 25*, 267–284.

Ferry, J.D. (1970). In *Viscoelastic properties of polymers* (2nd ed., p.12). New York: John Wiley.

Foegeding, E.A., Bowland, E.L., & Hardin, C.C. (1995). Factors that determine the fracture properties and microstructure of globular protein gels. *Food Hydrocolloids, 9*, 237–249.

Friedman, H.H., Whitney, J.E., & Szczesniak, A.S. (1963). The texturometer: A new instrument for objective texture measurement. *Journal of Food Science, 28*, 390 396.

Gelatin Manufacturers Institute of America. (1993). *Standard methods for the sampling and testing of gelatins*. Author: New York.

Glicksman, M. (1969). *Gum technology in the food industry*. New York: Academic Press.

Guiseley, K.B., Stanley, N.F., & Whitehouse, P.A. (1980). Carrageenan. In R.Z. Davidson (Ed.), *Handbook of water soluble gums and resins* (pp. 5–11). New York: McGraw-Hill.

Hamann, D.D. (1983). Structural failure in solid foods. In M. Peleg & E.B. Bagley (Eds.), *Physical properties of foods* (pp. 351–383). New York: Van Nostrand Reinhold/AVI.

Harvey, H.G. (1960). Gels—with special reference to pectin gels. In *Texture in Foods* (pp. 29–63). Society of Chemical Industry.

Institute of Food Technology. (1959). Pectin standardization. *Food Technology, 13*, 496.

Izydorczyk, M.S., Biliaderis, C.G., & Bushuk, W. (1990). Oxidative gelation studies of water-soluble pentosans of wheat. *Journal of Cereal Science, 11*, 153–169.

Jones, N.R. (1977). Uses of gelatin in edible products. In A.G. Ward & A. Courts (Eds.), *The science and technology of gelatin* (pp. 366–392). New York: Academic Press.

Kamata, Y., & Kinsella, J.E. (1989). A comparison of creep phenomena in food protein gels. *Journal of Food Science, 54*, 170–172.

Kang, K.S., & Pettitt, D.J. (1993). Xanthan, gellan, welan and rhamsan. In R.L. Whistler & J.N. BeMiller (Eds.), *Industrial gums* (3rd ed., pp. 341–398). New York: Academic Press.

Keetels, C.J.A.M., & van Vliet, T. (1993). Mechanical properties of concentrated starch systems during heating, cooling, and storage. In E. Dickinson & P. Walstra (Eds.), *Food colloids and polymers* (pp. 266–271). London: Royal Society of Chemistry.

Keogh, K.M., Laine, K.I., & O'Connor, J.F. (1995). Rheology of sodium caseinate-carrageenan mixtures. *Journal of Texture Studies, 26*, 635–652.

Kinsella, J.E., & Whitehead, D.M. (1989). Proteins in whey: chemical, physical and functional properties. *Advances in Food and Nutrition Research, 33*, 343–438.

Konstance, R. P. (1993). Axial compression properties of calcium caseinate gels. *Journal of Dairy Science, 76*, 3317–3326.

Langley, K.R., Millard, D., & Evans, E.W. (1986). Determination of tensile strength of gels prepared from fractionated whey proteins. *Journal of Dairy Research, 53*, 285–292.

Langton, M., Astrom, A., Stading, M., & Hermansson, A.M. (1997). Effect of microstructure on sensory perception of particulate gels. In E. Dickinson & B. Bergenstahl (Eds.), *Food colloids* (pp. 18–28). London: Royal Society of Chemistry.

Lelievre, J., Mirza, I.A., & Tang, J. (1992a). Measurement of the tensile failure of gels. *Journal of Texture Studies, 23*, 349–358.

Lelievre, J., Mirza, I.A., & Tang, J. (1992b). Failure testing of gellan gels. *Journal of Food Engineering, 16*, 25–37.

Luyten, H. (1988). *The rheological and fracture properties of Gouda cheese*. Unpublished doctoral dissertation, Wageningen Agricultural University, Wageningen, the Netherlands.

Luyten, H., & van Vliet, T. (1995). Fracture properties of starch gels and their rate dependency. *Journal of Texture Studies, 26*, 281–298.

Luyten, H., & van Vliet, T., & Walstra, P. (1992). Comparison of various methods to evaluate fracture phenomena on food materials. *Journal of Texture Studies, 23*, 245–266.

Maekaji, K., & Kawamura, D. (1985). Variation in rheological properties of konjac gel with the corn cultivar. *Agricultural and Biological Chemistry, 49*, 2483–2484.

Marin, G. (1988). In A.A. Collyer & D.W. Clegg (Eds.), *Rheological measurement* (pp. 297–343). London: Elsevier Applied Science.

Marrs, M. (1997, June). Melt-in-mouth gels. *World of Ingredients*, June, pp. 39–40.

McClements, D.J., & Keogh, M.K. (1995). Physical properties of cold-setting gels formed from heat-denatured whey protein isolate. *Journal of the Science of Food and Agriculture, 69*, 7–14.

Mitchell, J.R. (1976). Rheology of gels. *Journal of Texture Studies, 7*, 313–339.

Montejano, J.G., Hamann, D.D., & Lanier, T.C. (1985). Comparison of two instrumental methods with sensory texture of protein gels. *Journal of Texture Studies, 16*, 403–424.

Morris, E.R. (1990). Industrial hydrocolloids. In D.M. Bloor & E. WynJones (Eds.), *The structure, dynamics and equilibrium properties of colloidal systems* (pp. 449–470). Dordrecht, the Netherlands: Kluwer.

Moss, T., Mohri, M., Artik, N., & Matsumura, Y. (1989). Rheological properties of heat-induced gel of soy bean 11S globulin under high ionic strength (mu = 0.5). *Journal of Texture Studies, 19*, 361–371.

Munoz, A.M., Pangborn, R.M., & Noble, A.C. (1986). Sensory and mechanical attributes of gel texture. II. Gelatin, sodium alginate and kappa-carrageenan gels. *Journal of Texture Studies, 17*, 17–36.

Murayama, A., Ichikawa, Y., & Kawabata, A. (1995). Sensory and rheological properties of κ-carrageenan gels mixed with locust bean gum, tara gum or guar gum. *Journal of Texture Studies, 26*, 239–254.

Nussinovitch, A., Ak, M.M., Normand, M.D., & Peleg, M. (1990). Characterization of gellan gels by uniaxial compression, stress relaxation and creep. *Journal of Texture Studies, 21*, 37–49.

Nussinovitch, A., Peleg, A., & Normand, M.D. (1989). A modified Maxwell and nonexponential model for characterization of the stress relaxation of agar and alginate gels. *Journal of Food Science, 54*, 1013–1016.

Ortega, D., Rakitsky, W.G., & Sanderson, G.R. (1994). In G.O. Phillips, D.J. Wedlock, & P.A. Williams (Eds.), *Gums and stabilisers for the food industry 7* Oxford, UK: IRL.

Papageorgiou, M., Kasapis, S., & Richardson, R.K. (1994). Steric exclusion phenomena in gellan/gelatin systems I. Physical properties of single and binary gels. *Food Hydrocolloids, 8*, 97–112.

Peleg, M. (1983). The semantics of rheology and texture. *Food Technology, 37*(11), 54–61.

Pomeranz, Y., & Meloan, C.E. (1994). Rheology. In *Food analysis: Theory and practice* (pp. 449–487). New York: Chapman & Hall.

Pons, M., & Fiszman, S.M. (1996). Instrumental texture profile analysis with particular reference to gelled systems. *Journal of Texture Studies, 27*, 597–624.

Purkayastha, P., & Peleg, M. (1987). Comparison between projected mechanical equilibrium conditions of selected food materials in stress relaxation and creep. *Journal of Texture Studies, 17*, 433–444.

Rector, D.J., Kella, N.K., & Kinsella, J.E. (1989). Reversible gelation of whey proteins: Melting, thermodynamics and viscoelastic behaviour. *Journal of Texture Studies, 20*, 457–471.

Rees, D.A. (1972). Polysaccharide gels: A molecular view. *Chemical Industry*, 630–636.

Rees, D.A., Morris, E.R., Thom, D., & Madden, J.K. (1982). Shapes and interactions of carbohydrate chains. In G.O. Aspinall (Ed.), *The polysaccharides* (Vol. 1, pp. 195–290). New York: Academic Press.

Richardson, R.K., Morris, E.R., Ross-Murphy, S.B., Taylor, L.J., & Dea, I.C.M. (1989). Characterization of perceived texture of thickened systems by dynamic viscosity measurements. *Food Hydrocolloids, 3*, 175–191.

Rohm, H., Jaros, D., & Benedikt, J. (1995). Constant strain rate compression of biopolymer gels. *Journal of Texture Studies, 26*, 665–674.

Rolin, C. (1993). Pectin. In R.L. Whistler & J.N. BeMiller (Eds.), *Industrial gums* (3rd ed., pp. 257–295). New York: Academic Press.

Ross-Murphy, S.B. (1984). Rheological methods. *Critical Reports in Applied Chemistry, 5*, 103–137.

Ross-Murphy, S.B. (1988). In J.R. Mitchell & J.M.V. Blanshard (Eds.), *Food structure* (pp. 387–400). London: Butterworths.

Ross-Murphy, S.B. (1995). Rheological characterisation of gels. *Journal of Texture Studies, 26*, 391–400.

Sanderson, G.R., Bell, V.L., Burgum, D.R., Clark, R.C., & Ortega, D. (1988). The texture of gelatin gels. In G.O. Phillips, D.J. Wedlock, & P.A. Williams (Eds.), *Gums and stabilisers for the food industry 4* (pp. 219–229). Oxford, UK: IRL.

Sanderson, G.R., & Ortega, D. (1993). In K. Nishinari (Ed.), *Food hydrocolloids* (pp.). New York: Plenum.

Savoie, V.J., & Arntfield, S.D. (1996). Effect of pH and cations on the thermally induced gelation of ovalbumin. *Journal of Texture Studies, 27*, 287–306.

Selby, H.H., & Whistler, R.L. (1993). Agar. In R.L. Whistler & J.N. BeMiller (Eds.), *Industrial gums* (3rd ed., pp. 87–104). New York: Academic Press.

Shama, F., & Sherman, P. (1973). Identification of stimuli controlling the sensory evaluation of viscosity. II Oral methods. *Journal of Texture Studies, 4*, 111–118.

Shim, J.L. (1985). Gellan gum/Gelatin blends. U.S. Patent 4,517,216.

Stading, M., & Hermansson, A.M. (1991). Large deformation properties of β-lactoglobulin gel structures. *Food Hydrocolloids, 5*, 339–352.

Steinhauser, J., Otterbach, G., & Gierschner, K. (1979). *Ind. Obst.-Gemüseverwert, 64*, 179.

Szczesniak, A.S. (1963). Classification of textural characteristics. *Journal of Food Science, 28*, 385–389.

Szczesniak, A.S. (1983). Physical properties of foods: What they are and their relation to other food properties. In M. Peleg & E.B. Bagley (Eds.), *Physical properties of foods* (pp. 1–42). New York: Van Nostrand Reinhold/AVI.

Szczesniak, A.S., Brandt, M.A., & Friedman, H.H. (1963). *Journal of Food Science, 28*, 397–403.

Szczesniak, A.S., & Farkas, E. (1962). *Journal of Food Science, 27*, 381.

Tang, Q., McCarthy, O.J., & Munro, P.A. (1995). Effects of pH on whey protein concentrate gel properties: Comparisons between small deformation (dynamic) and large deformation (failure) testing. *Journal of Texture Studies, 26*, 255–272.

Therkelsen, G.H. (1993). Carrageenan. In R.L. Whistler & J.N. BeMiller (Eds.), *Industrial gums* (3rd ed., pp. 145–181). New York: Academic Press.

Trzeciecki, M., Swiderski, F., & Fedorowicz, E. (1992). Studies on selection of methods for comparative estimation of gelling properties of vegetable hydrocolloids and gelatine. *Fruit Processing, 2*, 173–175.

Tschoegl, N.W., Rinde, J.A., & Smith, T.L. (1970). *Rheologica Acta, 9*, 223.

Uhlenbrock, W. (1983). *Gordian, 83*, 148.

van Vliet, T. (1995). Mechanical properties of concentrated food gels. In E. Dickinson (Ed.), *Food macromolecules and colloids* (Special Pub. No. 82, pp. 447–455). Cambridge, UK: Royal Society of Chemistry.

van Vliet, T., Luyten, H., & Walstra, P. (1991). In E. Dickinson (Ed.), *Food polymers, gels and colloids*, (Special Pub. No. 82, pp. 392–403). Cambridge, UK: Royal Society of Chemistry.

van Vliet, T., Luyten, H., & Walstra, P. (1993). In E. Dickinson & P. Walstra (Eds.), *Food colloids and polymers* (Special Pub. No. 113, p. 175). Cambridge, UK: Royal Society of Chemistry.

Whistler, R.L. (1993). Introduction to industrial gums. In R.L. Whistler & J.N. BeMiller (Eds.), *Industrial gums* (3rd ed., pp. 1–19). New York: Academic Press.

Wood, F.W. (1979). Psychophysical studies on liquid foods and gels. In P. Sherman (Ed.), *Food texture and rheology* (pp. 21–31). London: Academic Press.

Yano, T., & Shiinoki, Y. (1986). Viscoelastic behaviour of an agar-gelatin mixture gel as a function of its composition. *Food Hydrocolloids, 1*, 153–161.

Yano, T., Shiinoki, Y., Miyawaki, O., & Sakiyama, T. (1987). Instantaneous elastic deformation of a porous gel analysed by the finite element method. *Journal of Food Engineering, 6*, 217–230.

Zoon, P., van Vliet, T., & Walstra, P. (1989a). Rheological properties of rennet-induced skim milk gels. 4. The effect of pH and NaCl. *Netherlands Milk and Dairy Journal, 43*, 17–34.

Zoon, P., van Vliet, T., & Walstra, P. (1989b). Rheological properties of rennet-induced skim milk gels. 5. Behaviour at large deformation. *Netherlands Milk and Dairy Journal, 43*, 35–52.

Index

T

U